Mathematics of Planet Earth

Volume 13

This series provides a variety of well-written books of a variety of levels and styles, highlighting the fundamental role played by mathematics in a huge range of planetary contexts on a global scale. Climate, ecology, sustainability, public health, diseases and epidemics, management of resources and risk analysis are important elements. The mathematical sciences play a key role in these and many other processes relevant to Planet Earth, both as a fundamental discipline and as a key component of cross-disciplinary research. This creates the need, both in education and research, for books that are introductory to and abreast of these developments.

Springer's MoPE series will provide a variety of such books, including monographs, textbooks, contributed volumes and briefs suitable for users of mathematics, mathematicians doing research in related applications, and students interested in how mathematics interacts with the world around us. The series welcomes submissions on any topic of current relevance to the international Mathematics of Planet Earth effort, and particularly encourages surveys, tutorials and shorter communications in a lively tutorial style, offering a clear exposition of broad appeal.

Responsible Editors:
Martin Peters, Heidelberg (martin.peters@springer.com)
Robinson dos Santos, São Paulo (robinson.dossantos@springer.com)
Additional Editorial Contacts:
Donna Chernyk, New York (donna.chernyk@springer.com)
Masayuki Nakamura, Tokyo (masayuki.nakamura@springer.com)

Bertrand Chapron • Dan Crisan • Darryl D. Holm •
Etienne Mémin • Jane-Lisa Coughlan

Editors

Stochastic Transport in Upper Ocean Dynamics III

STUOD 2023 Workshop, Plouzané, France,
September 25–28

 Springer

Editors

Bertrand Chapron
Ifremer - Institut Français de Recherche
pour l'Exploitation de la Mer
Plouzané, France

Dan Crisan
South Kensington Campus
Imperial College London
London, UK

Darryl D. Holm
South Kensington Campus
Imperial College London
London, UK

Etienne Mémin
Campus Universitaire de Beaulieu
Inria - Institut National de Recherche en
Sciences et Technologies du Numérique
Rennes, France

Jane-Lisa Coughlan
South Kensington Campus
Imperial College London
London, UK

ISSN 2524-4264 ISSN 2524-4272 (electronic)
Mathematics of Planet Earth
ISBN 978-3-031-70659-2 ISBN 978-3-031-70660-8 (eBook)
https://doi.org/10.1007/978-3-031-70660-8

This work was supported by Horizon 2020 Framework Programme (856408).

This Springer imprint is published by the registered company Springer Nature Switzerland AG
The registered company address is: Gewerbestrasse 11, 6330 Cham, Switzerland

If disposing of this product, please recycle the paper.

Preface

This volume contains the Proceedings of the 4th **Stochastic Transport in Upper Ocean Dynamics Annual Workshop** held on **25–28 September 2023**. The workshop is a core part of the Stochastic Transport in Upper Ocean Dynamics (STUOD) project, which is supported by a European Research Council (ERC) Synergy Grant and is led by four Principal Investigators that bring together three world-class institutions: Prof. Bertrand Chapron, French Research Institute for Exploitation of the SEA (IFREMER), Prof. Dan Crisan, Imperial College London (ICL), Prof. Darryl D. Holm, Imperial College London (ICL) and Prof. Etienne Mémin, National Institute for Research in Digital Science and Technology (INRIA).

The project aims to deliver new capabilities for assessing variability and uncertainty in upper ocean dynamics and provide decision makers a means of quantifying the effects of local patterns of sea level rise, heat uptake, carbon storage and change of oxygen content and pH in the ocean. The project will make use of multimodal data and will enhance the scientific understanding of marine debris transport, tracking of oil spills and accumulation of plastic in the sea.

As in previous years, the 4th STUOD Annual Workshop 2023 focused on a range of fundamental topic areas, including:

1. Observations at high resolution of upper ocean properties such as temperature, salinity, topography, wind, waves and velocity
2. Large-scale numerical simulations
3. Data-based stochastic equations for upper ocean dynamics that quantify simulation errors
4. Stochastic data assimilation to reduce uncertainty

Each chapter in the present volume illustrates one or several of these topic areas. Many chapters offer new mathematical frameworks that are intended to enhance future research in the STUOD project. The workshop was held in the hybrid mode and brought together early-career academics, postgraduate students, senior members of the community and other invited guests from across the world that included the UK, France, Netherlands, Italy, Germany and the USA.

The scientific programme of the four-day workshop featured an inspirational range of talks, theoretical and applied sessions, as well as networking opportunities. In particular, it showcased topics on: Data assimilation and stochastic modelling, Data models, Data, Numerics for ocean models, Physics models, Theoretical analysis as well as covering new and future STUOD developments. Several members of the STUOD External Advisory Board gave invited talks: Prof. Sebastian Reich (University of Potsdam), Prof. Jeroen Molemaker (UCLA), Prof. Rosemary Morrow (Laboratoire d'Études en Géophysique et Océanographie Spatiales), Prof. Baylor Fox-Kemper (Brown University) and Prof. Peter Korn (Max Planck Institute for Meterology). The programme also included individual presentations by the STUOD Principal Investigators and postdoctoral researchers that overall provided opportunities for investigators, at both early and established stages of their career, to foster future research collaborations and enable the next generation of researchers.

Workshop attendees on 26 September 2023

The following is a brief description of the **14** contributions included in the proceedings:

The submitted manuscripts include the chapter by **Alexander Lobbe, Dan Crisan and Oana Lang** entitled **"Generative Modelling of Stochastic Rotating Shallow Water Noise"**. They present their recent work in developing a generic methodology for calibrating the noise in fluid dynamics stochastic partial differential equations where the stochasticity was introduced to parametrize subgrid-scale processes. The stochastic parametrization of sub-grid scale processes is required in the estimation of uncertainty in weather and climate predictions, to represent systematic model errors arising from subgrid-scale fluctuations. The methodology uses a principal component analysis (PCA) technique based on the ansatz that the increments of the stochastic parametrization are normally distributed. In this chapter, the PCA technique is replaced by a generative model technique. This enables them to avoid imposing additional constraints on the increments. The methodology is tested on a stochastic rotating shallow water model with the elevation variable of the model used as input data. The numerical simulations show that the noise is indeed non-Gaussian. The generative modelling technology gives good RMSE, CRPS score and forecast rank histogram results.

Albert Dombret, Darryl D. Holm, Ruiao Hu, Oliver D. Street and Hanchun Wang in their chapter entitled "**Collisions of Burgers Bores with Nonlinear Waves**" treat nonlinear wave-current interactions in their simplest form—as an overtaking collision. In one spatial dimension, the chapter investigates the collision interaction formulated as an initial value problem of a Burgers bore overtaking solutions of two types of nonlinear wave equations—Korteweg-de Vries (KdV) and nonlinear Schrodinger (NLS). The bore-wave state arising after the overtaking Burgers-KdV collision in numerical simulations is found to depend qualitatively on the balance between nonlinearity and dispersion in the KdV equation. The Burgers-KdV system is also made stochastic by following the stochastic advection by Lie transport approach (SALT).

The work of **Franco Flandoli, Andrea Papini and Marco Rehmeier** entitled "**Average Dissipation for Stochastic Transport Equations with Lévy Noise**" shows that, in one spatial and arbitrary jump dimension, the averaged solution of a Marcus-type SPDE with pure jump Lévy transport noise satisfies a dissipative deterministic equation involving a fractional Laplace-type operator. To this end, they identify the correct associated Lévy measure for the driving noise. They consider this a first step in the direction of a non-local version of enhanced dissipation, a phenomenon recently proven to occur for Brownian transport noise and the associated local parabolic PDE by the first author. Moreover, they present numerical simulations, supporting the fact that dissipation occurs for the averaged solution, with a behaviour akin to the diffusion due to a fractional Laplacian, but not in a pathwise sense.

Daniel Goodair demonstrates in his chapter entitled "**General Solution Theory for the Stochastic Navier-Stokes Equations**" how solutions to the incompressible Navier-Stokes Equations with transport and advection noise can be recovered through recent developments in the solution theory for stochastic partial differential equations (SPDEs). Local-in-time and global-in-time results are presented. Applications to the Stochastic Navier-Stokes Equations posed on the torus and a smooth bounded domain are detailed; in the latter case, both the no-slip and Navier boundary conditions are considered. Martingale weak solutions in 3D and weak solutions in 2D are proven in all cases. In 2D, strong solutions for the torus and Navier boundary are shown, whilst local strong solutions on the torus in 3D are also retrieved.

The work of **Darryl D. Holm, Ruiao Hu and Oliver D. Street** entitled "**Geometric Theory of Perturbation Dynamics Around Non-equilibrium Fluid Flows**" investigates the evolution of linear perturbations of time-dependent ideal fluid flows with advected quantities, expressed in terms of the second order variations of the action corresponding to a Lagrangian defined on a semidirect product space. This approach is related to Jacobi fields along geodesics and several examples are given explicitly to elucidate their approach. Numerical simulations of the perturbation dynamics are also presented.

Jin Won Kim and Sebastian Reich in their chapter entitled **"On Forward-Backward SDE Approaches to Conditional Estimation"** investigate the representation of conditional expectation values for partially observed diffusion processes in terms of appropriate estimators. They cite the work of Kalman and Bucy who have established a duality between filtering and estimation in the context of time-continuous linear systems. This duality has recently been extended to time-continuous nonlinear systems in terms of an optimization problem constrained by a backward stochastic partial differential equation. They revisit this problem from the perspective of appropriate forward-backward stochastic differential equations. Their approach sheds new light on the conditional estimation problem and provides a unifying perspective. It is also demonstrated that certain formulations of the estimation problem lead to deterministic formulations similar to the linear Gaussian case as originally investigated by Kalman and Bucy. Finally, they discuss application of the proposed formulation to optimal control problem on partially observed diffusion processes.

The work of **Colin J. Cotter, Dan Crisan and Maneesh Kumar Singh** entitled **"Data Assimilation for the Stochastic Camassa-Holm Equation Using Particle Filtering: A Numerical Investigation"** explores data assimilation for the Stochastic Camassa-Holm equation through the application of the particle filtering framework. Specifically, their approach integrates adaptive tempering, jittering and nudging techniques to construct an advanced particle filtering system. All filtering processes are executed utilizing ensemble parallelism. They conduct extensive numerical experiments across various scenarios of the Stochastic Camassa-Holm model with transport noise and viscosity to examine the impact of different filtering procedures on the performance of the data assimilation process. Their analysis focuses on how observational data and the data assimilation step influence the stability of the obtained results.

Arnaud Debussche, Etienne Mémin and Antoine Moneyron in their chapter entitled **"Some Properties of a Non-hydrostatic Stochastic Oceanic Primitive Equations Model"** study how relaxing the classical hydrostatic balance hypothesis affects theoretical aspects of the LU primitive equations well-posedness. They focus on models that sit between incompressible 3D LU Navier-Stokes equations and standard LU primitive equations, aiming for numerical manageability while capturing non-hydrostatic phenomena. Their main result concerns the well-posedness of a specific stochastic interpretation of the LU primitive equations. This holds with rigid-lid type boundary conditions and when the horizontal component of noise is independent of depth. These conditions can be related to the dynamical regime in which the primitive equations remain valid. Moreover, under these conditions, they show that the LU primitive equations solution tends towards the one of the deterministic primitive equations for a vanishing noise, thus providing a physical coherence to the LU stochastic model.

The work of **Arnaud Debussche, Etienne Mémin and Antoine Moneyron** entitled **"Derivation of Stochastic Models for Coastal Waves"** considers a stochastic nonlinear formulation of classical coastal waves models under location uncertainty (LU). In the formal setting investigated here, stochastic versions of the Serre-Green-Naghdi, Boussinesq and classical shallow water wave models are obtained through an asymptotic expansion, which is similar to the one operated in the deterministic setting. However, modified advection terms emerge, together with advection noise terms. These terms are well-known features arising from the LU formalism, based on momentum conservation principle.

Paul Platzer and Bertrand Chapron in their chapter entitled **"The Effects of Unresolved Scales on Analogue Forecasting Ensembles"** apply similar states in a database called "catalogue" such as a reanalysis, analogues provide simple yet efficient ensemble forecasts in atmospheric and ocean sciences. Typically performed on low-resolution images of large-scale atmospheric or ocean circulation, analogue forecasting encounters uncertainties due to unresolved small spatial scales, as the latter contribute to the time-evolution of the circulation but not to the similarity criterion used to search for analogues. Another source of uncertainty are the finite distances between the analogues and the initial target large-scale variables. They disentangle these two sources of uncertainty using a modified version of the Lorenz system, where stochastic terms account for unresolved small spatial scales. For large enough catalogue size and forecast horizon, we show that the analogue forecasting ensemble spread is dominated by the effect of stochastic terms, with only little influence of the initial analogue-to-target distances. Conversely, for short-term forecast and small catalogue size, the analogue ensemble is mostly influenced by initial analogue-to-target distances and not by the effects of unresolved scales. This result calls for adjustments of the classical analogue method for small forecast horizons.

The work of **Sebastian Reich** entitled **"Particle-Based Algorithm for Stochastic Optimal Control"** posits that the solution to a stochastic optimal control problem can be determined by computing the value function from a discretization of the associated Hamilton-Jacobi-Bellman equation. Alternatively, the problem can be reformulated in terms of a pair of forward-backward SDEs, which makes Monte-Carlo techniques applicable. More recently, the problem has also been viewed from the perspective of forward and reverse time SDEs and their associated Fokker-Planck equations. This approach is closely related to techniques used in diffusion-based generative models. Forward and reverse time formulations express the value function as the ratio of two probability density functions, one stemming from a forward McKean-Vlasov SDE and another one from a reverse McKean-Vlasov SDE. In this chapter, they extend this approach to a more general class of stochastic optimal control problems and combine it with ensemble Kalman filter type and diffusion map approximation techniques in order to obtain efficient and robust particle-based algorithms.

Valentin Resseguier presents in his chapter entitled **"Maximum Likelihood Estimation of Subgrid Flows from Tracer Image Sequences"** a sequence of tracer satellite images, several methods (e.g. optical flow) that exist to successfully estimate the main advecting current. Yet, this estimate is limited in resolution. To go beyond, they propose a new parametric estimation method to estimate second-order statistics of the residual small-scale velocity. They first express stochastic transport in a discrete setting to apply standard MLE techniques. Then they propose an efficient method to solve the MLE optimization problem through a fast log-likelihood gradient evaluation algorithm.

The work of **Francesco L. Tucciarone, Long Li, Etienne Mémin and Louis Thiry** entitled **"Transport Noise Defined from Wavelet Transform for Model-Based Stochastic Ocean Models"** introduces the simulation of planetary flows at all the scales that have a significant impact on the climate system, which is unachievable with nowadays computational resources. Parametrization of the scales smaller than the simulation resolution is thus crucial to correctly resolve the ocean dynamics. In this work, a novel parametrization of the subgrid scales by means of the wavelet transform is introduced in the shallow water and primitive models within the so-called Location Uncertainty framework.

James Woodfield in his chapter entitled **"Stochastic Fluids with Transport Noise: Approximating Diffusion from Data Using SVD and Ensemble Forecast Back-propagation"** introduces and tests methods for the calibration of the diffusion term in Stochastic Partial Differential Equations (SPDEs) describing fluids. They take two approaches: one uses ideas from the singular value decom-position and the Biot-Savart law. The other backpropagates through an ensemble forecast, with respect to diffusion parameters, to minimize a probabilistic ensemble forecasting metric. They describe the approaches in the specific context of solutions to SPDEs describing the evolution of fluid particles, sometimes called inviscid vortex methods. The methods are tested in an idealized setting in which the reference data is a known realization of the parametrized SPDE, and also using a forecast verification metric known as the Continuous Rank Probability Score (CRPS).

Finally, the STUOD Organizing Committee would again like to acknowledge the financial support received from the European Research Council (ERC) under the European Union's Horizon 2020 Research and Innovation Programme (ERC, Grant Agreement No 856408) for providing funds to cover the travel expenses of the invited speakers, catering costs and administrative support as well as the in-kind support from IFREMER for hosting at the Salon de l'Océan in Plouzané, France.

STUOD Organizing Committee
Prof. Bertrand Chapron (IFREMER)
Prof. Dan Crisan (ICL)
Prof. Darryl D. Holm (ICL)
Prof. Etienne Mémin (INRIA)
Dr Jane-Lisa Coughlan (ICL)

Plouzané, France	Bertrand Chapron
London, UK	Dan Crisan
London, UK	Darryl D. Holm
Rennes, France	Etienne Mémin
London, UK	Jane-Lisa Coughlan
May 2024	

Contents

Generative Modelling of Stochastic Rotating Shallow Water Noise

Alexander Lobbe, Dan Crisan, and Oana Lang

1 Introduction

Stochastic parameterizations address the uncertainty stemming from unaccounted for or neglected physical effects, as well as inaccurate observational data and imperfect theoretical models. Over the past two decades, there has been significant research in the area of stochastic parameterizations, largely driven by their application in quantifying uncertainty generated by downsampling high-resolution solutions to lower resolutions. More recently, numerous stochastic parameterization approaches have emerged to tackle such challenges, see for instance [2–4, 11, 15].

The accurate calibration of the stochastic model parameters can be used in the application of stochastic models, for example, in data assimilation and forecasting processes. Recently, several numerical techniques for calibration ([2–4, 17]) have been developed to demonstrate the effective integration of data-driven models and advanced data assimilation methods. In such studies, the calibration algorithms typically involve computing the full trajectories of the corresponding fluid parcels, which is often expensive numerically. The approach we introduced in [6] operates with entire solution fields instead. The methodology accounts for small-scale effects which are unresolved as a result of working with models run at coarse resolution, and it uses a principal component analysis (PCA) technique that relies on the ansatz that the data is Gaussian. However, the Gaussian assumption may not be exactly fulfilled in practice.

In this chapter, we replace the PCA technique with a generative model one, a technical change which allows us to model closer to the data by relaxing the Gaussian assumption. As with the previous work, we aim to design data-

A. Lobbe · D. Crisan · O. Lang (✉)
Department of Mathematics, Imperial College London, London, UK
e-mail: alex.lobbe@imperial.ac.uk; d.crisan@imperial.ac.uk; o.lang15@imperial.ac.uk

© The Author(s) 2025
B. Chapron et al. (eds.), *Stochastic Transport in Upper Ocean Dynamics III*,
Mathematics of Planet Earth 13, https://doi.org/10.1007/978-3-031-70660-8_1

1

driven models in which real uncertainty is accounted for, based on input from measurements and statistically-informed initial data.

We give next a brief description of the stochastic parametrization framework and the calibration methodology introduced in [6]. We denote by m^f the deterministic model state and assume that the evolution of m^f is governed by a partial differential equation of the following form

$$\frac{dm^f}{dt} = \mathcal{A}(m^f), \quad t \geq 0, \tag{1}$$

where $\mathcal{A}$ is the model operator. Given that we implement the equation numerically, let us assume that the partial differential Eq. (1) is discretised in time and space and that the evolution of m^f satisfies

$$m^f_{t_{n+1}} = m^f_{t_n} + \mathcal{A}\left(m^f_{t_n}\right)\Delta, \tag{2}$$

where $0 \leq t_1 \leq t_2 \leq \ldots$ is an equidistant time grid with mesh Δ. Higher order numerical schemes are possible: the same procedure will apply to those. We will denote by m^c the (discretised) stochastic model thought of as being modelled on a coarser spatial grid than that on which m^f is simulated, hence the superscript c. The aim of the stochastic parametrization is to compensate for the loss of scales when moving from a fine grid to a coarse grid. The effect of the *unresolved scales* can be mathematically modelled by a term of the form

$$\sum_{i=1}^{M} \mathcal{M}(m^c_{t_n})\xi_i \sqrt{\Delta} W^i_{t_n}, \tag{3}$$

where $\mathcal{M}$ is a suitably chosen operator and M is the number of sources of noise, $(\xi_k)_{k=1}^{M}$ are (space dependent but time independent) vector fields and W_{t_n} are independent normally distributed random variables $W^k_{t_n} \sim N(0, 1)$.[1] In other words, we have

$$m^c_{t_{n+1}} = m^c_{t_n} + \mathcal{A}\left(m^c_{t_n}\right)\Delta + \sum_{k=1}^{M} \mathcal{M}(m^c_{t_n})\xi_k W^k_{t_n}\sqrt{\Delta} \tag{4}$$

The choice of the stochastic parametrization (3) is such that asymptotically, as Δ tends to 0, one deduces that the model run on the coarse grid approximates the stochastic partial differential equation (SPDE)

$$dm^c = \mathcal{A}(m^c)dt + \mathcal{M}(m^c)dW_t, \quad t \geq 0, \tag{5}$$

---

[1] It is this assumption that will be removed in the current study.

where $W_t = W(t, x)$ is a space-time Brownian motion. We note that, theoretically, the solutions of both the deterministic Eq. (1) and its stochastic counter-part (5) live on the same physical domain (such as $\mathbb{R}^n$, the torus, a horizontal strip, etc), however their time *discretisations* (2) and (4) are approximated on different space grids and the space discretisation for (2) is finer than the one for (4). Therefore, in the numerical resolution for (2) and (4), then we could distinguish between the model operator for (2), and call it, say $\mathcal{A}^f$ and that for (4), and call it, say, $\mathcal{A}^c$.

In [6] we estimated the number of sources of noise M and the space dependent vector fields $(\xi_k)_{k=1}^N$ by using a principal component analysis methodology. Obviously, this hinged on the assumption that the stochastic parametrization that models the small scale dynamics has Gaussian increments. In practice this may not always be the case. In this chapter we lift this assumption and assume that effect of the *unresolved scales* is mathematically modelled by a term of the form

$$\mathcal{M}(m_{t_n}^c)N_n \qquad (6)$$

where $(N_n)_{n\geq 1}$ are independent indentically distributed random variables, but not necessarily with a Gaussian distribution. Again, just as in [6], we estimate the distribution of the independent noises from data and the calibration procedure introduced below is agnostic to the source of the input data. The *data* can be real data, such as satellite observations of e.g. ocean sea-surface height, data from re-analysis such as ERA5 ([9]), or synthetic data from a model run of (3) computed on a sufficiently large time window $[0, T]$. The use of a coarser grid computation in subsequent data assimilation of model reduction will lead to a significant reduction of computational effort.

Generative models are a class of machine learning models designed to generate new data samples from an unknown distribution. They are trained on a given dataset of samples from the same distribution. An important class of generative models are *diffusion models* which have gained more attention recently due to their ability to generate high-quality and diverse samples. The core idea behind *diffusion models* is to iteratively transform the training data through a diffusion-like mechanism into samples from a known distribution (a Gaussian distribution for example). In the process, the forward and the backward diffusions are learned using a neural network. Once the learning is complete, samples from the unknown distribution are obtained by running the backward diffusion initiated from samples from the Gaussian distribution. We give details of the methodology we use which is based on [7] in Section 3, specifically tailored to calibrate a stochastic rotating shallow water model.

In this chapter we use synthetic data coming from a realization of the (deterministic) rotating shallow water model, for which we keep the same notation m^f for now. The model run is then mollified using a procedure that will eliminate the small/fast scales effects, for example by using a low-pass filter, Gaussian mollifier, Helmholtz projection, subsampling, etc, or combinations thereof. We will denote by $C(m^f)$ the resulting mollification of the data. Note that both m^f and $C(m^f)$ live on the same space. We emphasise that $C(m^f)$ is not the solution of (5) and its time discretisation will not satisfy (4). However, we make the ansatz that the difference between the two processes $\hat{m} := m^f - C(m^f)$ has a stochastic representation given by (6), in other words, we will have

$$\hat{m}_{t_{n+1}} - \hat{m}_{t_n} \approx \mathcal{M}(m_{t_n}^c)N_n, \tag{7}$$

where N_n has an unknown distribution to be modeled by a certain score-based generative model known as a diffusion Schrödinger bridge, following [7]. Details of our implementation are included in the Sect. 3.

In the context of stochastic modeling in fluid dynamics, uncertainty plays a significant role, and accurately calibrating these models to real-world scenarios is crucial for reliable predictions. In this chapter we show that diffusion models can be used to quantify the uncertainty due to *unresolved scales*. The end result is that we can generate ensembles of fluid states. These fluid models can be affected by uncertainty coming from other sources not just from unresolved scale, see [1] for details. For example, one may want to model fast scales through a stochastic parametrization. This is in line with the Hasselmann paradigm (see [5]) where a stochastic model of climate variability entails slow changes of climate that are explained as the integral response to continuous random excitation by short period "weather" disturbances. Therefore the model will incorporate a rapidly varying "weather" system (essentially the atmosphere) modelled *stochastically*, and a slowly responding "climate" system (the ocean, cryosphere, land vegetation, etc.) modelled *deterministically*. The essential feature of stochastic climate models is that the non-averaged "weather" components are also retained. They appear formally as stochastic forcing terms. Calibrating stochasticity that models fast scales is different. In this case, the "truth" is already stochastic (the stochasticity is part of the model) and the data is made out of increments of the truth - minus the drift term. The low pass filter is not used here as the stochasticity is not a result of the coarsening procedure. However, we can still apply the generative model approach to infer the stochastic terms. In this case, the original model is in fact stochastic:

$$dm = \mathcal{A}(m)dt + \mathcal{M}(m)dW_t, \quad t \geq 0. \tag{8}$$

As a result

$$m_{t_{n+1}} - m_{t_n} - \mathcal{A}\left(m_{t_n}\right)\Delta \approx \mathcal{M}(m_{t_n})N_n. \tag{9}$$

In other words, the data consists of the increments $m_{t_{n+1}} - m_{t_n} - \mathcal{A}\left(m_{t_n}\right)\Delta$ out of which we compute the samples from the distribution of N_n.

. In the following subsection we provide an overview of the contents of the chapter.

1.1 Outline of the Chapter

In Sect. 2 we describe the particular fluid dynamics model we will be working with throughout the chapter. Specifically this is a rotating shallow water (RSW) model, similar to the model used in the earlier works ([6], ...). The novelty here compared to previous works is that we use a non-dimensionalised version of the rotating shallow water model, whose derivation is briefly outlined in Sect. 2. We choose to use a non-dimnensionalised model in order to be able to change the physical properties of the flow based on the selection of the non-dimensional numbers: the Rossby and Froude numbers in this case. Given that we are modelling subgrid scale effects by noise, it is instrumental that the coarse and fine scale deterministic model evolutions are sufficiently different, to ensure that there is a clear target for the noise term.

In Sect. 3 we introduce the generative model used in the numerical studies of this chapter. The theoretical framework allows for different types of generative models, some of which we mention in the beginning of Sect. 3 below. We choose to use the *diffusion Schrödinger bridge* model because it is a promising candidate for the fluid modelling studies we perform for several reasons. Specifically, the model is relatively transparent from a mathematical point of view, due to the form of the mathematically derived diffusion model. This is very expressive as a machine learning model, because of the underlying parametric model which is a neural network. Finally, we think that the *Schrödinger bridge* is useful due to the iterative nature that should make the calibration of the number of diffusion steps less critical.

In Sect. 4 we present the numerical study and results we have obtained based on the non-dimensionalised rotating shallow water equations. We verify that the evolution of the fluid we simulate indeed exhibits a loss of scales. Next, we show that the non-standard dataset we use to train the diffusion Schrödinger bridge is indeed representable by the parametric model. Further, the stochastic ensemble run with the generative model is shown to have an advantage compared to Gaussian noise in terms of different forecast metrics. The RMSE and CRPS scores significantly improve when the generative noise is used in the low initial uncertainty setting.

In Sect. 5 we summarise the conclusions of our study and identify directions for future studies.

2 Rotating Shallow Water Model

In this chapter, we base our study on a stochastic approximation of the nondimensionalised rotating shallow water model

$$d_t \mathbf{u} + (\mathbf{u} \cdot \nabla)\mathbf{u} + \frac{f}{\text{Ro}} \hat{\mathbf{z}} \times \mathbf{u} + \frac{1}{\text{Fr}^2} \nabla(\eta - b) = 0$$

$$d_t \eta + \nabla(\eta \mathbf{u}) = 0$$

(10)

where

- $\mathbf{u}(x, t) \doteq (u(x, t), v(x, t))$ is the horizontal fluid velocity vector field
- $\eta(x, t)$ is the height of the fluid column
- $f \in \mathbb{R}$ is the Coriolis parameter, $f = 2\Theta \sin \varphi$ where Θ is the rotation rate of the Earth and φ is the latitude; $f\hat{z} \times \mathbf{u} = (-fv, fu)^T$, where $\hat{z}$ is a unit vector pointing away from the centre of the Earth
- $\mathrm{Fr} = \frac{U}{\sqrt{gH}}$ is the Froude number (dimensionless) which is connected to the stratification of the fluid flow. Here U is a typical scale for horizontal speed and H is the typical vertical scale, while g is the gravitational acceleration.
- $\mathrm{Ro} = \frac{U}{f_0 L}$ is the Rossby number (also dimensionless) which describes the effects of rotation on the fluid flow: a small Rossby number ($\mathrm{Ro} \ll 1$) suggests that the rotation term dominates over the advective terms.
- $b(x, t)$ is the bottom topography function.

The initial condition (Fig. 1) is computed from an initial η-field from a geostrophic balance assumption (see details in Sect. 4.1). We work with the corresponding discrete version of (10), that is

$$\mathbf{u}_{n+1} - \mathbf{u}_n + (\mathbf{u}_n \cdot \nabla)\mathbf{u}_n \Delta + \frac{f}{\mathrm{Ro}}\hat{z} \times \mathbf{u}_n \Delta + \frac{1}{\mathrm{Fr}^2}\nabla(\eta_n - b)\Delta = 0$$

$$\eta_{n+1} - \eta_n + \nabla \cdot (\eta_n \mathbf{u}_n)\Delta = 0$$

(11)

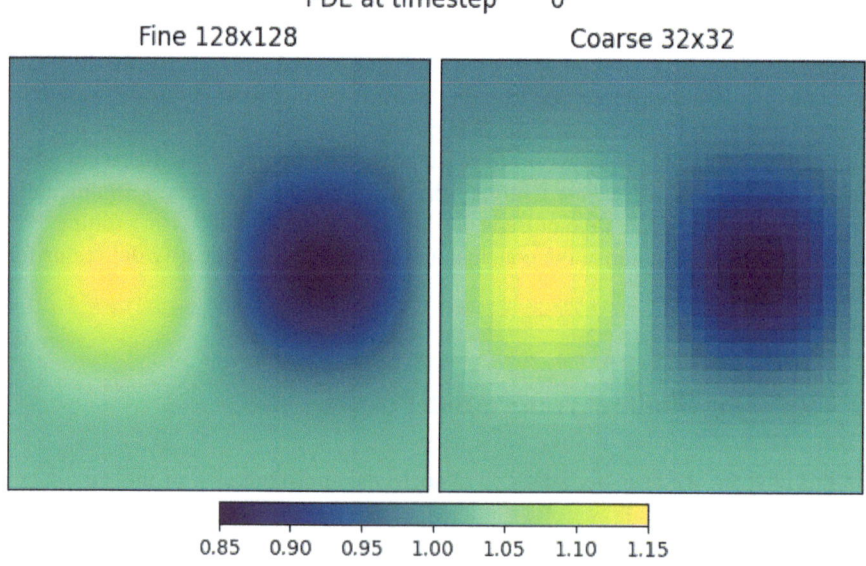

Fig. 1 Initial Condition for the non-dimensional height variable on the fine (128×128) and coarse (32×32) grids

In [6] we perturbed the iteration corresponding to (11) with spatial Gaussian noise of the form

$$W_n(x) = \sqrt{\Delta} \sum_{i=1}^{M} \boldsymbol{\xi}_i(x) W_n^i \tag{12}$$

where $(\boldsymbol{\xi}_i)_i$ are divergence-free elements of the covariance basis functions of the SALT ([11]) noise parametrisation and $W_n^i \sim N(0, 1)$ are independent i.i.d. random variables. When we do this, we obtain the following recurrence formula

$$\mathbf{u}_{n+1} - \mathbf{u}_n + (\tilde{\mathbf{u}}_n \cdot \boldsymbol{\nabla})\mathbf{u}_n + \mathbf{u}_n \cdot \boldsymbol{\nabla} W_n(x) + \frac{f}{\mathrm{Ro}}\hat{\mathbf{z}} \times \tilde{\mathbf{u}}_n + \frac{1}{\mathrm{Fr}^2}\boldsymbol{\nabla}(\eta_n - b)\Delta = 0$$

$$\eta_{n+1} - \eta_n + \boldsymbol{\nabla} \cdot (\eta_n \tilde{\mathbf{u}}_n) = 0 \tag{13}$$

where[2]

$$\tilde{\mathbf{u}}_n = \mathbf{u}_n \Delta + W_n(x) \tag{14}$$

The choice of the perturbation (14) is such that the iteration (13) is an approximation of the stochastic partial differential equation

$$\mathrm{d}\mathbf{u} + \left[(\mathbf{u} \cdot \boldsymbol{\nabla})\mathbf{u} + \frac{f}{\mathrm{Ro}}\hat{\mathbf{z}} \times \mathbf{u}\right]\mathrm{d}t + \sum_i \left[(\boldsymbol{\xi}_i \cdot \boldsymbol{\nabla})\mathbf{u} + \boldsymbol{\nabla}\boldsymbol{\xi}_i \cdot \mathbf{u} + \frac{f}{\mathrm{Ro}}\hat{\mathbf{z}} \times \boldsymbol{\xi}_i\right] \circ \mathrm{d}W_t^i$$

$$= -\frac{1}{\mathrm{Fr}^2}\boldsymbol{\nabla}(\eta - b)\,\mathrm{d}t$$

$$\mathrm{d}\eta + \boldsymbol{\nabla} \cdot (\eta\mathbf{u})\,\mathrm{d}t + \sum_i \boldsymbol{\nabla} \cdot (\eta\boldsymbol{\xi}_i) \circ \mathrm{d}W_t^i = 0$$

$$\tag{15}$$

where $\circ$ denotes Stratonovich integration and W^i are standard i.i.d. Brownian motions as before. The Stratonovich stochastic term generates a second order correction when writing the system in Itô form, but this is dealt with using the intrinsic properties of the numerical scheme. We have explained this part in detail in the Appendix of [6].

To bring the rotating shallow water example in line with the general notation presented in the introduction, observe that m^f is represented here by the pair $(\mathbf{u}, \eta)$ which solves the partial differential Eq. (10). Since we will be working only with discrete approximations, we can directly identify m^f with the solution of (11). Then m^c is the solution of (13). In other words:

---

[2] The term $\tilde{\mathbf{u}}$ is a velocity perturbation which is specific for this stochastic version of the RSW model.

$$m_{t_n}^c := \begin{pmatrix} \mathbf{u}_{t_n}^c \\ \eta_{t_n}^c \end{pmatrix}$$

where $(\mathbf{u}_{t_n}^c, \eta_{t_n}^c)$ solves (13). Then

$$\mathcal{M}(m_{t_n}^c)(\zeta) = \mathcal{M} \begin{pmatrix} \mathbf{u}_{t_n}^c \\ \eta_{t_n} \end{pmatrix} (\zeta) := \begin{pmatrix} \nabla \mathbf{u}_{t_n}^c \cdot \zeta + u_{t_n}^c \cdot \nabla \zeta \\ \nabla \eta_{t_n}^c \cdot \zeta. \end{pmatrix} \tag{16}$$

with ζ typically corresponding to ξ_i,[3] and

$$\mathcal{A}(m_{t_n}^c) = \mathcal{A} \begin{pmatrix} \mathbf{u}_{t_n}^c \\ \eta_{t_n}^c \end{pmatrix} = \begin{pmatrix} (\mathbf{u}_{t_n}^c \cdot \nabla)\mathbf{u}_{t_n}^c + \frac{f}{\mathrm{Ro}}\hat{\mathbf{z}} \times \mathbf{u}_{t_n}^c + \frac{1}{\mathrm{Fr}^2}\nabla(\eta_{t_n}^c - b) \\ \nabla \cdot (\eta_{t_n}^c \mathbf{u}_{t_n}^c) \end{pmatrix}. \tag{17}$$

Based on [6] we have

$$\hat{m}_{t_{n+1}} - \hat{m}_{t_n} \approx \sum_{i=1}^{M} \mathcal{M}(m_{t_n}^c)\xi_i(x)\sqrt{\Delta}W_{t_n}^i = \mathcal{M}(m_{t_n}^c)W_{t_n}(x) \tag{18}$$

with $\mathcal{M}$ given above in (16) and

$$W_{t_n}(x) = \sqrt{\Delta} \sum_{i=1}^{M} \xi_i(x) W_{t_n}^i \tag{19}$$

for the particular case of the RSW model. In practice, however, we first generate the increments in (18) for η only and then we use them together with a geostrophic balance assumption to compute the corresponding noise increments for the two components of $\mathbf{u}_{t_n}^c$. That is, for the RSW model we work mainly with $\mathcal{M}(\eta_{t_n}^c)(\zeta) = (\nabla \eta_{t_n}^c \cdot \zeta)$ which corresponds to a *transport noise*. The novelty of the current work is that we replace the spatial Gaussian noise $W_{t_n}(x)$ in (7) with a general noise $N_n(x)$ such that

$$\hat{m}_{t_{n+1}} - \hat{m}_{t_n} \approx \mathcal{M}(m_{t_n}^c)N_n \tag{20}$$

where N_n has an unknown distribution which is modelled using a diffusion Schrödinger bridge.

To ensure that the noise N_n is divergence-free, we generate a scalar random field $\tilde{N}_n$ such that $N_n = \nabla^\perp \tilde{N}_n$. In other words, $\tilde{N}_n$ sastisfies a hyperbolic partial

---

[3] Here we can observe one more time the challenges posed by transport noise in general (and SALT noise in this particular case) as this always involves calculating derivatives corresponding to both the model variable m and the (noise) variable ζ. In other words, the operator $\mathcal{M}$ and the variable ζ are inherently intertwined.

differential equation

$$\hat{m}_{t_{n+1}} - \hat{m}_{t_n} = -C(m_{t_n}^f)\nabla^{\perp}\tilde{N}_n \tag{21}$$

where C is a low-pass filter as defined in [6] Section 3. We refer to Section 3 in [6] for the detailed procedure we use to solve this partial differential equation.

3 Score-Based Generative Models: Diffusion Schrödinger Bridge

Score-based generative models are a recent trend in generative modelling and have achieved state-of-the-art results in several benchmark tasks in Machine Learning (cite some). In general, generative models are used to generate previously unseen samples from an underlying probability distribution which is typically available only through a dataset of samples. The generation of new samples is performed by a learned model which represents the unknown distribution of the training data either implicitly or explicitly. Classical examples of generative models in the literature include, among many others, Gaussian Mixture Models (GMMs), Hidden Markov Models (HMMs), Autoregressive Models [21], Variational Autoencoders (VAEs) [13], Generative Adversarial Networks (GANs) [8], Energy-Based Models (EBMs) [14], Normalising Flows [16] and, most recently, Diffusion Generative Models [7, 10, 18–20]. A comprehensive review of the field of generative modelling with diffusion models is provided in [22]. Note that, since the data distribution is typically a distribution over a very high-dimensional state-space, neural networks are the standard choice for the underlying parametric model in nearly all generative models used nowadays.

Although there are several different types of diffusion models in the literature, they all follow a common principle. That is, the data is being gradually diffused by successively adding noise (we call this below a *noising process*) until it becomes essentially a sample from a pure noise distribution, such as a Gaussian with known mean and covariance. Then, one can reverse the noising process to generate samples from the unknown data distribution by drawing a pure noise sample and running the reverse diffusion process. This is called the *denoising* process.

In this work, we are focusing on a specific type of score-based diffusion known as the Diffusion Schrödinger Bridge (DSB) model, developed in [7]. Broadly speaking, the DSB model is an extension of the classical score-based diffusion model by an optimal transport procedure known as iterative proportional fitting (IPF) which allows to iterate the score-based diffusion training and can work with shorter noising and denoising processes as a result. In the following subsection we outline the basics of the DSB model following [7].

3.1 Learning Diffusion Schrödinger Bridge

The noising process is modelled by a forward Markov Chain $\{X_k\}_{k=0}^N$ on $\mathbb{R}^d$ such that

$$X_{k+1} = X_k + \gamma_{k+1} f(X_k) + 2\gamma_{k+1} V_{k+1}, \quad k = 0, \ldots, N,$$

where $\{V_{k+1}\}_k \sim \mathcal{N}(0, \mathbf{1})$ are i.i.d. Gaussian random variables, $f : \mathbb{R}^d \to \mathbb{R}^d$ is a drift function and $\{\gamma_k\}_k$ are typically small stepsize parameters. The symbol $\mathbf{1}$ denotes the identity matrix. We assume the initial density $X_0 \sim p_0 = p_{data}$. The joint density p of the Markov Chain $X_{0:N} = (X_0, \ldots, X_N)$ can be decomposed into the corresponding forward transition densities $\{p_{k+1|k}\}_k$ as follows

$$p(x_{0:N} = \{x_k\}_{k=0}^N \in (\mathbb{R}^d)^{N+1}) = p_0(x_0) \prod_{k=0}^{N-1} p_{k+1|k}(x_{k+1}|x_k).$$

Similarly we can write down the backward decomposition

$$p(x_{0:N}) = p_N(x_N) \prod_{k=0}^{N-1} p_{k|k+1}(x_k|x_{k+1}) = p_N(x_N) \prod_{k=0}^{N-1} \frac{p_k(x_k) p_{k+1|k}(x_{k+1}|x_k)}{p_{k+1}(x_{k+1})},$$

where $\{p_k\}_k$ are the marginal densities and $\{p_{k|k+1}\}_k$ are the reverse transition densities. The methodology is based on sampling from p_{data} using the reverse decomposition initialized at $p_N = p_{prior}$. To achieve this, we need to approximate the reverse transition densities. To this end note the earlier assumption that the backward transitions are normally distributed as

$$p_{k+1|k}(x_{k+1}|x_k) = \mathcal{N}(x_{k+1}; x_k + \gamma_{k+1} f(x_k), 2\gamma_{k+1}\mathbf{1})$$

Then we apply a Taylor approximation (see [7]) to get

$$p_{k|k+1}(x_k \mid x_{k+1}) = p_{k+1|k}(x_{k+1} \mid x_k) \exp\left[\log p_k(x_k) - \log p_{k+1}(x_{k+1})\right]$$
$$\approx \mathcal{N}(x_k; x_{k+1} - \gamma_{k+1} f(x_{k+1})$$
$$+ 2\gamma_{k+1} \nabla \log p_{k+1}(x_{k+1}), 2\gamma_{k+1}\mathbf{1})$$

Then, the backward transitions are Gaussian, with a drift depending on the parameters f and $\{\gamma_k\}_k$ and on the *score functions* $\{\nabla \log p_k\}_k$. Note that we can integrate out the initial density from the marginals such that

$$p_{k+1}(x_{k+1}) = \int p_0(x_0) p_{k+1|0}(x_{k+1}|x_0) dx_0$$

and thus

$$\nabla \log p_{k+1}(x_{k+1}) = \mathbb{E}_{p_{0|k+1}}[\nabla_{x_{k+1}} \log p_{k+1|0}(x_{k+1}|X_0)].$$

The conditional expectation above is intractable, but the joint distribution is available through samples, so we can use regression to find it

$$s_{k+1} = \arg\min_{s} \mathbb{E}_{p_{0,k+1}}\left[\left\|s(X_{k+1}) - \nabla_{x_{k+1}} \log p_{k+1|0}(X_{k+1} \mid X_0)\right\|^2\right],$$

where $\|\cdot\|$ denotes the L_2-norm on $\mathbb{R}^d$. We can thus learn a parametrised approximation of the score (all scores simultaneously)

$$s_{\theta^\star}(k, x_k) \approx \nabla \log p_k(x_k)$$

via Denoising Score Matching (Vincent 2011) as

$$\theta^\star = \arg\min_{\theta} \sum_{k=1}^{N} \mathbb{E}_{p_{0,k}}[\|s_\theta(k, X_k) - \nabla_{x_k} \log p_{k|0}(X_k|X_0)\|^2].$$

So we estimate the score function and then sample $X_0 \overset{approx}{\sim} p_{data}$ using the diffusion started at $p_N \approx p_{prior}$ such that

$$X_k = X_{k+1} - \gamma_{k+1} f(X_{k+1}) + 2\gamma_{k+1} s_{\theta^\star}(k+1, X_{k+1}) + \sqrt{2\gamma_{k+1}}\mathcal{N}(0, 1).$$

Let $\mathcal{P}_{N+1}$ be the space of sequences of probability densities of length $N + 1$. In the Schrödinger Bridge framework, we consider the joint density $p \in \mathcal{P}_{N+1}$ of the Markov Chain X and we want to find a density $\pi^\star \in \mathcal{P}_{N+1}$ such that

$$\pi^\star = \operatorname{argmin}\left\{\mathrm{KL}(\pi \mid p) : \pi \in \mathcal{P}_{N+1}, \pi_0 = p_{\text{data}}, \pi_N = p_{\text{prior}}\right\}, \tag{22}$$

where for any two probability densities p and q over a space $\mathcal{X}$,

$$KL(p\|q) = \int_{\mathcal{X}} p(x) \log\left(\frac{p(x)}{q(x)}\right) dx$$

denotes the Kullback-Leibler divergence[4] between probability distributions. Assuming $\pi^\star$ is available, a generative model can be obtained by sampling $X_N \sim p_{\text{prior}}$, followed by the reverse-time dynamics $X_k \sim \pi^\star_{k|k+1}(\cdot \mid X_{k+1})$ for $k \in \{N - 1, \dots, 0\}$.

---

[4] Note that the Kullback-Leibler divergence is not a distance in a strict mathematical sense because it is not symmetric. Hence the name divergence.

A well-known solution method to find a minimum of 22 is iterative proportional fitting (IPF). Initialised at $\pi^0 = p(x_{0:N})$ this method defines the following iterative process

$$\pi^{2n+1} = \operatorname{argmin} \left\{ \mathrm{KL}\left(\pi \mid \pi^{2n}\right) : \pi \in \mathcal{P}_{N+1}, \pi_N = p_{\mathrm{prior}} \right\}$$

$$\pi^{2n+2} = \operatorname{argmin} \left\{ \mathrm{KL}\left(\pi \mid \pi^{2n+1}\right) : \pi \in \mathcal{P}_{N+1}, \pi_0 = p_{\mathrm{data}} \right\}$$

A positive result for the feasibility of IPF in our setting is provided in Proposition 1 below.

Proposition 1 (Proposition 2 in [7]) *Assume that* $\mathrm{KL}\left(p_{data} \otimes p_{prior} \mid p_{0,N}\right) < +\infty$. *Then for any* $n \in \mathbb{N}$, π^{2n} *and* π^{2n+1} *admit positive densities w.r.t. the Lebesgue measure denoted as* p^n *resp.* q^n *and for any* $x_{0:N} \in \mathcal{X}$, *we have* $p^0(x_{0:N}) = p(x_{0:N})$ *and*

$$q^n(x_{0:N}) = p_{prior}(x_N) \prod_{k=0}^{N-1} p_{k|k+1}^n (x_k \mid x_{k+1}), \, p^{n+1}(x_{0:N})$$

$$= p_{data}(x_0) \prod_{k=0}^{N-1} q_{k+1|k}^n (x_{k+1} \mid x_k)$$

In practice we have access to $p_{k+1|k}^n$ and $q_{k|k+1}^n$. Hence, to compute $p_{k|k+1}^n$ and $q_{k+1|k}^n$ we use

$$p_{k|k+1}^n (x_k \mid x_{k+1}) = \frac{p_{k+1|k}^n (x_{k+1} \mid x_k) \, p_k^n(x_k)}{p_{k+1}^n(x_{k+1})}, \, q_{k+1|k}^n (x_{k+1} \mid x_k)$$

$$= \frac{q_{k|k+1}^n (x_k \mid x_{k+1}) \, q_{k+1}^n (x_{k+1})}{q_k^n(x_k)}.$$

The following Proposition 2 details a possible loss function to use in the training of the DSB model, known as Mean Matching. Different variations to be used for training can be found in [7].

Proposition 2 (Proposition 3 in [7]) *Assume that for any* $n \in \mathbb{N}$ *and* $k \in \{0, \ldots, N-1\}$,

$$q_{k|k+1}^n (x_k \mid x_{k+1}) = \mathcal{N}\left(x_k; B_{k+1}^n(x_{k+1}), 2\gamma_{k+1}\mathbf{I}\right), \, p_{k+1|k}^n (x_{k+1} \mid x_k)$$

$$= \mathcal{N}\left(x_{k+1}; F_k^n(x_k), 2\gamma_{k+1}\mathbf{I}\right),$$

with $B_{k+1}^n(x) = x + \gamma_{k+1} b_{k+1}^n(x)$, $F_k^n(x) = x + \gamma_{k+1} f_k^n(x)$ *for any* $x \in \mathbb{R}^d$. *Then we have for any* $n \in \mathbb{N}$ *and* $k \in \{0, \ldots, N-1\}$

$$B_{k+1}^n = \underset{B \in L^2(\mathbb{R}^d, \mathbb{R}^d)}{\text{argmin}} \ \mathbb{E}_{p_{k,k+1}^n} \left[\left\| B(X_{k+1}) - \left(X_{k+1} + F_k^n(X_k) - F_k^n(X_{k+1}) \right) \right\|^2 \right],$$

$$F_k^{n+1} = \underset{F \in L^2(\mathbb{R}^d, \mathbb{R}^d)}{\text{argmin}} \ \mathbb{E}_{q_{k,k+1}^n} \left[\left\| F(X_k) - \left(X_k + B_{k+1}^n(X_{k+1}) - B_{k+1}^n(X_k) \right) \right\|^2 \right].$$

Note that, here, we use neural networks $B_{\beta^n}(k, x) \approx B_k^n(x)$ and $F_{\alpha^n}(k, x) \approx F_k^n(x)$ to parametrise the unknown drifts in the transition densities. We describe the DSB training process according to the chapter [7]. For pseudocode of this algorithm, the reader may refer to [7]. Roughly speaking, the transition densities (drifts) for both the forward and backwards noising and denoising processes are modelled by a collection of neural networks. The initial joint density $p(x_{0:N})$ is given from a random initialisation of the networks. Each DSB iteration consists of a forward and backward run. The first step is a forward iteration, where we take samples from the dataset and diffuse them according to the dynamics given by the noising process with the parameterised forward net. The losses collected during the forward iterations are applied to the *backward* nets. In the inner backward iterations we sample from the prespecified prior density (Gaussian) and run the denoising process according to the backward net. In the backward iterations we apply the gradient descent steps for the *forward* nets. The inner iterations are each run until convergence. In practise, a prespecified number of iterations that we tune.

4 Numerical Results

In the following we describe our numerical results using the DSB method on the non-dimensionalised rotating shallow water equations. On a general note, we found that the method of generating noise from a generative model is generally stable and we obtain suitable new noise samples from the trained model. In our setup we generate the noise as an integral of the velocity perturbations used in the SPDE model, akin to a streamfunction in classical GFD. Therefore, the output from the generative model is subject to the application of a gradient. Thus, a small percentage of the data produces large gradients which can lead to instabilities in the numerical evolution of the SPDE. We mitigate this problem by clipping large gradients. Note that this is not expected to be problematic for two reasons. Firstly, the amount of clipped data locations is small and secondly, we are interested in the overall spatial correlations in the generated noise, which remains intact after clipping the gradients because large gradients primarily occur in the boundary regions.

4.1   Fine vs Coarse Scale

We use an initial height field for η given by

$$\eta_0(x, y) = 1 - \frac{a}{2}\operatorname{atan}(\frac{y}{L_y} - \frac{1}{2}) + a\sin\left(\frac{2\pi x}{L_x}\right) + \frac{a}{2}\sin\left(\frac{2\pi x}{L_x}\right)\sin\left(\frac{\pi y}{L_y}\right)^4,$$

where the domain is the rectangle $[0, L_x] \times [0, L_y]$. We use the domain $[0, 1] \times [0, 1]$ in our simulations. Moreover, $a \in \mathbb{R}$ is a parameter. We chose $a = 0.1$. The initial condition for the nondimensional simulation is computed from the initial η-field using the geostrophic balance condition. This condition in dimensional form for the dimensional velocity component v^d in y-direction reads as

$$f^d v^d = g \frac{\partial \eta^d}{\partial x^d},$$

where $g = 9.81 \frac{m}{s^2}$ is the gravitational constant and f^d is a parameter associated with the Coriolis force. The superscript d indicates dimensional variables throughout. In non-dimensional form, we have

$$f_0 f U v = g \frac{\Delta \eta}{L \Delta x} \quad \leftrightarrow \quad v = \frac{1}{f} \frac{\mathrm{Ro}}{\mathrm{Fr}^2} \frac{\partial \eta}{\partial x}$$

where U is a typical velocity scaling, f_0 is a typical Coriolis scaling and L is a typical length scale. Here, Ro denotes the Rossby number and Fr denotes the Froude number. The simulations are run at Ro $= 0.2$ and Fr $= 1.1$. Thus also for the non-dimensional velocity component in x-direction we have the geostrophic balance condition

$$u = -\frac{1}{f} \frac{\mathrm{Ro}}{\mathrm{Fr}^2} \frac{\partial \eta}{\partial y}.$$

The finally used initial condition is scaled so that the initial u and v variables are $O(1)$ on the domain.

We use the intermediate fields during the spin-up of the system as a qualitative sanity check to verify that the solutions on the fine and coarse grids diverge as the flow progresses. Specifically, we expect to observe fine-scale features (waves) to develop in the fine-grid simulation, which are not resolved in the coarse scale run. Indeed, the plot in Fig. 2 shows that this is indeed occurring.

4.2 Training Data

We generate the training data as solutions of the hyperbolic calibration equation through a forward run of the fine-scale PDE. The collected solutions are thought of as stream functions for the velocity perturbations of the SPDE and are assumed to be sampled from a fixed (in time) probability distribution that we aim to model through the generative model. To ease the training, we perform a nonlinear transformation

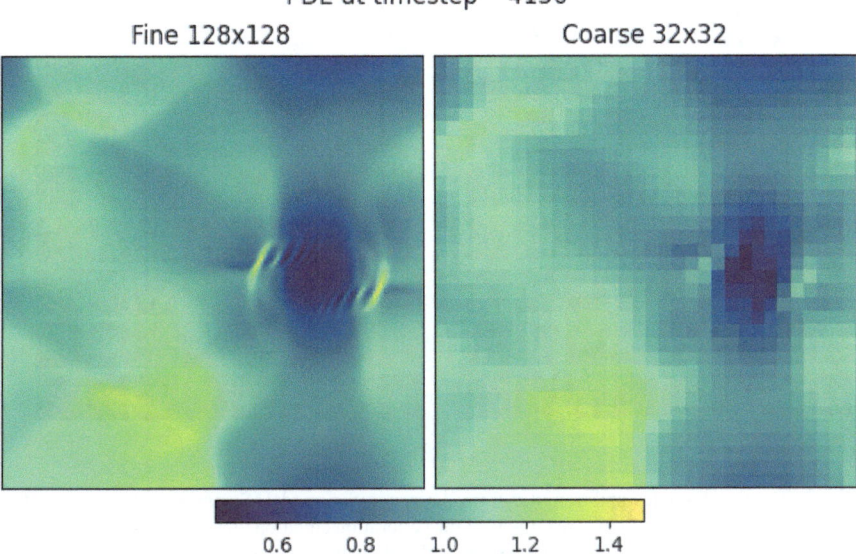

Fig. 2 Snapshot of non-dimensional height variable at fine and coarse run after 4150 fine scale timesteps. We can observe the development of fine scale features which are absent (unresolved) in the coarse field

($arcsinh$-transform) [12] of the obtained data and normalise the values globally to the interval [0, 1]. This is standard practise in Machine Learning and Statistics. Specifically, let $\Psi : \Omega \to \mathbb{R}$ denote a solution of the calibration equation. Then we transform it using

$$\hat{\Psi} = \frac{1}{\vartheta} \operatorname{arcsinh}(\vartheta \Psi)$$

with parameter $\vartheta = 2e5$. The transformed dataset $\{\hat{\Psi}_i\}_i$ is then normalised to the range [0, 1] by

$$\psi_i = \frac{\hat{\Psi}_i - \min_{i,x,y} \hat{\Psi}_i(x, y)}{\max_{i,x,y}(\hat{\Psi}_i(x, y) - \min_{i,x,y} \hat{\Psi}_i(x, y))}.$$

Samples from the training set are displayed in Fig. 3a and the pixel distribution across all samples after transformation and normalisation is depicted in the histogram in Fig. 4a.

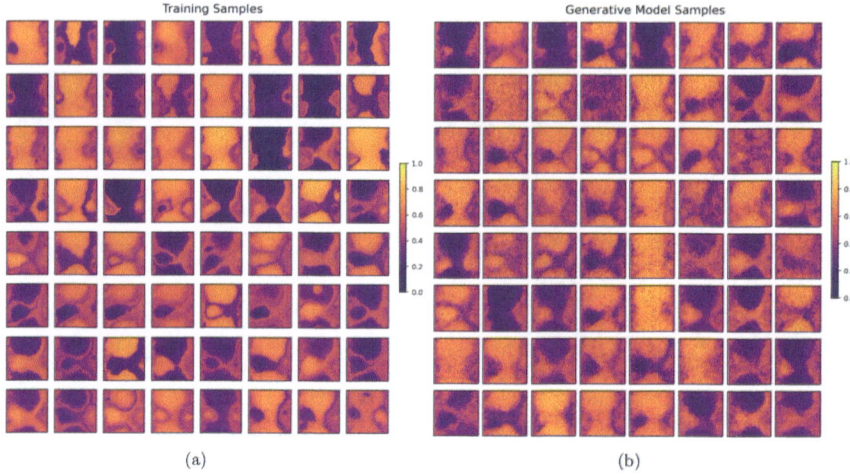

Fig. 3 Training Samples and Samples from the generative model. Samples of the training data after transformation. The fields are outputs of the calibration equation thought of a stream functions for the velocity perturbations in the SPDE. The data have been transformed by a *arcsinh* transformation and normalized to the interval [0, 1]. Samples from the generative model. (**a**) Training samples. (**b**) Generated samples

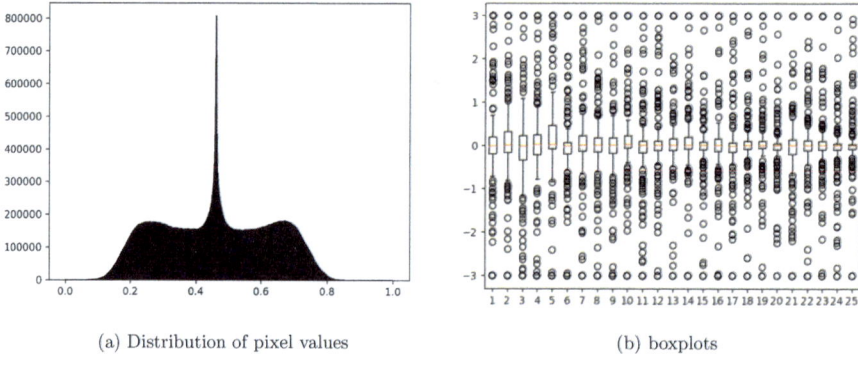

(a) Distribution of pixel values   (b) boxplots

Fig. 4 (**a**) Distribution of pixelvalues throughout the whole training data set. The data transformation has been choosen s.th. the pixel values achieve a good coverage of the data range [0, 1]. This way the model can better distinguish between variations in shade.(**b**) Distribution of pixelvalues in central grid locations in the generated dataset, i.e. the output from the generative model

4.3 Generative Model Output

Samples from the generative model output after training are shown in Fig. 3b. The model has been trained for 5 DSB Iterations using 30 diffusion steps. These samples are subsequently reverse-transformed and the gradients are applied to convert the stream function information into velocity perturbations. The distribution of velocity perturbations (here for the velocity component u) at some central grid locations is depicted in the boxplots in Fig. 4b. It shows that the median of the generated noise distributions at each of the grid locations is close to zero and their magnitude rarely exceeds 1, as we expect in a non-dimensional simulation. The boxplots also reflect the fact that we clip the noise data at a magnitude of 3 to avoid instabilities due to outliers. The plot shows that the amount of data clipped is negligible as it is well within the outlier regime at all gridpoints. Moreover, the generative noise is not Gaussian. We perform a one-sample Kolmogorov-Smirnov Test to show that the generated noise is not Gaussian. Specifically, we use the velocity perturbations computed from the generative model noise to inspect the distribution in different spatial locations on the computational grid ($8 \times 8 = 64$ total locations). For each location, we perform independent KS-tests for normality, which show that at none of the tested locations could we detect a normal distribution of the generated noise (p-value less than 0.05) (Fig. 5).

4.4   Forecast Studies

The aim of the generative model is to produce a distribution for the noise in the rotating shallow water model that is advantageous for the modelling of the effects of unrepresented small scales (see e.g. Fig. 8). To this end, we use the established ensemble and forecast metrics root mean square error (RMSE) and continuous ranked probability score (CRPS). We also use rank histograms. The forecasts are produced from a forward ensemble run of the SPDE with different noises for comparison. The first noise is the generative model noise, which we compare to two different typed of Gaussian noises. One is a Gaussian with the same overall mean and variance as the generative model, i.e. a Gaussian with covariance $\sigma 1$, where $\sigma \in \mathbb{R}$ is the standard deviation of the dataset obtained as the output from the generative model. The second Gaussian noise has a diagonal covariance that is varies in space given as diag(σ), where $\sigma \in \mathbb{R}^{mn \times mn}$ is the vector of the standard deviations of the generated noise at all individual spatial locations. The forecasts are run for a lead time of 200 calibration time steps and then reset to the fine PDE value from which the ensemble is relaunched launched. Additionally, we apply a normally distributed perturbation to each ensemble, to represent initial uncertainty. We chose three different scenarios here, one is a scenario of no initial incertainty, then small initial uncertainty with variance 0.001^2 and a large initial uncertainty scenario with variance 0.05^2.

The results of the CRPS score are depicted in Fig. 6 below and the results of the RMSE metric are shown in Fig. 7. Both metrics show better forecasts for the generative model in case of no and low initial uncertainty in the ensemble. Especially the forecast results for the height variable are significantly better in the generative noise setting. The results in the u and v variables are somewhat less pronounced (Fig. 8).

Noise data at different points in space compared to Gaussian

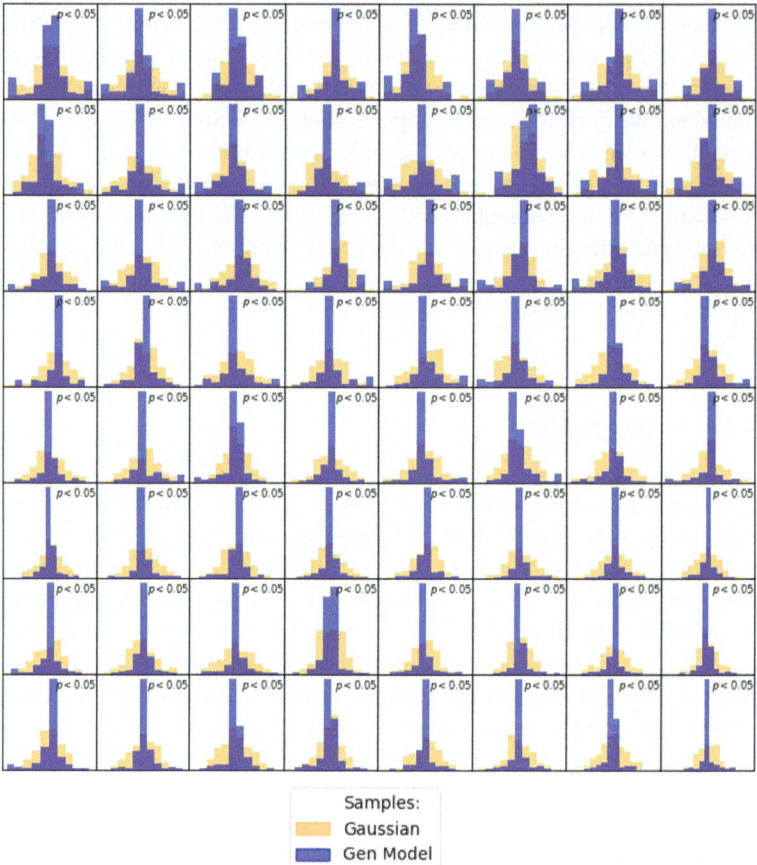

Fig. 5 Distribution of the generated noise values compared to Gaussians at every pixel in a central region. A Kolmogorov-Smirnov one-sample test has been performed to check if the data come from a normal distribution. The hypothesis was rejected for all locations indicating that the generated noise does not come from a simple normal distribution

The forward run ensembles are also assessed for a longer duration with ensembles of 10 particles for 1000 calibration steps, without resetting to the truth. We produce rank histograms from those runs using repetitions with different noise samples. Here, we compare the generative model noise to a Gaussian with the same overall mean and variance as the generative model. The results are shown in Fig. 9. Figure 9a shows the rank histograms for the Gaussian noise which are overall more uniform than the clearly overdispersed Gaussian ensembles in Fig. 9b.

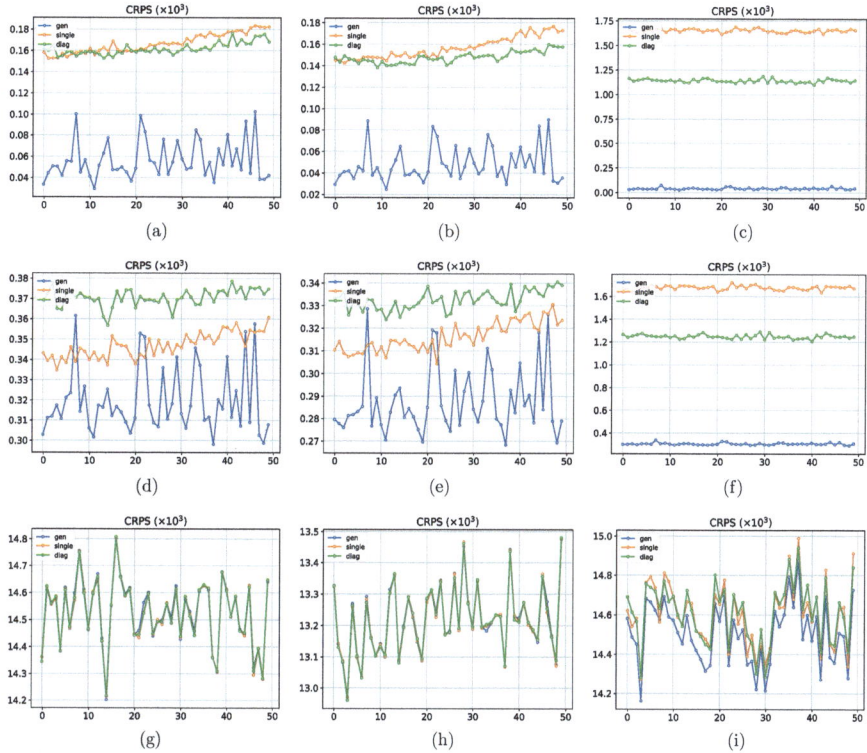

Fig. 6 CRPS Scores at forecast times for a lead time of 200 calibration timesteps for all three shallow water variables (u,v, and e). We plot them for different initial noise standard deviations: no noise (0.0), small noise (0.001) and large noise (0.05). The graphs show the results for three different velocity perturbation distributions. Blue lines show the generative model noise, green lines are diagonal gaussian noise with spatially independent variance, and orange is gaussian noise with the same variance in all locations. We observe that small initial noise shows an advantage of the generative model, which fades away as initial uncertainty becomes large. Also, the generative model performs significantly better in the implicit variable e. (**a**) Variable u, Std 0.0. (**b**) Variable v, Std 0.0. (**c**) Variable e, Std 0.0. (**d**) Variable u, Std 0.001. (**e**) Variable v, Std 0.001. (**f**) Variable e, Std 0.001. (**g**) Variable u, Std 0.05. (**h**) Variable v, Std 0.05. (**i**) Variable e, Std 0.05

5 Conclusions and Future Work

In this work the feasibility of using modern generative models for the generation of appropriate noise distributions in stochastic models for subgridscale effects in fluid dynamics has been investigated. To this end we implemented a Diffusion Schrödinger Bridge model for the generation of Rotating Shallow Water noise and performed a comparative study of the generated ensemble in terms of established forecast metrics, RMSE and CRPS. The results show that the generative model samples display an advantage over the gaussian noise ensemble in the case of low initial uncertainty. This result indicates that the generative model is more effective

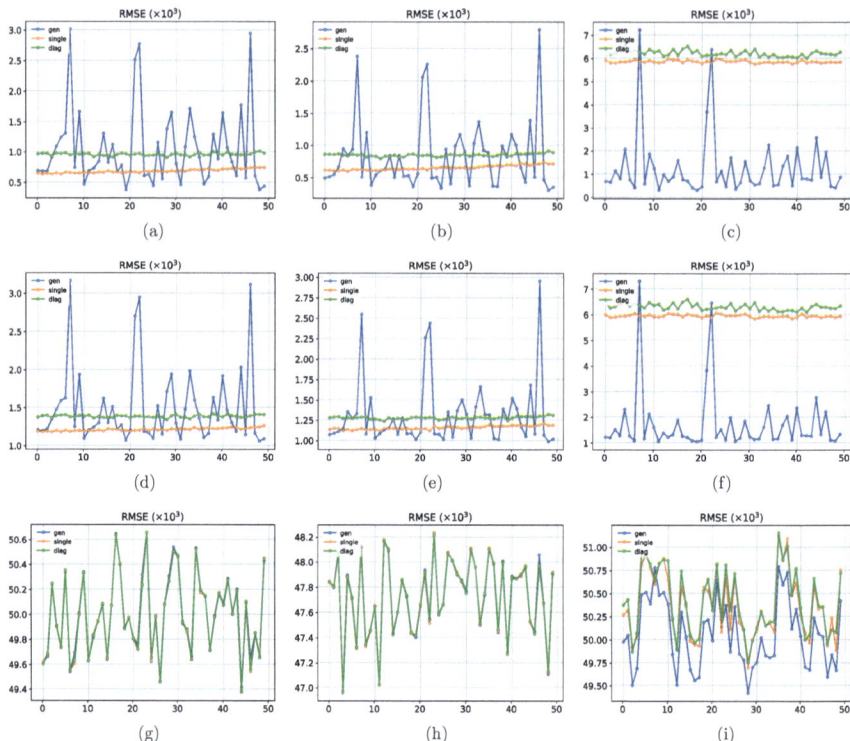

Fig. 7 RMSE scores at forecast times for a lead time of 200 calibration timesteps for all three shallow water variables (u,v, and e). We plot them for different initial noise standard deviations: no noise (0.0), small noise (0.001) and large noise (0.05). The graphs show the results for three different velocity perturbation distributions. Blue lines show the generative model noise, green lines are diagonal gaussian noise with spatially independent variance, and orange is gaussian noise with the same variance in all locations. We observe that small initial noise shows an advantage of the generative model, which fades away as initial uncertainty becomes large. Also, the generative model performs significantly better in the implicit variable e. (**a**) Variable u, Std 0.0. (**b**) Variable v, Std 0.0. (**c**) Variable e, Std 0.0. (**d**) Variable u, Std 0.001. (**e**) Variable v, Std 0.001. (**f**) Variable e, Std 0.001. (**g**) Variable u, Std 0.05. (**h**) Variable v, Std 0.05. (**i**) Variable e, Std 0.05

at capturing the fine scale effects on the coarse dynamics than an Gaussian noise ensemble.

In future work, we will compare the generative model noise against a model using a Karhunen-Loeve decomposition of the dataset, according to the previously developed method in [6]. Moreover, studies on different underlying fluid models need to be performed in addition to this initial study on the rotating shallow water equations.

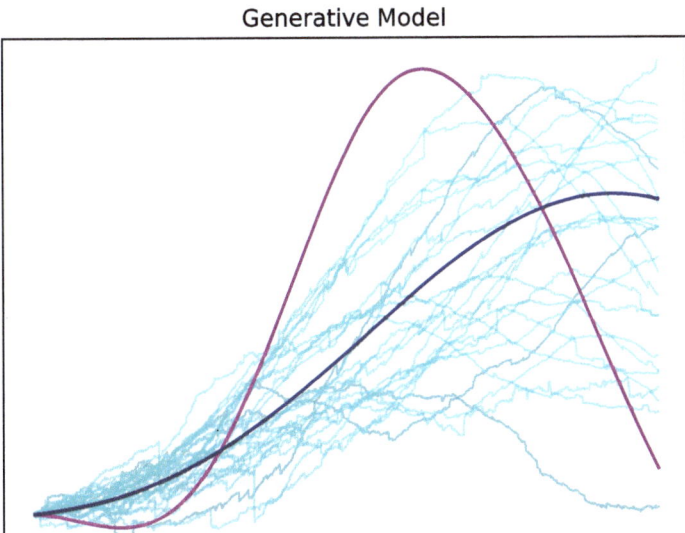

Fig. 8 Ensemble plot for forward runs of the SPDE with generative model noise. We compare the evolution of the fine scale PDE projected onto the coarse grid with the evolution of the PDE run on the coarse grid and an ensemble with generative noise at a central grid location. Horizontal axis is time

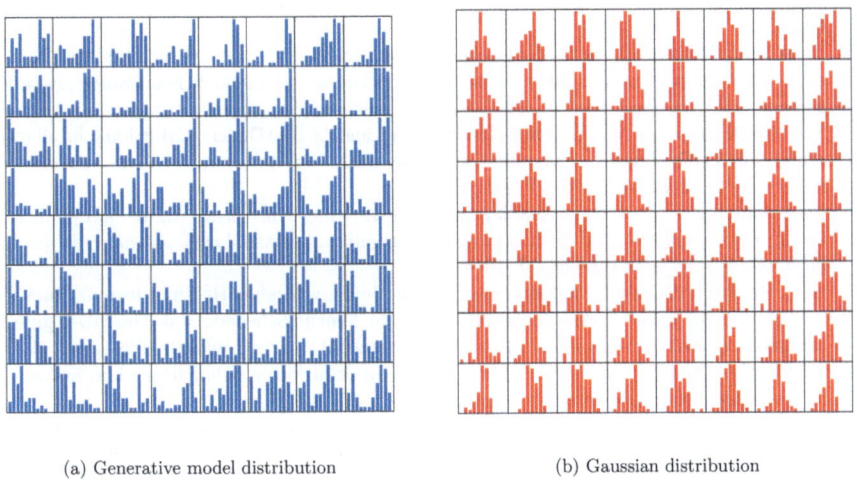

(a) Generative model distribution   (b) Gaussian distribution

Fig. 9 Rank histograms. (**a**) Rank Histograms of the generative noise ensemble with a given forecast horizon. (**b**) Rank Histograms of the Gaussian noise ensemble with a given forecast horizon

Funding
All three authors have been supported by the European Research Council (ERC) under the European Union's Horizon 2020 Research and Innovation Programme (ERC, Grant Agreement No 856408).

Data Availability Statement
Data sharing not applicable to this article as no datasets were generated or analysed during the current study.

Conflict of Interest Statement
On behalf of all authors, the corresponding author states that there is no conflict of interest.

References

1. J. Berner and et al. Stochastic parameterization: Towards a new view of weather and climate models. *Bulletin of the American Meteorological Society*, 98(3):565–588, 2017.
2. Rüdiger Brecht, Long Li, Werner Bauer, and Etienne Mémin. Rotating shallow water flow under location uncertainty with a structure-preserving discretization. *Journal of Advances in Modeling Earth Systems*, 13(12), 2021.
3. Colin Cotter, Dan Crisan, Darryl Holm, Wei Pan, and Igor Shevchenko. Modelling uncertainty using stochastic transport noise in a 2-layer quasi-geostrophic model. *Foundations of Data Science*, 2(2):173, 2020.
4. Colin Cotter, Dan Crisan, Darryl D Holm, Wei Pan, and Igor Shevchenko. Numerically modeling stochastic lie transport in fluid dynamics. *Multiscale Modeling & Simulation*, 17(1):192–232, 2019.
5. Dan Crisan, D Holm, and Peter Korn. An implementation of hasselmann's paradigm for stochastic climate modelling based on stochastic lie transport *. *Nonlinearity*, 36:4862–4903, 08 2023.
6. Dan Crisan, Oana Lang, Alexander Lobbe, Peter Jan van Leeuwen, and Roland Potthast. Noise calibration for spdes: a case study for the rotating shallow water model. *Foundations on Data Science*, 2023.
7. Valentin De Bortoli, James Thornton, Jeremy Heng, and Arnaud Doucet. Diffusion schrödinger bridge with applications to score-based generative modeling. *Advances in Neural Information Processing Systems*, 34:17695–17709, 2021.
8. Ian Goodfellow, Jean Pouget-Abadie, Mehdi Mirza, Bing Xu, David Warde-Farley, Sherjil Ozair, Aaron Courville, and Yoshua Bengio. Generative adversarial nets. *Advances in neural information processing systems*, 27, 2014.
9. Hans Hersbach, Bill Bell, Paul Berrisford, Shoji Hirahara, András Horányi, Joaquín Muñoz-Sabater, Julien Nicolas, Carole Peubey, Raluca Radu, Dinand Schepers, et al. The ERA5 global reanalysis. *Quarterly Journal of the Royal Meteorological Society*, 146(730):1999–2049, 2020.
10. Jonathan Ho, Ajay Jain, and Pieter Abbeel. Denoising diffusion probabilistic models. *Advances in neural information processing systems*, 33:6840–6851, 2020.
11. Darryl D Holm. Variational principles for stochastic fluid dynamics. *Proceedings of the Royal Society A: Mathematical, Physical and Engineering Sciences*, 471(2176):20140963, 2015.
12. Lonnie Magee John B. Burbidge and A. Leslie Robb. Alternative transformations to handle extreme values of the dependent variable. *Journal of the American Statistical Association*, 83(401):123–127, 1988.
13. Diederik P Kingma and Max Welling. Auto-encoding variational bayes. *arXiv preprint arXiv:1312.6114*, 2013.

14. Yann LeCun, Sumit Chopra, Raia Hadsell, M Ranzato, and Fujie Huang. A tutorial on energy-based learning. *Predicting structured data*, 1(0), 2006.
15. Etienne Mémin. Fluid flow dynamics under location uncertainty. *Geophysical & Astrophysical Fluid Dynamics*, 108(2):119–146, 2014.
16. George Papamakarios, Eric Nalisnick, Danilo Jimenez Rezende, Shakir Mohamed, and Balaji Lakshminarayanan. Normalizing flows for probabilistic modeling and inference. *The Journal of Machine Learning Research*, 22(1):2617–2680, 2021.
17. Valentin Resseguier, Long Li, Gabriel Jouan, Pierre Dérian, Etienne Mémin, and Bertrand Chapron. New trends in ensemble forecast strategy: uncertainty quantification for coarse-grid computational fluid dynamics. *Archives of Computational Methods in Engineering*, 28(1):215–261, 2021.
18. Jascha Sohl-Dickstein, Eric Weiss, Niru Maheswaranathan, and Surya Ganguli. Deep unsupervised learning using nonequilibrium thermodynamics. In *International conference on machine learning*, pages 2256–2265. PMLR, 2015.
19. Yang Song and Stefano Ermon. Generative modeling by estimating gradients of the data distribution. *Advances in neural information processing systems*, 32, 2019.
20. Yang Song, Jascha Sohl-Dickstein, Diederik P Kingma, Abhishek Kumar, Stefano Ermon, and Ben Poole. Score-based generative modeling through stochastic differential equations. In *International Conference on Learning Representations*, 2021.
21. Aäron van den Oord, Nal Kalchbrenner, and Koray Kavukcuoglu. Pixel recurrent neural networks. In Maria Florina Balcan and Kilian Q. Weinberger, editors, *Proceedings of The 33rd International Conference on Machine Learning*, volume 48 of *Proceedings of Machine Learning Research*, pages 1747–1756, New York, New York, USA, 20–22 Jun 2016. PMLR.
22. Ling Yang, Zhilong Zhang, Yang Song, Shenda Hong, Runsheng Xu, Yue Zhao, Wentao Zhang, Bin Cui, and Ming-Hsuan Yang. Diffusion models: A comprehensive survey of methods and applications. *ACM Computing Surveys*, 56(4):1–39, 2023.

Collisions of Burgers Bores with Nonlinear Waves

Albert Dombret, Darryl D. Holm, Ruiao Hu, Oliver D. Street, and Hanchun Wang

1 Introduction

Our topic is the nonlinear momentum exchange between surface waves and the currents which carry them. For example, the wind stress creates waves and swells on the sea surface. Those sea surface waves may then exchange momentum with the fluid flow at the surface. We model the last step via a composition-of-maps approach in which waves are taken as a degree of freedom which exchanges momentum and energy with the fluid current that carries them.

Our composition-of-maps approach models nonlinear surface waves as propagating in the reference frame of the fluid velocity at the surface. The total momentum of the system may be written as the sum of wave momentum and fluid momentum in the fixed Eulerian frame. By Newton's 2nd Law, the time rate of change of this total momentum equals the true force acting on the fluid flow in an inertial frame. However, the acceleration of the fluid velocity is only part of the rate of change of this total force. Thus, the acceleration of the fluid velocity appears to acquire an additional fictitious force (such as the Coriolis force, or Craik-Leibovich force) when experienced in the non-inertial reference frame of the fluid velocity.

The non-inertial force in the fluid frame arising from the shift of the fluid momentum by the addition of the wave momentum in the Eulerian frame can be regarded as an additional source of circulation around a Lagrangian loop carried by the fluid velocity. As with the Coriolis force or the Craik-Leibovich vortex

A. Dombret
Physics, École Normale Supérieure, Paris, France
e-mail: albert.dombret@ens.psl.eu

D. D. Holm · R. Hu · O. D. Street · H. Wang (✉)
Mathematics, Imperial College London, London, UK
e-mail: d.holm@ic.ac.uk; ruiao.hu15@imperial.ac.uk; o.street18@imperial.ac.uk; hanchun.wang21@imperial.ac.uk

© The Author(s) 2025
B. Chapron et al. (eds.), *Stochastic Transport in Upper Ocean Dynamics III*,
Mathematics of Planet Earth 13, https://doi.org/10.1007/978-3-031-70660-8_2

force, the difference in fluid velocity circulation dynamics induced by measuring the fluid velocity circulation relative to the wave velocity circulation can be exhibited by calculating the Kelvin theorem for the full system and subtracting out the wave velocity contribution to the circulation integral. To treat this wave-current momentum interaction dynamics, a theory based on the composition of the fluid flow map and the wave dynamics map has been developed recently in [13, 14].

This wave-current momentum interaction dynamics can be illustrated in 1D by setting up an initial condition for a wave-current 'collision' in which, for example, a Burgers ramp/cliff solution (an advancing B-bore velocity profile of the fluid current) overtakes a set of KdV nonlinear dispersive wave packets in its path. The numerical simulations in the present work show for these initial conditions that when the B-bore overtakes the KdV wave packets, the large fluid velocity gradient at the leading edge of the B-bore can rapidly feed the amplitude of the KdV waves so much that—in the frame of motion of the Burgers leading edge—the KdV solution can incorporate part of the Burgers velocity and carry it forward ahead of the new Burgers leading edge as a compound wave.

Physically, though, real bores driven by the tide and advancing up the Severn river for example tend not to show this numerically simulated formation of compound waves. Instead, they tend to show a train of small amplitude surface waves which have been swept up and embedded in the shallow ramp profile behind the advancing front of the bore [5]. This qualitative difference in behaviour is found in numerical simulations in the present work to depend on the balance between nonlinearity and dispersion in the KdV equation. Namely, as the dispersion coefficient is raised at fixed nonlinearity coefficient the behaviour of the KdV waves in the B-KdV system can switch their behaviour from being passively incorporated into the Burgers velocity profile to actively incorporating part of the Burgers momentum and breaking away run ahead of the Burgers front as a compound wave. See Fig. 4 for a comparison of simulation results.

Thus, from the viewpoint of the present work, this different behaviour in our simulations of the B-KdV system arises because of a bifurcation depending on the balance between the KdV nonlinearity and the KdV dispersion and perhaps also with the Burgers nonlinearity. This type of bifurcation study is in progress also for the wave-current interaction of Burgers currents and waves governed by the nonlinear Schrodinger (NLS) equation. However, a discussion of the study for B-NLS collisions will be deferred to future work. For the treatment of wave-current interaction between NLS waves carried by Euler fluid motion in two-dimensions, see [13, 14].

2 Modelling Considerations

Previous work has shown that in models of wave mean flow interaction (WMFI), although the mean flow may not itself create waves, the interaction of the mean flow with existing waves can have strong effects both on the mean flow and on the waves, [14]. In this chapter, we will exhibit a geometric approach to wave-current interaction based upon composition of maps introduced in [14], which we illustrate by considering two examples of a bore—whose dynamics is governed by the Burgers equation—overtaking a set of water waves governed either by the Korteweg-de Vries (KdV) equation in one example, or by the nonlinear Schrödinger (NLS) equation in the other. The dynamics of all three of these types of water waves are well known. However, the application of the method of composition of flow maps to the collision interaction of a Burgers bore overtaking a set of nonlinear shallow water waves seems to be new.

The present work aims to investigate nonlinear wave-current interactions in their simplest forms, in one dimension (1D) on the real line $\mathbb{R}$. Even in 1D these interactions of different types of waves can be profound. In particular, we investigate overtaking interactions of ramps-and-cliffs shaped bore solutions of the inviscid Burgers equation,[1]

$$u_t + 3uu_x = 0,$$

interacting with:

1. Korteweg-de Vries (KdV) soliton solutions governed by

$$v_t + 6vv_x + \gamma v_{xxx} = 0.$$

   As for the Burgers equation, the KdV equation is Galilean invariant in the sense that a given solution $v(x, t)$ remains a solution when 'boosted' into a moving frame by replacing x with $x + ct$ everywhere in $v(x, t)$ so that

$$v(x, t) \mapsto v_{[c]}(x, t) = v(x + vt, t).$$

2. The Nonlinear Schrödinger (NLS)

$$i\hbar\psi_t = -\frac{1}{2}\psi_{xx} + \kappa|\psi|^2\psi,$$

   describes wave packets for a complex variable (wave function) $\psi(x, t)$. It has two types of solution known as focusing ($\kappa < 0$) and de-focusing ($\kappa > 0$). The amplitude of the solution is given by $|\psi|^2 = \psi^*\psi$.

---

[1] The factor of 3 in the PDE form of the inviscid Burgers equation here signals the geometric notation to be used later for Lie transport by vector field $u^\sharp$ acting on a 1-form-density, e.g, $\mathcal{L}_{u^\sharp}(mdx \otimes dx) \simeq (mu_x + (mu)_x)dx^2$ with $m = u$ and $u^\sharp$.

In the KdV equation, though, the factor of 6 in the nonlinearity is traditional, following [4, 10, 22]. The relevance of the ratio of coefficients of the nonlinearity and dispersion in the KdV equation for the qualitative result of collisions of the Burgers bore with KdV waves is demonstrated in Fig. 4 of Sect. 4.

The nonlinear Schrödinger equation is Galilean invariant in the sense that a given solution $\psi(x, t)$ remains a solution when 'boosted' into a moving frame by replacing x with $x + ct$ everywhere in $\psi(x, t)$ while also multiplying by a phase factor of $e^{-iv(x+ct/2)}$ so that

$$\psi(x, t) \mapsto \psi_{[c]}(x, t) = \psi(x + ct, t)\, e^{-ic(x+ct/2)}.$$

Remark 2.1 Although one might expect that the Burgers bores would either 'snowplow' the waves ahead or run them over whenever it encounters them, the coupled nonlinear wave equations will tell a different story. In fact, as we shall discuss, the interactions of Burgers bores with KdV and NLS solutions can be much more profound than a simple 'snowplow' effect.

Approach The equations governing wave-current dynamics in two dimensions have been derived and exemplified in [14]. The equations governing the coupling dynamics exemplified here between the Burgers ramps-and-cliffs[2] and the KdV and NLS solitons will be derived following the work of [14]. Namely, the *Bore-Soliton* equations treated here will be derived by Hamilton's principle in a variational framework which couples the sum of two Lagrangians for the separate bore and soliton degrees of freedom via insertion of a vector field representing the Burgers current velocity into a 1-form density representing the momentum map for each type of soliton. A variant of this general approach was introduced by Dirac and Frenkel [9] in coupling the Schrödinger equation probability current density J with the electromagnetic field vector potential A to study linear nonrelativistic quantum electrodynamics (QED). The quantum-classical $(J \cdot A)$ coupling of Schrödinger wave functions and Maxwell fields is known to produce profound cooperative effects, such as stimulated emission of radiation. The present work investigates what effects may occur when the Burgers current velocity is coupled to the dynamics of two well-known nonlinear wave soliton equations KdV and NLS via their momentum maps arising in their corresponding phase-space variational principles.

2.1 Examples

2.1.1 Burgers-KdV (B-KdV) Dynamics

For 1D wave-current interaction in the B-KdV case, the approach discussed here amounts to the composition of the two smooth invertible maps that govern the dynamics of the two continuum variables comprising the respective solutions for the bore momentum 1-form density $(u\, dx^2 \in \Lambda^1(\mathbb{R}) \otimes \mathrm{Den}(\mathbb{R}))$ and the KdV soliton

---

[2] Although the Euler-Poincaré variational derivation produces the inviscid Burgers equation, both global well-posedness of solutions and stability of the numerical simulations of the Burgers ramp-and-cliff dynamics require viscous regularisation.

density ($v\,dx \in \mathrm{Den}(\mathbb{R})$). The Burgers velocity vector field $u^{\sharp} := u\partial_x \in \mathbb{R}$ is tangent to the right action of smooth invertible maps of the real line $\mathbb{R}$ onto itself by the diffeomorphism Lie group. The same map governs KdV dynamics, except it is augmented by the Gel'fand-Fuchs 2-cocycle, which introduces the third-order dispersion term in KdV and which together with the diffeomorphisms comprises the Bott-Virasoro Lie group. Thus, the interaction of these two types of *coherent structures* for Burgers ramps-and-cliffs and KdV solitons will be governed by the composition of two time-dependent, smooth, Lie-group transformations of the real line $\mathbb{R}$ onto itself, in which one of these maps is extended by the Gel'fand-Fuchs 2-cocycle.

As discussed below in Sect. 4, Hamilton's principle for the sum of Lagrangians for Burgers and KdV equations coupled via the product of the fluid velocity and the wave momentum map yields the Burgers-KdV system of partial differential equations,[3]

$$\partial_t u + 3uu_x = -v\,\partial_x\!\left(\gamma v_{xx} + 3v^2\right),$$
$$\partial_t v + \partial_x(uv) = -\,\partial_x\!\left(\gamma v_{xx} + 3v^2\right). \tag{2.1}$$

The right-hand sides of these equations arise represent the coupling between the two individual equations. The KdV dynamics for its potential velocity $v\,dx = \phi_x\,dx = d\phi$ can be regarded as being 'swept' or 'advected' as a density by the velocity vector field, $u^{\sharp}$, of the Burgers fluid current as,

$$(\partial_t + \mathcal{L}_{u^{\sharp}})u\,dx^2 = -d\!\left(\gamma v_{xx} + 3v^2\right) \otimes v\,dx\,,$$
$$(\partial_t + \mathcal{L}_{u^{\sharp}})v\,dx = -d\!\left(\gamma v_{xx} + 3v^2\right). \tag{2.2}$$

However, this 'advection' is not passive. Numerical simulations presented in Figs. 1 and 2 demonstrate that the solutions of the B-KdV dynamics in Eq. (2.1), and their geometric equivalents in Eq. (2.2), have a significant impact on the Burgers velocity vector field $u^{\sharp}$, which 'sweeps' these solutions.

The figure below shows the typical Burgers-KdV overtaking collision.

2.1.2 Burgers-NLS

The Burgers Ramp/Cliff solution and the NLS solitons interact quite differently from the interactions of Burgers Ramp/Cliff solutions and the KdV solitons. The Lie-Poisson form and the canonical Hamilton's canonical equations represented by

---

[3] The Burgers-KdV *system* in (2.1) is not in the same category as the Burgers-KdV *equation* in [23].

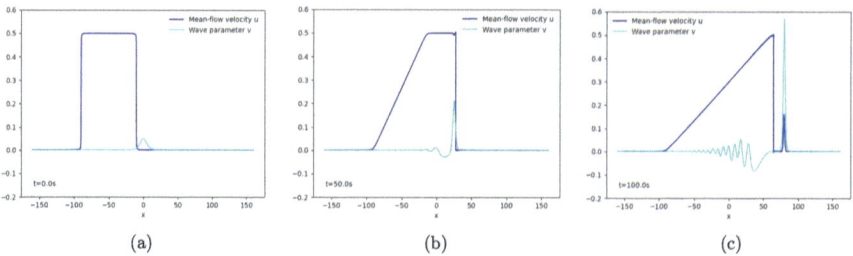

(a)   (b)   (c)

Fig. 1 The plots show the evolution of the mean-flow Burgers rightward velocity u and the wave parameter v in the coupled Burgers-KdV system of equations in (2.1) from for a small viscosity of $v = 0.01$ in the Burgers equation to stabilise the numerical simulation. At time $t = 0$, the Burgers bore overtakes the KdV soliton. At time $t = 50s$ the Burgers bore has started transferring momentum to the KdV wave and a wave moving rightward in the Burgers frame is developing. At time $t = 100s$ one sees the compound Burgers-KdV wave advancing ahead of the Burgers bore and leaving behind a leftward moving KdV wave train as viewed from the leading edge of the instantaneous Burgers motion. (**a**) t = 0s, (**b**) t = 50s, (**c**) t = 100s,

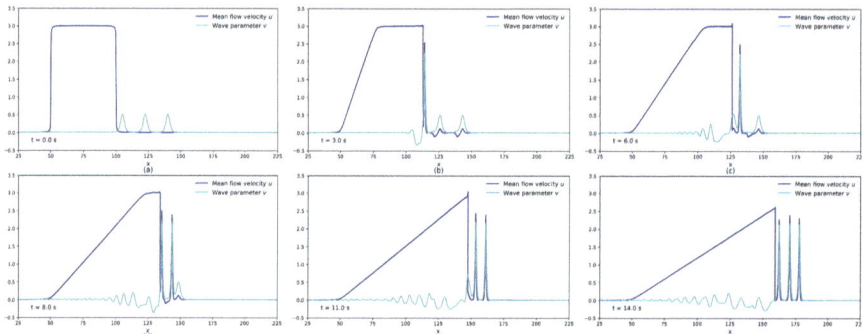

Fig. 2 (**a**) At time $t = 0$, the Burgers bore overtakes the 1st of three identical rightward moving KdV waves in the Burgers frame. (**b**) At time $t = 3s$, the Burgers bore has started transferring momentum to the first KdV wave, a KdV wave moving leftward in the frame of the bore is developing and the leading 2nd and 3rd KdV waves are beginning to create small Burgers waves. (**c**) At time 6 s, the 1st KdV wave has transferred most of its momentum to the 2nd (middle) KdV wave and a KdV wave train is moving leftward. (**d**) At time 8 s, momentum transfer from the bore has restored the amplitude of the 1st KdV wave and the 2nd KdV wave is overtaking the 3rd (rightmost) one. (**e**) At time t=11 s, the 2nd KdV wave has transferred its momentum to the rightmost 3rd wave and both of them have entrained part of the bore in becoming compound travelling waves. (**f**) At t=14s, all three KdV waves have become rightward moving compound Burgers-KdV travelling waves. The middle wave will eventually overtake and transfer momentum to the leading wave, so that the heights of the compound waves will be ordered in velocity

the polar decomposition $\psi = \sqrt{N}\exp{i\phi}$ for the Burgers-NLS interaction are given by

$$(\partial_t + \mathcal{L}_{u^\sharp})(u - N\partial_x\phi)(dx \otimes dx) = 0,$$

$$(\partial_t + \mathcal{L}_{u^\sharp + \phi_x^\sharp})(N\,dx),$$

$$\partial_t\phi + u\phi_x = -\frac{1}{2}\phi_x^2 - \frac{(\sqrt{N})_{xx}}{2\sqrt{N}} + F'(N). \tag{2.3}$$

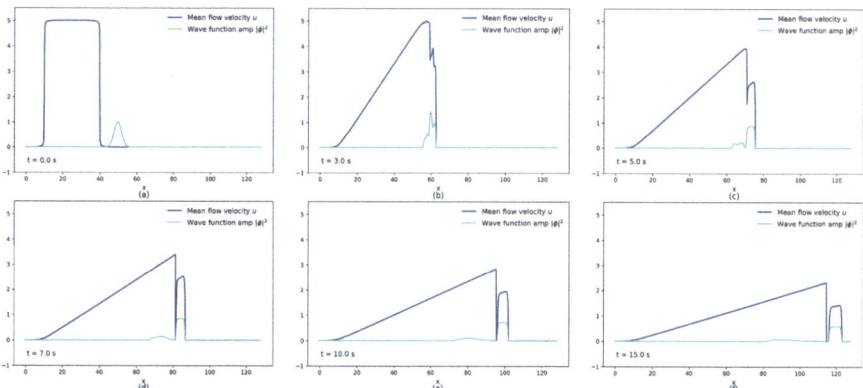

Fig. 3 This plot illustrates the evolution of the mean-flow velocity u and the wave function amplitude $|\phi|^2$ in the coupled Burgers-Nonlinear Schödinger (Burgers-NLS) system of equations treated in section 5. The Burgers equation ramp/cliff solution is regularised by viscosity of $\nu_1 = 0.1$ in this case and dissipation $\nu_2 = 0.1$ has been added to the NLS equation to reduce the frequency of its phase oscillations. At time $t = 0$, figure (**a**) shows that the Burgers bore starts to overtake the NLS Gaussian wave packet. Between $t = 3$ and $t = 10$, figure (**b**)-(**e**) shows the bore is clearly transferring momentum to the NLS wave package. At $t = 15$, figure (**f**) shows that a compound Burgers-NLS wave advances ahead of the ramp/cliff formation of the bore

The Fig. 3 shows an solution of the Burgers-NLS interaction Eq. (2.3) which is derived and discussed in Sect. 5.

Plan of the Chapter

- Section 3 provides the background materials for shallow water waves and in one spatial dimensions. The examples of inviscid Burgers', Korteweg–De Vries (KdV) and Camassa Holm (CH) equations are discussed.
- Section 4 discusses the Burgers-Korteweg-de Vries (B-KdV) results. In particular, Fig. 4 demonstrates the sensitivity of these results to the balance between nonlinearity and dispersion in the KdV nonlinear wave subsystem of the B-KdV collision. Additionally, we briefly consider the introduction of stochasticity into the B-KdV equations via the approach of Stochastic Advection by Lie Transport (SALT).
- Section 5 derives the one dimensional coupling of Burgers dynamics to the NLS equations. Note that this is a 1D analogue of the 2D coupling of Euler to NLS considered in [14].
- Section 6 provides a brief summary of the results in this chapter and an outlook for further developments.

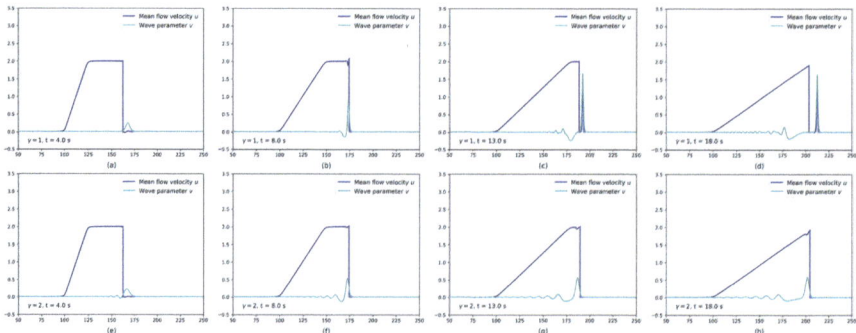

Fig. 4 These figures represents the wave-current interaction with different dispersion strengths $\gamma = 1$, and $\gamma = 2$ at a fixed value of 6 for the KdV nonlinearity coefficient. Figure (**a**)–(**d**) shows that for $\gamma = 1$ a compound wave is excited and propagates ahead away from the bore. In contrast, figure (**e**)–(**h**) shows that for $\gamma = 2$, the KdV wave moves along with the bore front

3 EPDiff and Shallow Water Waves

3.1 Introduction to Wave Equations

Wave equations are evolutionary equations for time dependent curves in a space of smooth maps $C^\infty(\mathbb{R}^n, V)$ for solutions, $u \in V$, taking values in a vector space V.

$$\partial_t u = f(u), \quad \text{or} \quad \partial_t u_i(\mathbf{x}, t) = f_i(u_i, \, u_{i,j}, \, u_{i,jk}, \, u_{i,jkl}, \, \dots). \tag{3.1}$$

Typically, V is $\mathbb{R}$ or $\mathbb{C}$, $n = 1$. We are interested in the Cauchy problem. Namely, solve (3.1) for $u(x, t)$, given the initial condition $u(x, 0)$ and boundary conditions $u(x|_{\partial D}, t)$.

Travelling Waves The simplest wave solution is called a travelling wave. This solution is a function u of the form

$$u(x, t) = F(x - ct),$$

where $F : \mathbb{R} \to V$ is a function defining the wave shape, and c is a real number defining the propagation speed of the wave. Thus, travelling waves preserve their shape and simply translate to the right at a constant speed, c.

Plane Waves A complex-valued travelling wave, called a plane wave, plays a fundamental role in the theory of linear wave equations. The general form of a plane wave velocity is

$$u(x, t) = \Re e(Ae^{i(kx - \omega t)}),$$

where $|A|$ is the wave amplitude, k is the wave number, ω is the wave frequency, and $c_p = \omega/k$ is the speed along the oscillating waveform.

3.2 Conservation Laws

Conservation laws for evolutionary equations of the form $u_t = f(u)$ satisfy

$$\frac{d}{dt} \int F(u)\, dx = \int \frac{\delta F}{\delta u} u_t\, dx = \int \frac{\delta F}{\delta u} f(u)\, dx = \int dG(u) = 0\,,$$

for some functions F and G of u and its derivatives, and for suitable boundary conditions.

For example, the inviscid Burgers equation

$$u_t + u u_x = 0\,, \tag{3.2}$$

has an infinite number of conservation laws, given by $C_n = \int \frac{u^n}{n}\, dx$

$$
\begin{aligned}
\frac{dC_n}{dt} &= \frac{d}{dt} \int \frac{u^n}{n}\, dx = \int u^{n-1} u_t\, dx = - \int u^n u_x\, dx \\
&= - \int \frac{1}{n+1} \partial_x u^{n+1}\, dx = - \frac{1}{n+1} \int d(u^{n+1}) = 0\,,
\end{aligned}
\tag{3.3}
$$

for homogeneous boundary conditions and any integer n.

Even so, the solutions of the inviscid Burgers equation carry the seeds of their own destruction, since they exhibit wave breaking in finite time. That is, without dissipation or dispersion their velocity profile would develop a negative vertical slope in finite time. This is shown in the proof of the following Lemma.

Lemma 3.1 (Steepening Lemma for the Inviscid Burgers Equation) *Suppose the initial profile of velocity $u(0, x)$ for the inviscid Burgers equation (3.2) has an inflection point of negative slope $u_x(0, \overline{x}(0)) < 0$ located at $x = \overline{x}(0)$ to the right of its maximum, and otherwise it decays to zero in each direction sufficiently rapidly for all of its conservation laws in Eq. (3.3) to be finite. Then the negative slope at the inflection point will become vertical in finite time.*

Proof Consider the evolution of the slope at the inflection point, defined by $s(t) = u_x(\overline{x}(t), t)$. Then the inviscid Burgers equation (3.2) yields an evolution equation for the slope, $s(t)$. Namely, using $u_{xx}(\overline{x}(t), t) = 0$ the spatial derivative of Eq. (3.2) leads to

$$\frac{ds}{dt} = -s^2 \quad \Longrightarrow \quad s(t) = \frac{s(0)}{1 + s(0)t}\,. \tag{3.4}$$

Thus, if $s(0) < 0$, the slope at the inflection point $s(t)$ will become increasingly more negative, until it becomes vertical at time $t = -1/s(0)$. $\qquad\square$

3.3   Survey of Weakly Nonlinear Water Wave Equations: KdV and CH

The derivation of weakly nonlinear water wave equations starts with Laplace's equation for the velocity potential of an inviscid, incompressible, and irrotational fluid moving in a vertical plane under gravity with an upper free surface, as, e.g., in [24].

The equations are then expanded in the small parameters $\epsilon_1 = a/h$ and $\epsilon_2 = h^2/l^2$. Here $\epsilon_1 \geq \epsilon_2 > \epsilon_1^2$ and a, h, and l denote the wave amplitude, the mean water depth, and a typical horizontal length scale (e.g., a wavelength), respectively. Length is measured in terms of l, height in h and time in l/c_0. The elevation η is scaled with a and fluid velocity u is scaled with $c_0 a/h$. Here, $c_0 = \sqrt{gh}$ is the linear wave speed for undisturbed water at rest at spatial infinity, where u and its derivatives u_x and u_{xx} are taken to vanish.

The result of the expansion to quadratic order in ϵ_1 and ϵ_2 is the equation for the surface elevation η (see e.g. [24], p. 466), while higher order terms (HOT) can be found in e.g. [25],

$$0 = \eta_t + \eta_x + \frac{3}{2}\epsilon_1 \eta \, \eta_x + \frac{1}{6}\epsilon_2 \, \eta_{xxx} - \frac{3}{8}\epsilon_1^2 \, \eta^2 \, \eta_x$$

$$+ \epsilon_1\epsilon_2 \left(\frac{23}{24}\eta_x \, \eta_{xx} + \frac{5}{12}\eta \, \eta_{xxx} \right) + \epsilon_2^2 \frac{19}{360}\eta_{xxxxx} + \text{HOT} \qquad (3.5)$$

where partial derivatives are denoted by subscripts.

Next, following Kodama [17, 18] one applies the near-identity transformation,

$$\eta = u + \epsilon_1 \, f(u) + \epsilon_2 \, g(u) \,,$$

to the η-Eq. (3.5) and seeks functionals $f(u)$ and $g(u)$ that consolidate the terms of order $O(\epsilon_1^2)$ and $O(\epsilon_1\epsilon_2)$ in (3.5) into one order $O(\epsilon_2^2)$ term under normal form transformations. This procedure produces the following 1+1 quadratically nonlinear Camassa-Holm (CH) equation for unidirectional water waves with fluid velocity, $u(x, t)$ and momentum $m = u - \alpha^2 u_{xx}$, with constant $\alpha^2 = (19/60)\epsilon_2$, see [8],

$$m_t + c_0 u_x + \frac{\epsilon_1}{2}(um_x + 2mu_x) + \epsilon_2 \frac{3}{20}u_{xxx} = 0 \,. \qquad (3.6)$$

After these normal form transformations, Eq. (3.6) is equivalent to the shallow water wave Eq. (3.5) up to, and including, terms of order $\mathcal{O}(\epsilon_2^2)$. For α^2 positive, Eq. (3.6) becomes the Camassa-Holm equation derived and shown to be completely integrable in [3]. Hereafter, we will take $\alpha^2 = 0$ and leave the Burgers-CH interaction for later work.

For $\alpha^2 = 0$, Eq. (3.6) remains completely integrable, as it restricts to the Korteweg-de Vries (KdV) equation,

$$u_t + c_0 u_x + 3u\, u_x + \gamma\, u_{xxx} = 0\,, \tag{3.7}$$

which admits the soliton solution $u(x, t) = u_0 \operatorname{sech}^2((x-ct)\sqrt{u_0/\gamma}/2), c = c_0 + u_0$
see, e.g., [1].

Remark 3.1 (Interacting Solutions at Different Orders) The higher order terms in the asymptotic expansion of the nonlinear shallow water wave equations in Eq. (3.5) represent degrees of freedom that are not accessed by the lower order terms.[4]

This observation in combination with the Galilean invariance of the Burgers equation and the KdV equation at the next order of the expansion then raises the following question: What happens when a KdV solution is boosted into the time-dependent frame of motion of a Burgers solution? This is the question we address in the present work.

4   Burgers–Korteweg-de Vries (KdV) Collisions

In 1D, we consider Hamilton's principle for the formulation of the Burgers-KdV equations where the Lagrangian is given by the sum of the kinetic energy of the Burgers' solution and the Whitham Lagrangian for the KdV solution as follows,

$$0 = \delta S = \delta \int_0^T \ell(u, \phi)\, dt\,,$$

$$\ell(u, \phi) := \int_{\mathbb{R}} \frac{1}{2}|u|^2 + \frac{1}{2}\phi_x\,(\phi_t + u\phi_x) + \left(\phi_x^3 - \frac{\gamma}{2}\phi_{xx}^2\right)\, dx\,. \tag{4.1}$$

In Hamilton's principle (4.1), the variation in u is constrained to have the form $\delta u = \partial_t \xi - \mathrm{ad}_u\, \xi$ obtained from the Euler-Poincaré theory [16]. The arbitrary variation ξ and the variation $\delta\phi$ are assumed to be arbitrary and vanishing at endpoints $t = 0$ and $t = T$. As we will see, invoking Hamilton's principle with this Lagrangian yields KdV dynamics in the frame of motion of the Burgers equation, and the KdV dynamics acts back directly on the Burgers dynamics.

Computing the variations in u and ϕ in (4.1) yields

$$0 = \delta S = \int_0^T \left\langle u + \frac{\phi_x^2}{2}\,,\, \delta u\right\rangle - \left\langle \phi_{tx} + (u\phi_x)_x + \gamma\phi_{xxxx} + 3\phi_x\phi_{xx}\,,\, \delta\phi\right\rangle\, dt$$

---

[4] This remark also applies to the asymptotic expansion of the corresponding Lagrangian in Hamilton's principle for the underlying fluid theory. See Gjaja and Holm [11] for discussion of the further benefits of applying asymptotic expansions of Hamilton's principle in hierarchies of fluid dynamical approximations.

$$= \int_0^T \left\langle u + \frac{v^2}{2} , \, \partial_t \xi - \mathrm{ad}_u \, \xi \right\rangle - \langle \partial_t v + (uv)_x + \gamma v_{xxx} + 3vv_x , \, \delta \phi \rangle \, dt$$

$$= \int_0^T \left\langle \left(\partial_t + \mathrm{ad}_u^*\right) \left(u + \frac{v^2}{2}\right) , \, \xi \right\rangle$$

$$- \langle \partial_t v + (uv)_x \quad + \gamma v_{xxx} + 3vv_x , \, \delta \phi \rangle \, dt , \tag{4.2}$$

where in the second line we have inserted the constrained variations of u and introduced the one-form $v \, dx = \phi_x \, dx$. In Eq. (4.2), the angle bracket operation $\langle m , \xi \rangle : (\Lambda^1(\mathbb{R}) \otimes \mathrm{Den}(\mathbb{R})) \times \mathfrak{X}(\mathbb{R}) \to \mathbb{R}$ denotes the L^2 dual pairing

$$\langle m , \xi \rangle := \int_{\mathbb{R}} \xi m \, dx := \int_{\mathbb{R}} (\xi \partial_x) \lrcorner \, (m dx) dx , \tag{4.3}$$

where $(\xi \partial_x) \lrcorner \, (m dx) dx$ denotes insertion of a vector field $\xi \partial_x$ into a differential 1-form density $m \, dx \otimes dx$, which one may be abbreviated as $m \, dx^2$ without confusion.[5] The arbitrary variation in ϕ yields the equation

$$\partial_t v + \partial_x (uv + \gamma v_{xx} + 3v^2) = 0 , \tag{4.4}$$

where we see that v, the solution that corresponds to the KdV part of the flow, is swept along by the HB u-solution. That is, the one-form $v \, dx$ is Lie transported by the vector field $u \partial_x$. In terms of the Lie derivative $\mathcal{L}_{u^\sharp}$, one can write the v-equation (4.4) in coordinates on the real line as,

$$\left(\partial_t + \mathcal{L}_{u^\sharp}\right)(v \, dx) = \left(\partial_t v + \partial_x (uv)\right) dx = -d(\gamma v_{xx} + 3v^2) , \tag{4.5}$$

where the exterior derivative d represents the spatial differential.

The arbitrary variations in ξ yields dynamics for the total momentum 1-form density m, defined by

$$m := \delta \ell / \delta u = u + \frac{1}{2} v^2 . \tag{4.6}$$

Noting that $\mathrm{ad}_{u^\sharp}^* m = \mathcal{L}_{u^\sharp} m$ when m is a one-form density, the dynamics of m can be written as

$$\left(\partial_t + \mathcal{L}_{u^\sharp}\right)(m \, dx^2) = \left(m_t + (\partial_x m + m \partial_x) u\right) dx^2 = 0 , \quad \text{with} \quad m := u + \tfrac{1}{2} v^2 , \tag{4.7}$$

where we have used the coordinate expression of $\mathcal{L}_{u^\sharp}$ on one-form densities. Thus, we may collect the Lie derivative forms of the Burgers-KdV equations as

---

[5] For a review of differential form notation and usage in fluid dynamics see [12].

$$\left(\partial_t + \mathcal{L}_{u^\sharp}\right)(m\,dx^2) = 0 \quad \text{and} \quad \left(\partial_t + \mathcal{L}_{u^\sharp}\right)(v\,dx) = -d\left(\gamma v_{xx} + 3v^2\right). \quad (4.8)$$

A short calculation to eliminate $m = u + \frac{1}{2}v^2$ in favour of u in (4.7) using the v-equation in (4.4) finally shows that the results of Hamilton's principle in Eq. (4.2) yields the system in Eq. (2.1).

Hence, the velocity equation and the momentum density equation in (4.8) together imply via the product rule for the Lie derivative that

$$\left(\partial_t + \mathcal{L}_{u^\sharp}\right)(u\,dx^2) = -\frac{1}{2}\left(\partial_t + \mathcal{L}_{u^\sharp}\right)(v^2\,dx^2)$$

$$= -(v\,dx) \otimes \left(\partial_t + \mathcal{L}_{u^\sharp}\right)(v\,dx),$$

so that $\quad (u_t + 3uu_x)dx^2 = (v\,dx) \otimes d(v_{xx} + 3v^2) = \left(v\,\partial_x(\gamma v_{xx} + 3v^2)\right)dx^2.$

$$(4.9)$$

Thus, Eq. (4.8) provide the geometric forms of the (u, v) mutual interaction wave-current system in (4.8).

$$\partial_t u + 3uu_x = -v\,\partial_x\left(\gamma v_{xx} + 3v^2\right),$$
$$\partial_t v + \partial_x(uv) = -\partial_x\left(\gamma v_{xx} + 3v^2\right). \quad (4.10)$$

Remark 4.1 Homogeneous boundary conditions have been enforced for all spatial integrations by parts in the previous proof of the Euler-Poincaré equations arising from the Lagrangian in Eq. (4.1). The definition $v\,dx = d\phi$ implies that the quantity $\bar{v} = \int_{\mathbb{R}} v(x, t)\,dx$ is constant in time for vanishing boundary conditions as $|x| \to \infty$.

Remark 4.2 (Hamiltonian Formulation for the Burgers-KdV Interaction) Upon writing the KdV velocity as $v := \phi_x$ and using (4.6) to write the Burgers kinetic energy in terms of m and v, the natural Hamiltonian for the combined system of KdV 'waves' interacting dynamically with the Burgers 'current' may be taken as

$$h(m, v) = \int_{\mathbb{R}} \frac{1}{2}\left(m - \frac{1}{2}v^2\right)^2 + \frac{\gamma}{2}v_x^2 - v^3\,dx. \quad (4.11)$$

The corresponding variational derivatives are given by

$$\delta h(m, v) = \int_{\mathbb{R}} u\,\delta m - \left(uv + 3v^2 + \gamma v_{xx}\right)\delta v\,dx.$$

Consequently, one may express Hamilton's equations for the Burgers-KdV equations as

$$\partial_t m = -\left(\partial_x m + m\partial_x\right)\frac{\delta h}{\delta m} = -\left(\partial_x m + m\partial_x\right)u \,,$$

$$\partial_t v = \partial_x \frac{\delta h}{\delta v} = -\partial_x (uv + \gamma v_{xx} + 3v^2) \,. \tag{4.12}$$

The Hamiltonian equations for the Burgers-KdV dynamics in (4.12) may also be written in diagonal matrix Poisson operator form, as

$$\partial_t \begin{bmatrix} m \\ v \end{bmatrix} = -\begin{bmatrix} \partial_x m + m\partial_x & 0 \\ 0 & \partial_x \end{bmatrix} \begin{bmatrix} \frac{\delta h}{\delta m} = u \\ \frac{\delta h}{\delta v} = (uv + \gamma v_{xx} + 3v^2) \end{bmatrix} \,. \tag{4.13}$$

Here we see that the Hamiltonian structure of the m-equation in (4.13) is Lie-Poisson, as expected from the Euler-Poincaré reduction. We also see the *second* Hamiltonian structure for the KdV equation, whose Poisson operator is simply the spatial partial derive ∂_x. This Hamiltonian structure does not involve introducing the Bott-Virasoro Lie algebra, as needed for the other Hamiltonian structure with the Hamiltonian $\frac{1}{2}\int_{\mathbb{R}} v^2 dx$ in order to capture the dispersive term γv_{xxx} in the KdV equation [19].

Remark 4.3 (Casimirs for the Diagonal (Untangled) Poisson Operator in (4.12)) Casimirs are functionals that Poisson commute with any other functional of the Hamiltonian variables, in this case (m, v). In general, the variational derivative of a Casimir functional is a null eigenvector of the Poisson operator. In the present case, the Casimirs are

$$C_m = \int_{\mathbb{R}} \sqrt{m}\, dx \quad \text{and} \quad C_v = \int_{\mathbb{R}} v\, dx \,.$$

In addition, for the present case, the Poisson brackets among the moments $f_m(v) = \int_{\mathbb{R}} v^m\, dx$ commute among themselves,

$$\{f_m, f_n\} = 0 \,.$$

Remark 4.4 (The Tangled Poisson Operator (4.12)) Transforming variables in the Hamiltonian in (4.11) from $h(m, v)$ to $h(u, v)$, by substituting

$$u = m - \tfrac{1}{2}v^2 \,, \tag{4.14}$$

leads to the equivalent Hamiltonian,

$$h(u, v) = \int \frac{1}{2}u^2 - \frac{\gamma}{2}v_x^2 + v^3\, dx \,. \tag{4.15}$$

The corresponding equivalent equations in Hamiltonian form in the transformed variables (u, v) may be expressed in terms of the semidirect product Lie-Poisson equations with a generalised 2-cocycle as

$$
\partial_t \begin{bmatrix} u \\ v \end{bmatrix} = - \begin{bmatrix} u\partial_x + \partial_x u & v\partial_x \\ \partial_x v & \partial_x \end{bmatrix} \begin{bmatrix} \frac{\delta h}{\delta u} = u \\ \frac{\delta h}{\delta v} = \gamma v_{xx} + 3v^2 \end{bmatrix}
$$

$$
= - \begin{bmatrix} 3\,uu_x + v\partial_x(\gamma v_{xx} + 3v^2) \\ \partial_x(uv) + \partial_x(\gamma v_{xx} + 3v^2) \end{bmatrix}. \tag{4.16}
$$

Evolution under the Lie-Poisson bracket corresponding to the Poisson operator in Eq. (4.16) is given by

$$
\frac{df}{dt} = \{f, h\}(u, v) = - \int_{\mathbb{R}} \begin{bmatrix} \delta f/\delta u \\ \delta f/\delta v \end{bmatrix}^T \begin{bmatrix} u\partial_x + \partial_x u & v\partial_x \\ \partial_x v & \partial_x \end{bmatrix} \begin{bmatrix} \delta h/\delta u \\ \delta h/\delta v \end{bmatrix} dx. \tag{4.17}
$$

The Poisson bracket in Eq. (4.17) is the sum of a Lie-Poisson bracket dual to the semidirect-product Lie algebra $\mathfrak{X}(\mathbb{R})ⓈDen(\mathbb{R})$ of vector fields $\mathfrak{X}(\mathbb{R})$ acting on densities $Den(\mathbb{R})$ on the real line $\mathbb{R}$ with dual coordinates $u \in \mathfrak{X}^*(\mathbb{R})$ and $v \in Den^*(\mathbb{R})$ plus constant antisymmetric bracket with ∂_x in the $\{v, v\}$ position, inherited as a central extension from the bi-Hamiltonian structure of the KdV equation. By skew symmetry of the Poisson operator in this bracket operation under the L^2 pairing, the equations in (4.16) preserve the Hamiltonian $h(u, v)$ in Eq. (4.15) above, since of course $\{h,h\}$ vanishes identically.

Numerical Method In the numerical study, we use the pseudo-spectrum method supported by Dedalus Project [2]. The computational domain is discretized into 32,768 points over a length of 256 units, utilizing a 3/2 dealiasing factor to ensure accuracy in the Fourier spectral representation. We use a semi-implicit backward differentiation formula (SBDF4) scheme with a fixed timestep of 10^{-6}.

4.1 Introducing Stochasticity into Burgers-KdV via the SALT Approach

Ideal fluid dynamics in the Eulerian representation admits a Lie symmetry reduced Euler-Poincaré formulation [16]. In turn, the Euler-Poincaré formulation defines advective transport in ideal Eulerian fluid dynamics in terms of the Lie derivative operation. The Stochastic Advection by Lie Transport (SALT) approach introduces stochasticity into the Euler-Poincaré formulation of ideal fluid dynamics as a semimartingale for the transport velocity,

$$
dx_t = u(x, t)dt + \sum_{i=1}^{N} \xi_i(x) \circ dW_t^i, \tag{4.18}
$$

thereby preserving the geometric form of the deterministic equations. In fact, the drift velocity $u(x,t)$ is the deterministic fluid transport velocity and the ξ's are determined from principle component analysis of observed or simulated data, as discussed in [6, 7].

Since we have formulated the Burgers-KdV equations (4.8) in terms of deterministic Lie transport, we may reformulate them in SALT form as

$$\left(d + \mathcal{L}_{dx_t}\right)(m\, dx^2) = 0 \quad \text{and} \quad \left(d + \mathcal{L}_{dx_t}\right)(v\, dx) = -d(\gamma v_{xx} + 3v^2)dt,$$

$$\text{with} \quad m := u + \tfrac{1}{2}v^2. \tag{4.19}$$

An analysis of these SALT HB-KdV equations is deferred and will be discussed elsewhere.

5 Burgers: Nonlinear Schrödinger (B-NLS) Collisions

Consider the following application of Hamilton's principle for the Burgers-NLS dynamics, with $F(N) = \kappa N^2$,

$$0 = \delta S = \delta \int_0^T \int_{\mathbb{R}} \frac{u^2}{2} - N\left(\phi_t + u\phi_x\right) - \frac{N}{2}\phi_x^2 - \frac{1}{2}(\partial_x \sqrt{N})^2 + F(N)\, dx dt,$$

$$= \int_0^T \int_{\mathbb{R}} (u - N\phi_x)\delta u + \left(-(\phi_t + u\phi_x) - \frac{1}{2}\phi_x^2 + \frac{(\sqrt{N})_{xx}}{2\sqrt{N}} + F'(N)\right)\delta N$$

$$+ \left(N_t + \partial_x\left(N(u + \phi_x)\right)\right)\delta\phi\, dx dt. \tag{5.1}$$

To check these equations, consider the NLS Hamiltonian in the fixed *laboratory* frame:

$$H_{Lab}(N, \phi) = \int \frac{N}{2}\phi_x^2 + \frac{1}{2}(\partial_x \sqrt{N})^2 - F(N))\, dx, \tag{5.2}$$

with variations

$$\delta H_{Lab} = \int \left(\frac{1}{2}\phi_x^2 + \frac{(\sqrt{N})_{xx}}{2\sqrt{N}} - F'(N)\right)\delta N + (-\partial_x(N\partial_x\phi))\delta\phi\, dx. \tag{5.3}$$

In the laboratory frame one then has Hamilton's equations

$$\partial_t \phi = -\frac{\delta H_{Lab}}{\delta N} = -\frac{1}{2}\phi_x^2 - \frac{(\sqrt{N})_{xx}}{2\sqrt{N}} + F'(N),$$

$$\partial_t N = \frac{\delta H_{Lab}}{\delta \phi} = -\partial_x(N\phi_x). \tag{5.4}$$

In the Burgers frame of motion, we have $\delta u = \partial_t \xi - \mathrm{ad}_u \xi$ and the Hamiltonian $H(N, \phi)$ is boosted into the Burgers frame by adding the momentum map coupling term

$$H(N, \phi) = H_{Lab}(N, \phi) + \int u N \phi_x \, dx \,.$$

The Lie-Poisson form and the canonical Hamilton's canonical equations in the Burgers frame then become

$$(\partial_t + \mathcal{L}_{u^\sharp})\Big((u - N\partial_x\phi)(dx \otimes dx)\Big) = 0 \,,$$

$$(\partial_t + \mathcal{L}_{(u+\phi_x)^\sharp})(N \, dx) = 0 \,, \tag{5.5}$$

$$(\partial_t + \mathcal{L}_{u^\sharp})\phi = -\frac{1}{2}\phi_x^2 - \frac{(\sqrt{N})_{xx}}{2\sqrt{N}} + F'(N) \,.$$

These boosted canonical equations in geometric form reveal the transport operations in the B-NLS equations and they agree with the results of Hamilton's principle in Eq. (5.1) when their coefficients are collected. In particular, though, they reveal where stochastic transport may be properly added in the B-NLS system. Namely, the addition of stochastic transport to the vector fields $u^\sharp$ or $\phi_x^\sharp$ may be added separately or together, provided the stochastic transport vector fields are uncorrelated.

6 Conclusion and Outlook

This chapter has derived and simulated 1D self-consistent models of wave-current interaction equations modelled by Burgers motion transporting KdV nonlinear wave evolution special initial conditions modelling the overtaking collisions of Burgers bores with KdV and NLS waves. In each case, we have stressed the generality of the derivations of the wave-current interaction equations via the composition-of-maps variational approach by writing the wave-current collision equations in coordinate-free differential form to reveal their geometric structure. We have also simulated the B-KdV and B-NLS equations computationally in order to illustrate their fascinating solution behaviour.

B-NLS wave-current collisions can be generalised to higher dimensions. In fact, the composition-of-maps approach for the coupling of ideal Euler fluid dynamics to NLS waves via composition of maps in 2D has already been derived, investigated, simulated and discussed in detail in [14]. However, the complexity of the 2D interactions of fluid flow with NLS waves seen in [14] warrants 1D investigation to better illustrate the rich solution behaviour in a simpler context.

The coupled wave-current models studied here were also made stochastic using the SALT approach which preserves the variational derivations. The effects of stochasticity in other Burgers—nonlinear wave interactions also deserve further investigation.

Acknowledgments We are grateful to our friends, colleagues and collaborators for their advice and encouragement in the matters treated in this chapter. We especially thank C. Cotter, D. Crisan, E. Luesink, for many insightful discussions of corresponding results similar to the ones derived here for other wave-current interactions. DH and OS are also grateful for partial support during the present work by European Research Council (ERC) Synergy grant DLV-856408 (Stochastic Transport in Upper Ocean Dynamics—STUOD). The work of RH was supported by Office of Naval Research (ONR) grant N00014-22-1-2082 (Stochastic Parameterisation of Ocean Turbulence—SPOT).

References

1. M.J. Ablowitz and H. Segur, *Solitons and the Inverse Scattering Transform*, SIAM: Philadelphia (1981).
2. Burns, Keaton J. and Vasil, Geoffrey M. and Oishi, Jeffrey S. and Lecoanet, Daniel and Brown, Benjamin P. Dedalus: A flexible framework for numerical simulations with spectral methods *Phys. Rev. Research 2, 023068*, https://doi.org/10.1103/PhysRevResearch.2.023068
3. R. Camassa and D.D. Holm, Phys. Rev. Lett. **71**, 1661 (1993). https://doi.org/10.1103/PhysRevLett.71.1661
4. Cheviakov, A. and Zhao, P., 2024. Shallow Water Models and Their Analytical Properties. In Analytical Properties of Nonlinear Partial Differential Equations: with Applications to Shallow Water Models (pp. 79–267). Cham: Springer International Publishing.
5. Cotter, C. and Bokhove, O., 2010. Variational water-wave model with accurate dispersion and vertical vorticity. Journal of engineering mathematics, 67, pp.33–54.
6. Cotter, C.J., Crisan, D., Holm, D.D., Pan, W. and Shevchenko, I., 2019. Numerically Modelling Stochastic Lie Transport in Fluid Dynamics, SIAM Multiscale Model. Simul., 17(1), 192–232. https://doi.org/10.1137/18M1167929
7. Cotter, C.J., Crisan, D., Holm, D.D., Pan, W. and Shevchenko, I., 2020. Data Assimilation for a Quasi-Geostrophic Model with Circulation-Preserving Stochastic Transport Noise. J Stat Phys 179, 1186–1221. https://doi.org/10.1007/s10955-020-02524-0
8. Dullin, H.R., Gottwald, G. and Holm, D.D. (2004). On asymptotically equivalent shallow water wave equations. Physica D 190, 1–14. https://doi.org/10.1016/j.physd.2003.11.004
9. Frenkel, J. and Dirac, P.A.M., 1934. *Wave mechanics: advanced general theory*. Clarendon Press Oxford.
10. Gardner, C.S., Greene, J.M., Kruskal, M.D. and Miura, R.M., 1967. Method for solving the Korteweg-deVries equation. Physical review letters, 19(19), p.1095.
11. Gjaja, I. and Holm, D.D., 1996. Self-consistent wave-mean flow interaction dynamics and its Hamiltonian formulation for a rotating stratified incompressible fluid. Physica D, 98, 343–378. https://doi.org/10.1016/0167-2789(96)00104-2
12. Holm, D. D. (2011). *Geometric Mechanics, Part I*. World-Scientific.
13. Holm, D. D., Hu, R., & Street, O. D. (2022a). Coupling of waves to sea surface currents via horizontal density gradients. Retrieved from http://arxiv.org/abs/2202.04446
14. Holm, D. D., Hu, R., & Street, O. D. (2022b, 12). Lagrangian reduction and wave mean flow interaction. Physica D 454, Article 133847. Retrieved from https://doi.org/10.1016/j.physd.2023.133847
15. Holm, D.D., Marsden, J.E. and Ratiu, T.S., 1998. Euler–Poincaré models of ideal fluids with nonlinear dispersion, Phys. Rev. Lett., **80**, 41734177. https://doi.org/10.1103/PhysRevLett.80.4173
16. Holm, D.D., Marsden, J.E. and Ratiu, T.S., 1998. The Euler-Poincaré equations and semidirect products with applications to continuum theories. Advances in Mathematics, 137(1), pp.1–81. https://doi.org/10.1006/aima.1998.1721
17. Y. Kodama, Phys. Lett. A **107**, 245, **112**, 193 (1985); **123**, 276 (1987).

18. Y. Kodama and A. V. Mikhailov, in *Algebraic Aspects of Integrable Systems: In Memory of Irene Dorfman*, edited by A. S. Fokas and I. M. Gelfand, Birkhäuser, Boston, (1996) pp 173–204.
19. Marsden, J.E., 1999. Park City lectures on mechanics, dynamics, and symmetry. Symplectic Geometry and Topology, 7, pp.335–430.
20. J.E. Marsden and T.S. Ratiu, *Introduction to Mechanics and Symmetry*, 2nd Edition, Springer:New York (1999).
21. Misiolek, G., 1998. A shallow water equation as a geodesic flow on the Bott-Virasoro group. Journal of Geometry and Physics, 24(3), pp.203–208.
22. Miura, R.M., 1976. The Korteweg–deVries equation: a survey of results. SIAM review, 18(3), pp.412–459.
23. Wazwaz, A.-M., 2010. Partial Differential Equations and Solitary Waves Theory (Springer, Berlin,Heidelberg.
24. G.B. Whitham, *Linear and Nonlinear Waves*, Wiley Interscience:New York (1974).
25. Li Zhi and N. R. Sibgatullin, J. Appl. Maths. Mechs. **61** 177 (1997)

Average Dissipation for Stochastic Transport Equations with Lévy Noise

Franco Flandoli, Andrea Papini, and Marco Rehmeier

1 Introduction

Stochastic transport (and advection, not discussed in this chapter) attracts more and more attention for its potentialities to describe small scale turbulence in several models and applications, see for instance the volumes [7–9] or the application to raindrop formation [15] and turbulence in pipes [2], among many others. Small scale turbulence is described by stochastic processes, space-dependent, either given a priori or inferred from data. Most models deal with white noise or Ornstein–Uhlenbeck processes, the first basic paradigms for any investigation of this kind. However, turbulent signals may be more complex. Two classes of processes seem to be the first ones to be considered after Gaussian noise: Fractional Gaussian noise, and α-stable processes. The first one has been considered in [10], reporting some preliminary results. In this chapter we give some preliminary results on α-stable processes, which seem to be the first of its kind. We also include a list of questions for future work in this direction.

The property of turbulent fluids we want to emphasize is the additional dissipation produced by turbulent eddies. In the white noise case this has been widely

F. Flandoli (✉)
Faculty of Sciences, Scuola Normale Superiore Pisa, Pisa, Italy
e-mail: franco.flandoli@sns.it

A. Papini
Department of Mathematical Sciences, Chalmers University of Technology & University of Gothenburg, Gothenburg, Sweden
e-mail: andreapa@chalmers.se

M. Rehmeier
Faculty of Sciences, Scuola Normale Superiore Pisa, Pisa, Italy
e-mail: mrehmeier@math.uni-bielefeld.de

Faculty of Mathematics, Bielefeld University, Bielefeld, Germany

© The Author(s) 2025   45
B. Chapron et al. (eds.), *Stochastic Transport in Upper Ocean Dynamics III*,
Mathematics of Planet Earth 13, https://doi.org/10.1007/978-3-031-70660-8_3

investigated, see for instance [8] and the references therein. Here, we consider this property in the case of α-stable transport noise. The most interesting case is undoubtedly when this transport noise models the small scales acting on the large ones in nonlinear models, but such a case is still too difficult in the α-stable case.

We note that the significance of noise in modeling turbulence phenomena lies in its role as a surrogate for elements omitted by deterministic mathematical models. For example, in scenarios like the formation of vortices due to boundary irregularities noise plays a crucial role. While the intricate mechanisms underlying vortex generation are still unclear, incorporating this phenomenon into models on a phenomenological basis is imperative to capture its observable macroscopic effects. Rather than to search for an exhaustive physical portrayal of vortex creation, the approach is to embrace a phenomenological perspective. In particular, using the empirical observation of vortex emergence near obstacles and their effect, hence integrating this observation into the equations, despite the inherent complexity of the process. To do so, we use impulsive forces that are introduced in equations to emulate the sudden alterations in flow behavior caused by vortex emergence. These impulsive forces act as surrogates for the impact of irregularities on fluid motion. Acknowledging the random nature of these impulsive events, their generation time is often assumed to follow a random pattern as well. Mathematically, this randomness is often modeled using a family of Poisson processes, commonly employed to represent stochastic events happening over time. Depending on the behavior of such random pattern a limit to a continuous Brownian noise can be obtained, but in case of a real boundary lacking symmetry, it is not clear that such a limit can always be performed and as such it is mandatory to focus on more realistic jump-diffusion processes.

As a primer, in this chapter we limit ourselves to linear transport of a scalar quantity, for instance heat, by an α-stable noise. Even in this simplified setting, many difficult technical questions emerge. The first one is which notion of integral should be used. In the Gaussian white noise case the basic rule is to choose a noise providing the correct invariance (conservation) properties. This leads to Stratonovich noise. In the α-stable case the same invariance is given by the so-called Markus noise. This type of stochastic integral has already been developed in the literature, both in the finite- and infinite-dimensional case [1, 3, 11, 12]. Its main advantage is that it preserves the ordinary rules of calculus for general Lévy processes as integrators. In this sense, the Marcus integral can be considered a natural extension of the Stratonovich integral. Indeed, for diffusion integrators, both integral notions coincide. We briefly review its definition in Sect. 2.

Given the model, the noise and the meaning of stochastic calculus, we limit ourselves here to investigate the following question: whether the expected value of the solution is dissipated, and whether this expected value satisfies a closed equation, similarly to the Gaussian noise case. We give affirmative answers to these questions: the expected value of the solution to the Marcus-type SPDE (2.1), for a suitable choice of an essentially α-stable symmetric Lévy measure, satisfies the second-order parabolic deterministic Eq. (3.4), see Proposition 3.3. The precise shape of the operator in (3.4) depends, of course, on the chosen Lévy measure for the pure jump Lévy process Z from (2.1). With the aforementioned choice of an α-stable symmetric Lévy measure, the resulting operator is close (but not identical) to the α-fractional Laplacian.

In order to support our claims, we include numerical simulations, which underline the dissipative character of the expected value of the solution to (2.1). We limit ourselves to dimension $d = 1$, in which an explicit solution can be computed via the method of characteristics. We simulate the evolution of the averaged solution profile, obtained with Monte Carlo method, showing a decay in both time and space, as expected from our theoretical results. More so, starting with a compact supported initial condition, we analyze, in Figs. 2b and 3a and b, the decay in time in the origin $x = 0$, obtaining an asymptotic behavior $\sim \beta(\sigma, m, \alpha)t^{-1/\alpha}$, with β depending on the velocity field norm, the dimension of the Lévy process, and the parameter α of the Lévy measure ν. This decay corresponds to the one of nonlocal PDEs. We remark that the pathwise solution profile shows no dissipativity when σ is constant and $d = 1$. It will be interesting to devote future work to the question whether pathwise dissipative behavior can be observed as well, for instance by improving the mixing property of the velocity field.

The organization of this chapter is as follows. In Sect. 2, we introduce our model and recall the notion and basic properties of Marcus stochastic integral equations. In Sect. 3, we state our theoretical results, see in particular Proposition 3.3. We present and discuss numerical simulations in Sect. 4 and, finally, pose some open questions for future research in Sect. 5.

2   Stochastic Transport Equations of Marcus-Type

We consider the following transport Marcus-type SPDE on $\mathbb{R}_+ \times \mathbb{R}^d$

$$du(t, x) = (\sigma(x)\nabla u(t, x)) \diamond dZ_t, \quad u(0, x) = u_0(x), \tag{2.1}$$

where $\sigma : \mathbb{R}^d \to \mathbb{R}^{d \times m}$ and $u_0 : \mathbb{R}^d \to \mathbb{R}$. Z is an m-dimensional pure jump Lévy process on a filtered probability space $(\Omega, \mathcal{F}, (\mathcal{F}_t)_{t \geq 0}, \mathbb{P})$,

$$Z_t = \int_0^t \int_{\overline{B_1(0)}} z\tilde{N}(dz, ds) + \int_0^t \int_{\overline{B_1(0)}^c} zN(dz, ds), \quad t \geq 0,$$

with Poisson random measure N, Lévy measure ν (i.e. ν is a Borel probability measure on $\mathbb{R}^m$ with $\nu(\{0\}) = 0$ and $\int_{\mathbb{R}^m} \min(1, z^2)d\nu(z) < \infty$), and $\tilde{N}(dz, dt) = N(dz, dt) - \mathbb{1}_{\overline{B_1(0)}}\nu(dz)dt$. Here $B_1(0)$ denotes the Euclidean ball in $\mathbb{R}^m$ with radius 1 centered at 0, and $\overline{B_1(0)}$ its closure. We make specific choices for σ and ν below. The symbol $\diamond$ denotes the Marcus stochastic integral, i.e. (2.1) is understood in the following integral sense:

$$u(t, x) = u_0(x) + \int_0^t \int_{\overline{B_1(0)}} e^{\sigma z} u(s-, x) - u(s-, x) \, \tilde{N}(dz, ds) \tag{2.2}$$

$$+ \int_0^t \int_{\mathbb{R}^m \setminus \overline{B_1(0)}} e^{\sigma z} u(s-, x) - u(s-, x) \, N(dz, ds)$$

$$+ \int_0^t \int_{\overline{B_1(0)}} e^{\sigma z} u(s-, x) - u(s-, x) - \nabla u(s-, x)$$

$$\cdot (\sigma(x)z) \, d\nu(z) ds, \quad (t, x) \in \mathbb{R}_+ \times \mathbb{R}^d,$$

where for $z \in \mathbb{R}^m$ and $f : \mathbb{R}^d \to \mathbb{R}$, $e^{\sigma z} f$ denotes the solution g of

$$\partial_t g(t, x) = \nabla g(t, x) \cdot (\sigma(x)z), \quad g(0, x) = f(x), \tag{2.3}$$

evaluated at $t = 1$. Here $s-$ denotes the left limit of $s \in \mathbb{R}$. When σ and f are sufficiently regular, the solution to this first-order linear transport PDE is unique and given by $g(t, x) = f(\phi_{t,0}(x))$, where $(t, x) \mapsto \phi_{t,0}(x)$ is the inverse of the unique solution flow $(t, x) \mapsto \phi_{0,t}(x)$ for the ODE

$$\partial_t \phi_{0,t}(x) = -\sigma(\phi_{0,t}(x))z, \quad \phi_{0,0}(x) = x$$

on $\mathbb{R} \times \mathbb{R}^d$.

Remark 2.1 The choice of Marcus-type integral over the more commonly used Stratonovich or Itô-type integral stems from the physical property that we want to maintain when using jump-diffusion processes in our SPDE. In particular it is the only integral that passes to the limit without additional terms, maintaining the usual properties of calculus, when considering C^∞ approximations of the active noise in the problem. Note that this is exactly the same reason why most of this theory is developed with the Stratonovich-type integral when considering Brownian noise.

We recall the following definition and result from [11].

Definition 2.2 An $(\mathcal{F}_t)$-adapted random field $u : \mathbb{R}_+ \times \mathbb{R}^d \times \Omega \to \mathbb{R}$ is a solution to (2.1), if it is a càdlàg C^2-semimartingale and (2.2) is satisfied for $\mathbb{P}$-a.e. $\omega \in \Omega$.

Proposition 2.3 If $u_0 \in C_b^2(\mathbb{R}^d)$ and $\sigma \in C_b^4(\mathbb{R}^d, \mathbb{R}^{d \times m})$, then there is a unique solution to (2.1), and it is given by

$$u(t, x) = u_0(\varphi_{t,0}(x)),$$

where $(t, x) \mapsto \varphi_{t,0}(x)$ denotes the inverse of the stochastic flow of the Marcus-SDE

$$\varphi_{0,t}(x) = x - \int_0^t \sigma(\varphi_{0,s-}(x)) \diamond dZ_t, \quad t \geq 0, x \in \mathbb{R}^d.$$

The definition of solution to this SDE is similar to the infinite-dimensional case, precisely it is given by

$$\varphi_{0,t}(x) = x - \int_0^t \int_{B_1(0)} \Psi_z(\varphi_{0,s-}(x)) - \varphi_{0,s-}(x)\, d\tilde{N}(dz, ds)$$

$$- \int_0^t \int_{\mathbb{R}^m \setminus \overline{B_1(0)}} \Psi_z(\varphi_{0,s-}(x)) - \varphi_{0,s-}(x)\, dN(dz, ds)$$

$$- \int_0^t \int_{B_1(0)} \Psi_z(\varphi_{0,s-}(x)) - \varphi_{0,s-}(x) - z\sigma(\varphi_{0,s-}(x))\, \nu(z)ds,$$

where $\Psi_z(y)$ denotes the solution to

$$\partial_t f(t) = z\sigma(f(t)), \quad t \in \mathbb{R}, \qquad f(0) = y \in \mathbb{R}^d,$$

evaluated at $t = 1$, see [12] and [1, Ch.4,6].

2.1 Special Cases

We are specifically interested in the case $d = 1$, $\sigma(x) = \sigma \in \mathbb{R}^m$ constant and $\nu = \frac{C}{|z|^{m+\alpha}}dz$, where C is either a constant depending on m and $\alpha \in (0, 2)$, or a function of z. In this case, the solution to (2.3) is given by $g(t, x) = f(x + \sigma \cdot zt)$, and the last integral term in (2.2) simplifies to

$$\int_0^t \int_{B_1(0)} u(s-, x + \sigma \cdot z) - u(s-, x) - \nabla u(s-, x)\sigma z\, d\nu(z)ds.$$

Moreover, in this case we have $\varphi_{0,t}(x) = x - \sigma \cdot Z_t$. Since for any choice of σ and ν both stochastic integrals from (2.2) are martingales, taking expectation yields

$$\mathbb{E}[u(t, x)] - \mathbb{E}[u_0(x)] = \mathbb{E}\Bigg[\int_0^t \int_{B_1(0)} u(s-, x + \sigma \cdot z)$$

$$- u(s-, x) - \nabla u(s-, x)\sigma z\, d\nu(z)ds\Bigg], \qquad (2.4)$$

where we write $\mathbb{E}[X] = \int_\Omega X\, d\mathbb{P}$ for a random variable $X : \Omega \to \mathbb{R}$, provided the integral is defined. Also note that for $d = 1$, every divergence-free vector field is constant.

3 Averaged Enhanced Dissipation

We started our investigation with the following question: is the expected value of
the solution dissipated? If so, does it satisfy some closed equation?

In this section we give affirmative answers to these questions proving the
following:

Claim 3.1 Under suitable hypotheses on the initial condition and on the Lévy
noise, let u be the unique solution of (2.1) in the sense of Definition 2.2. Then,
$U(t, x) := \mathbb{E}[u(t, x)]$ satisfies the second-order parabolic deterministic equation
with an operator dependent on the Lévy measure and close (but not identical) to the
α-fractional Laplacian.

Let $d = 1$, σ be constant, and ν have a radially symmetric density (for instance,
the classical symmetric α-stable density $\frac{1}{|z|^{m+\alpha}}$), and set $\theta := |\sigma|$. Then, due to the
radial symmetry, the RHS of (2.4) without expectation, i.e. for each fixed $\omega \in \Omega$,
equals

$$\int_0^t \int_{B_1(0)} u(s-, x + \theta z_1) - u(s-, x) - \nabla u(s-, x)\theta z_1 \, d\nu(z) ds, \tag{3.1}$$

where we denote by z_i the i-th component of $z = (z_1, \ldots, z_m) \in \mathbb{R}^m$. In order to
further calculate this integral, we need the following lemma. Below, we denote by
π_1 the canonical projection $\pi_1 : \mathbb{R}^m \to \mathbb{R}$, $\pi_1(z) = z_1$.

Lemma 3.2 *Let $\alpha \in (0, 2)$.*

(i) Set $\nu_\alpha := \frac{1}{|z|^{m+\alpha}} dz$. Then

$$\nu_{\alpha,1} := \nu \circ \pi_1^{-1} = \frac{C(m, \alpha)}{|y|^{1+\alpha}} dy,$$

with

$$C(m, \alpha) = |\mathbb{S}^{m-2}| \int_0^\infty (1 + r^2)^{-\frac{m+\alpha}{2}} r^{m-2} dr < \infty,$$

where $|\mathbb{S}^{m-2}|$ denotes the surface area of $\mathbb{S}^{m-2}$, the unit sphere in $\mathbb{R}^{m-1}$.
(ii) Let $\nu_{\mathbb{1},\alpha} := \mathbb{1}_{B_1(0)}(z) \frac{1}{|z|^{m+\alpha}} dz$. Then

$$\nu_{\mathbb{1},\alpha,1} := \nu_{\mathbb{1},\alpha} \circ \pi_1^{-1} = \frac{C(y, m, \alpha)}{|y|^{1+\alpha}} dy,$$

with

$$C(y, m, \alpha) := \mathbb{1}_{[-1,1]}(y)|\mathbb{S}^{m-2}| \int_0^{\frac{\sqrt{1-y^2}}{|y|}} (1 + r^2)^{-\frac{m+\alpha}{2}} r^{m-2} dr, \quad y \in \mathbb{R}.$$

Proof

(i) Let $A \in \mathcal{B}(\mathbb{R})$, and for $z = (z_1, \ldots, z_m) \in \mathbb{R}^m$, write $z' = (z_2, \ldots, z_m)$. Then

$$
\begin{aligned}
\nu_{\alpha,1}(A) = \nu_\alpha(A \times \mathbb{R}^{m-1}) &= \int_A \int_{\mathbb{R}^{m-1}} (z_1^2 + |z'|^2)^{-\frac{m+\alpha}{2}} dz_1 dz' \\
&= \int_A |z_1|^{-m-\alpha} \int_{\mathbb{R}^{m-1}} \left(1 + \frac{|z'|^2}{z_1^2}\right)^{-\frac{m+\alpha}{2}} dz' dz_1 \\
&= \int_A |z_1|^{-1-\alpha} \int_{\mathbb{R}^{m-1}} (1 + |z'|)^{-\frac{m+\alpha}{2}} dz' dz_1 \\
&= \int_A C(m, \alpha) |z_1|^{-1-\alpha} dz_1,
\end{aligned}
$$

where the third equality follows from the transformation rule and the final one by calculating the inner integral by spherical coordinates.

(ii) Due to the definition of $\nu_{1,\alpha}$, the proof is similar to the first part.

$\square$

To convey our further procedure, first consider (3.1) with domain of integration $\mathbb{R}^m$ instead of $\overline{B_1(0)}$, and choose $\nu = C(m, \alpha)^{-1} \nu_\alpha$. Then, by Lemma 3.2 (i), (3.1) equals

$$
\int_0^t \int_{\mathbb{R}} \frac{u(s-, x + \theta y) - u(s-, x) - \nabla u(s-, x)\theta y}{|y|^{1+\alpha}} dy ds.
$$

Inserting in (2.4) and interchanging the expectation with the temporal and spatial integral as well as with the gradient shows that $U(t, x) := \mathbb{E}[u(t, x)]$ solves

$$
\partial_t U(t, x) = \mathcal{L}_\alpha U(t-, x), \tag{3.2}
$$

where the operator $\mathcal{L}_\alpha$ is defined by

$$
\mathcal{L}_\alpha f(x) := \int_{\mathbb{R}} \frac{f(x + \theta y) - f(x) - \nabla f(x)\theta y}{|y|^{1+\alpha}} dy.
$$

Note that (3.2) is a deterministic, nonlocal second-order parabolic equation for the expected value of $u(t, x)$, which itself solves (pathwise) (2.1). We point out the similarity of $\mathcal{L}_\alpha$ with the fractional Laplacian $(-\Delta)^\alpha$ on $\mathbb{R}$,

$$
(-\Delta)^\alpha f(x) = \int_{\mathbb{R}} \frac{f(x + y) - f(x) - \nabla f(x)y \mathbb{1}_{(-1,1)}(y)}{|y|^{1+\alpha}} dy.
$$

In fact, for $\theta = 1$ and $\alpha > 1$, $\mathcal{L}_\alpha = (-\Delta)^\alpha$, since in this case $\int_{B_1(0)^c} \frac{y}{|y|^{1+\alpha}} dy = 0$.

Now, in order to take into account the proper domain of integration $\overline{B_1(0)}$ in (3.1), we repeat the previous lines with the choice

$$\nu = |\mathbb{S}^{m-2}|^{-1} \nu_{\mathbb{1},\alpha}. \tag{3.3}$$

Then, similarly to (3.2), we arrive at

$$\partial_t U(t, x) = \mathcal{L}_{\mathbb{1},\alpha} U(t-, x), \quad (t, x) \in \mathbb{R}_+ \times \mathbb{R}, \tag{3.4}$$

satisfied by $U(t, x) = \mathbb{E}[u(t, x)]$, where we set

$$\mathcal{L}_{\mathbb{1},\alpha} f(x) := \int_{-1}^{1} f(x + \theta y) - f(x) - \nabla f(x)\theta y \, d\nu_{\mathbb{1},\alpha,1}(y)$$

$$= \int_{-1}^{1} c(m, \alpha, y) \frac{f(x + \theta y) - f(x) - \nabla f(x)\theta y}{|y|^{1+\alpha}} \, dy$$

with

$$c(y, m, \alpha) := \int_{0}^{\frac{\sqrt{1-y^2}}{|y|}} (1 + r^2)^{-\frac{m+\alpha}{2}} r^{m-2} dr.$$

For the finite positive weight $c(y, m, \alpha)$, we note $c(y, m, \alpha) \xrightarrow{|y| \to 1} 0$, and that $c(y, m, \alpha)$ is symmetric around $y = 0$. Since $c(y, m, \alpha)$ is bounded on $(-1, 1)$, by Taylor formula $\mathcal{L}_{\mathbb{1},\alpha} f(x)$ is well-defined and finite for any $\alpha \in (0, 2)$, $x \in \mathbb{R}$, and $f \in C^2(\mathbb{R})$. The latter is satisfied for $x \mapsto u(t-, x)$, for every $t \geq 0$ and $\mathbb{P}$-a.e. $\omega \in \Omega$, as well as for $x \mapsto U(t-, x)$, for every $t \geq 0$. Interchanging the expectation with the temporal and spatial integral in (3.1) is justified, since u is given as in Proposition (2.3) and since φ is a stochastic flow of smooth diffeomorphisms, see [1, Thm.6.10.10]. Therefore, we have arrived at the following result.

Proposition 3.3 *Consider* (2.1) *for* $d = 1$, $u_0 \in C_b^2(\mathbb{R})$, $\sigma \in \mathbb{R}^m$ *constant, Z with Lévy measure* $\nu_{\mathbb{1},\alpha}$ *as in* (3.3), *and let u be the unique solution in the sense of Definition 2.2. Then,* $U(t, x) := \mathbb{E}[u(t, x)]$ *solves* (3.4).

Remark 3.4 Note that Eq. (3.4) is dissipative since it involves a fractional Laplace-like operator. For such operators, there are results for the decay of solutions, yielding a link with our numerical results (see Sect. 4) and the operator $\mathcal{L}_{\mathbb{1},\alpha}$. More precisely, consider

$$\partial_t u(t, x) = (-\Delta)^{\alpha/2} u(t, x), \quad u(0, x) = u_0(x), \quad (t, x) \in \mathbb{R}_+ \times \mathbb{R}.$$

There exists a C_0-semigroup $(S_\alpha(t))_{t \geq 0}$ such that $u(t, x) = S_\alpha(t) * u_0(x)$. The following estimate in dimension $d = 1$ was obtained in [4].

$$\|u(t)\|_\infty = \|\mathcal{S}_\alpha(t) * u_0\|_\infty \le t^{-\alpha/2} \|u_0\|_1,$$

where $\| \cdot \|_\infty$ and $\| \cdot \|_1$ denote the usual L^∞- and L^1-norms, respectively, thus showing the dissipative behavior of the solution. This behavior is retrieved also numerically (see below), suggesting a similar behavior for $\mathcal{L}_{\mathbb{1},\alpha}$ due to is similarity with the fractional Laplacian.

So far, we were unable to prove similar results for individual paths of the solution to (2.1). In fact, the pathwise profile shows no sign of dissipativity (see Fig. 1) in our simple special case, where σ is constant and $d = 1$. So, a conclusion on the pathwise dissipative behavior and an Ito-Stratonovich diffusion limit-type result is yet to be reached. A main reason for this appears to be the absence of any mixing property of divergence-free vector fields in dimension $d = 1$, yielding the pathwise profile a simple translation in time. We expect that in dimension $d > 1$ and for suitable vector fields as in [5, 6], a pathwise dissipative behavior of the solution to (2.1) solution is possible.

4 Numerical Results

In order to support the results of the previous section, here we present numerical simulations of the SPDE (2.1) in our special case, thereby underlining the dissipative character of the expected value of its solution. The code used can be found in [14]. As in Sect. 3, we limit ourselves to dimension $d = 1$, we consider a constant vector field $\sigma \in \mathbb{R}^m$, and we choose the Lévy measure of the driving pure jump Lévy process to be $\nu_{\mathbb{1},\alpha}$, as defined in Lemma 3.2 (ii). The jump dimension $m \in \mathbb{N}$ of the Lévy process is arbitrary. Under these assumptions, an explicit solution to (2.1) can be computed via the method of characteristics, which can be exploited for simple numerical simulations, namely

$$u(t, x; \omega) := u_0(x + \sigma \cdot Z_t(\omega)). \tag{4.1}$$

For all simulations below we fixed a smooth bump function $u_0(x) := \exp(-\frac{0.01}{0.01-\min(0.01,x^2)})$ as initial condition. The choice of such an initial condition are twofold: the first one is to deal with the finite space domain in which the numerical simulations is performed, hence selection of a function with sufficiently small support. The other reason is to appreciate clearly the transport-like behavior of the solution as expressed in (4.1). Nonetheless, in future works it is fundamental to tackle a numerical analysis based on different and more physical initial condition.

To simulate our α-stable Lévy process trajectory, we need to take into account the fact that our choice of Lévy measure $\nu_{\mathbb{1},\alpha}$ neglects large jumps. To this end, using the independent increment and self-similarity property, we compute the next step of the trajectory by cutting away jumps larger than one and generating a new realization until the jump size is sufficiently small. To implement the α-stable distribution we used the R-package $Stabledif$ [13]. The time step is $dt = 10^{-4}$, in the range $[0, 2]$, while the space domain is selected to be the interval $[-1, 1]$. Note that the Eq. (2.1) is posed on $\mathbb{R}^d$, i.e. there is a slight discrepancy between the equation and our numerical simulations that needs to be taken into account when we interpreting the results. The support of suppu$_0$ is sufficiently small so that almost no mass escapes the system throughout the time evolution. The space discretization size is selected at $dx = 10^{-3}$.

In our numerical analysis, we set our parameters in the range $\theta = \|\sigma\| \in [0, 1]$, $\alpha \in (0, 2) \setminus \{1\}$ and $m = 1, 2, 5, 10$. Our results are qualitatively consistent, and we are here analyzing and presenting figures for $\theta = 0.5$, $\alpha = 1.5$ and $m = 2$. In future works, we expect to give a precise quantification of the dependence on these parameters or the decay rate in time of the averaged solution. Our results, as of now, focus on the qualitative behavior in time for the averaged solution. This behavior depends only on the parameter α of the stable distribution.

Figure 1 presents a solution trajectory to (2.1) at different times. Precisely, the black curve represents the initial condition, while the blue ones show the solution profile at times $t = 1/2$, $t = 1$. Similarly, the grey curves represent the times $t = 3/2$, $t = 2$. As expected, the solution trajectories are translations, with no mixing property arising from the constant vector field σ, thus not showing any dissipation. The L^2 norm in space is preserved, and even for large times the solution preserves energy with an error of 10^{-14}, which is only due to the space domain being finite in the simulations. Nonetheless it is clear that the energy is preserved, as expected, due to the transport character of (2.1).

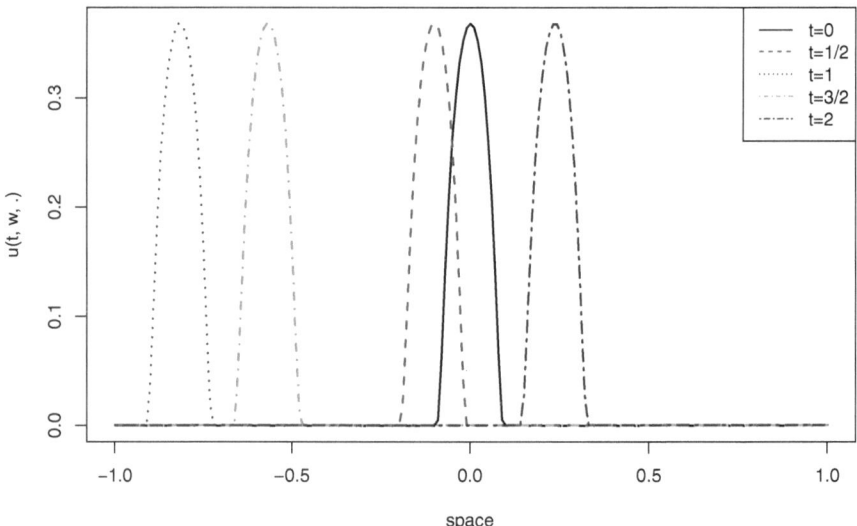

Fig. 1 Solution trajectory at several times

To show consistency with the theoretical results, we simulate the evolution of the averaged solution profile, obtained with Monte Carlo methods over averaging 5000 samples, as pictured in Fig. 2a and b, showing a decay in both time and space. In

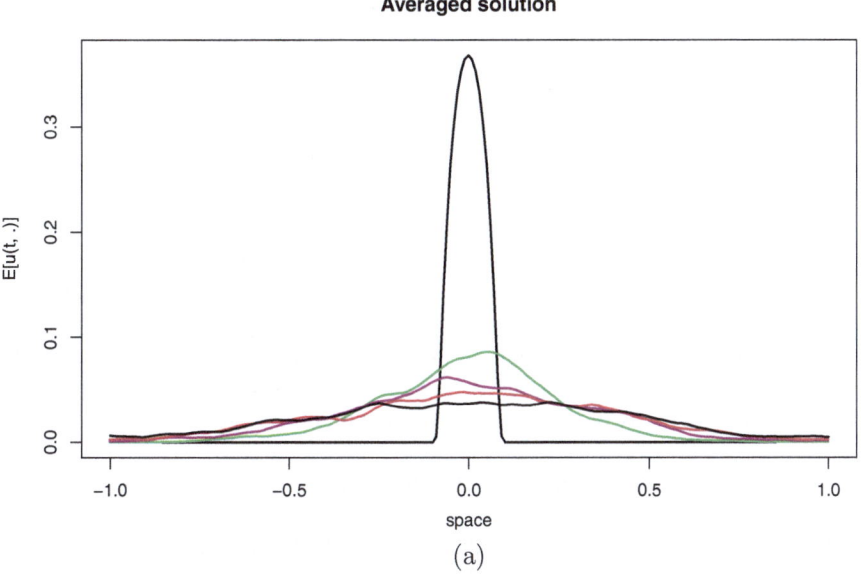

(a)

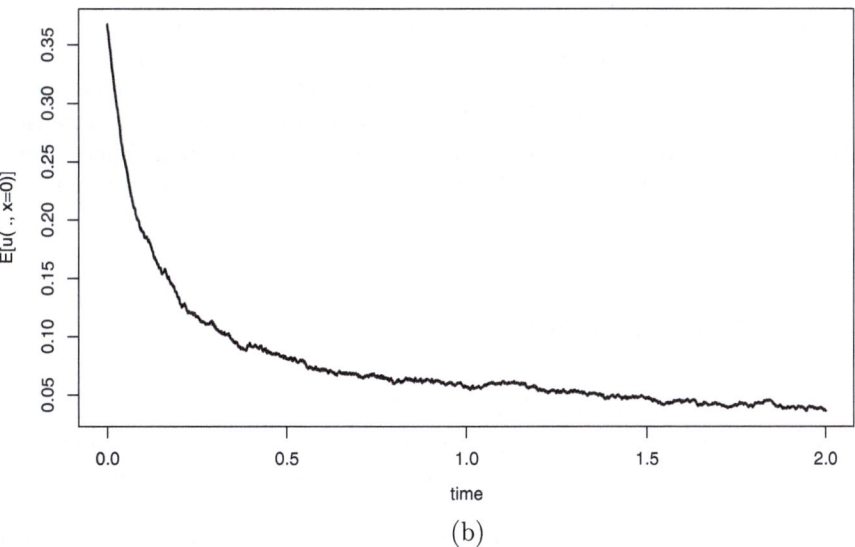

(b)

Fig. 2 Averaged solution to Eq. (2.1). (**a**) Decay in space of averaged solution at several times. Lines from top to bottom, $x = 0$ as reference: u_0, $\mathbb{E}[u_t] at t = 1/2, t = 1, t = 3/2 and t = 2$. (**b**) Decay in time of averaged solution at $x = 0$

the first figure, analogously to the pathwise result, we present the averaged solution at different times. The black graph represents the initial condition, while the green and purple ones show the average solution at times $t = 1/2$, $t = 1$, respectively, and at times $t = 3/2$, $t = 2$ for the red and blue ones. In this case, a diffusive behavior is present and a decay in space and time is observed. Concerning the space behavior, the profiles are still not smooth. The reason is twofold, one being the averaging procedure, the second one arising from the fractional operator obtained for the equation modeling the averaged system (i.e. (3.4)) and the corresponding small jumps.

Concerning the decay, particularly care for $x = 0$, in which the initial condition has its maximum, and note the decay of the averaged solution in time with a power law-like asymptotic behavior.

More so, starting with a compact supported initial condition, we show, in Fig. 3a and b, the time decay in the origin $x = 0$, with a nonlinear regression to estimate the asymptotic and power law-like behavior, linking it to the operator $\mathcal{L}_{1,\alpha}$ proposed in the theoretical section. In particular, in Fig. 3a, we performed a regression using the insight of Remark 3.4 numerically to show that, with a residual error of less than 0.003, we have a decay of the averaged solution in time of the following form:

$$\mathbb{E}[u(t, x = 0)] \sim \beta(\sigma, m, \alpha)t^{-1/\alpha},$$

with β depending on $\theta := \|\sigma\|$, the dimension of the Lévy process $m \in \mathbb{N}$, and the parameter α of the Lévy measure ν. Here, we have not delved into an analysis of the behavior of β, which could be theoretically examined, as discussed in Remark 3.4.

Therefore, it is crucial for future research to explore how the strength of β, and consequently the velocity field, interact with the decay of the profile.

In Fig. 3b, we plot the log-log version of the decay in time, showing the inverse asymptotic behavior of the averaged solution profile in space and time and its concordance with the results of the theoretical section and the nonlinear regression. More so, on the right in Fig. 3b, the tail was analyzed in the time frame $z[500 : 1001] := [0.5, 1]$, showing a rough behavior, but with respect to the regression line, the error is in the range 0.003–0.006, which validates our results.

5 Conclusions and Open Questions

In this work we have answered positively the following question: is the expected value of the solution dissipated? In fact, under suitable hypothesis on the initial condition and on the Lévy noise, the unique solution of (2.1) satisfies a second-order parabolic deterministic equation with an operator dependent on the Lévy measure and close (but not identical) to the α-fractional Laplacian. As such this is a first step to understand the effect of jump-diffusion processes into fluid motion and the effect of modelling turbulence with a noise term.

We conclude this work with a few questions for future work in this direction.

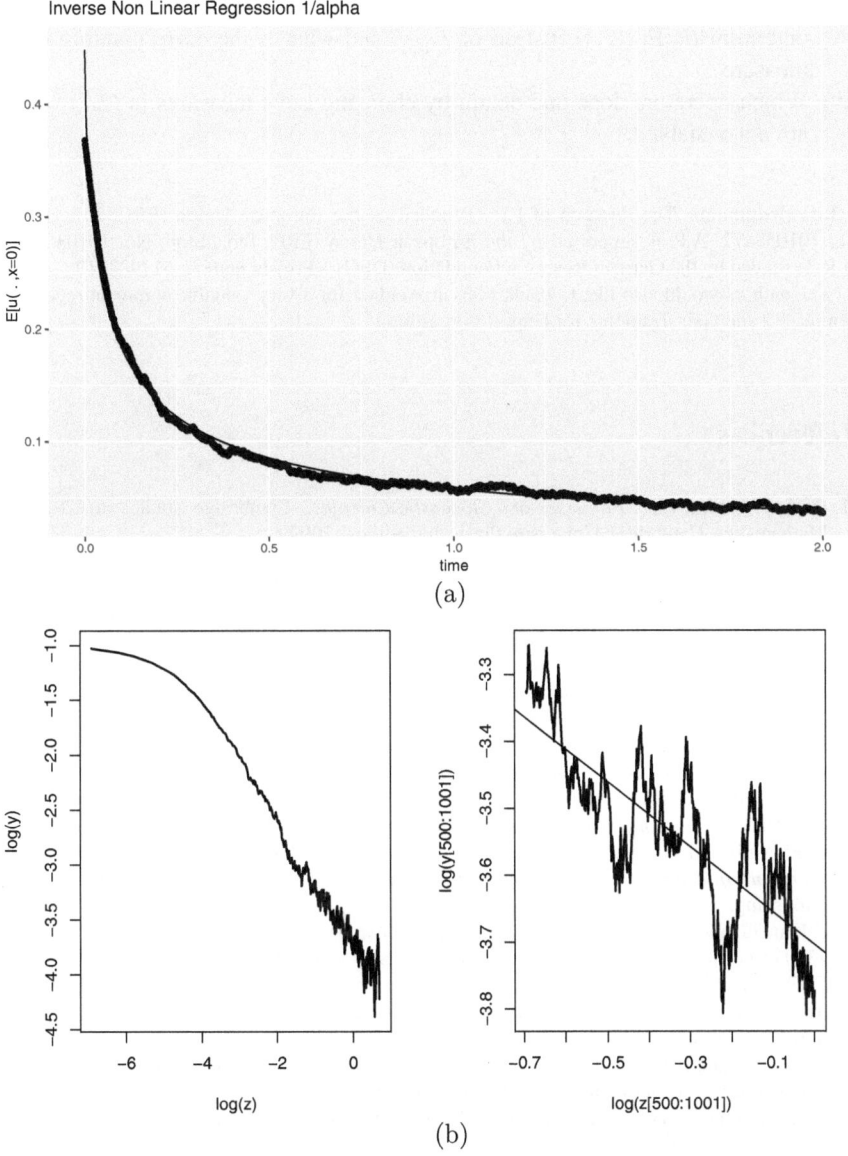

Fig. 3 Nonlinear regression on (3.4). (**a**) Nonlinear regression $\sim t^{-1/alpha}$, $R.E. = 0.003$. (**b**) Log-Log Plot and tail behavior. $z := t$ (*time*), $y := \mathbb{E}[u(t, x = 0)]$. Left: full decay. Right: zoom on tail for larger times

(i) Can similar results be obtained for non-constant σ (in dimension $d \geq 2$)?

(ii) Related to (i), do vector fields with suitable mixing properties lead to a pathwise dissipation result for solutions to (2.1)?

(iii) Is it possible to obtain precisely the fractional Laplacian as the operator of the deterministic Eq. (3.4) instead of $\mathcal{L}_{\mathbb{1},\alpha}$, and what is the corresponding Lévy measure?

(iv) Which operators does one obtain by choosing Lévy measures in (2.1) which are not α-stable?

Acknowledgments The research of F.F. is funded by the European Union (ERC, NoisyFluid, No. 101053472. A.P. is supported by the European Union (ERC, StochMan, No. 101088589). M.R. is funded by the German Research Foundation (DFG)—Project number 517982119.

The authors would also like to thank Marvin Weidner for a very valuable comment regarding Lemma 3.2 and Gaia Tramonte for helpful suggestions.

References

1. D. Applebaum. *Lévy Processes and Stochastic Calculus*. Cambridge Studies in Advanced Mathematics. Cambridge University Press, 2nd edition, 2009.
2. A. Cavalieri, É. Mémin, and G. Tissot. Input-output analysis of the stochastic Navier–Stokes equations: Application to turbulent channel flow. *Phys. Rev. Fluids*, 8:033904, 2023.
3. A. Chechkin and I. Pavlyukevich. Marcus versus Stratonovich for systems with jump noise. *Journal of Physics A: Mathematical and Theoretical*, 47(34):342001, 2014.
4. A. Fino and M. Kirane. The Cauchy problem for heat equation with fractional Laplacian and exponential nonlinearity. *Communications on Pure and Applied Analysis*, 19(7):3625–3650, 2020.
5. F. Flandoli, L. Galeati, and D. Luo. Quantitative convergence rates for scaling limit of SPDEs with transport noise. *arXiv preprint 2104.01740*, 2021.
6. F. Flandoli, L. Galeati, and D. Luo. Eddy heat exchange at the boundary under white noise turbulence. *Philosophical Transactions of the Royal Society A*, 380(2219):20210096, 2022.
7. F. Flandoli and E. Luongo. The dissipation properties of transport noise. In *Stochastic Transport in Upper Ocean Dynamics*, pages 69–85, Cham, 2023. Springer International Publishing.
8. F. Flandoli and E. Luongo. *Stochastic partial differential equations in fluid mechanics*, volume 2330 of *Lecture Notes in Mathematics*. Springer, Singapore, 2023.
9. F. Flandoli, S. Morlacchi, and A. Papini. Effect of transport noise on Kelvin–Helmholtz instability. In *Stochastic Transport in Upper Ocean Dynamics II*, pages 29–52, Cham, 2024. Springer Nature Switzerland.
10. F. Flandoli and F. Russo. Reduced dissipation effect in stochastic transport by Gaussian noise with regularity greater than 1/2. *arXiv preprint 2305.19293*, 2023.
11. L.-S. Hartmann and I. Pavlyukevich. First order linear Marcus SPDEs. *arXiv preprint 2303.00674*, 2023.
12. S. Marcus. Modeling and analysis of stochastic differential equations driven by point processes. *IEEE Transactions on Information Theory*, 24(2):164–172, 1978.
13. M. Mächler, D. Würtz, and Rmetrics core team members. stabledist: Stable distribution functions. 2016-09-12.
14. A. Papini, F. Flandoli, and M. Rehmeier. Average dissipation for stochastic transport equations with Lévy noise. *Zenodo*, 2024.
15. A. Papini, R.Huang, and F. Flandoli. Turbulence enhancement of coagulation: The role of eddy diffusion in velocity. *Physica D: Nonlinear Phenomena*, 448:133726, 2023.

General Solution Theory for the Stochastic Navier-Stokes Equations

Daniel Goodair

1 Introduction

The theoretical analysis of Stochastic Navier-Stokes Equations dates back to the work of Bensoussan and Temam [5] in 1973, where the problem of existence of solutions is addressed in the presence of a random forcing term. The well-posedness question for additive and multiplicative noise has since seen significant developments, for example through the works [4, 6, 7], [8–10, 18, 23, 37, 39, 44] and references therein. The choice of noise to best encapsulate physical properties of this fluid equation is in constant study, recently yielding a strong argument for transport type stochastic perturbations (where the stochastic integral depends on the gradient of the solution). The chapter of Brzeźniak, Capinski and Flandoli [7] in 1992 was one of the first to bring attention to the significance of fluid dynamics equations with transport noise, whilst such ideas have only recently been cemented through the specific stochastic transport schemes of [32] and [43]. In these chapters Holm and Mémin establish a new class of stochastic equations driven by transport type noise which serve as fluid dynamics models by adding uncertainty in the transport of fluid parcels to reflect the unresolved scales. The physical significance of such equations in modelling, numerical analysis and data assimilation continues to be well documented, see [1, 13–17, 20, 21, 33, 34, 36, 38, 50] and in particular [19] for a comprehensive account of the topic. With this motivation our main object of study is the Navier-Stokes Equation under Stochastic Advection by Lie Transport (SALT) introduced in [32], given by

D. Goodair (✉)
Imperial College London, Mathematics, London, UK
e-mail: djg116@ic.ac.uk

B. Chapron et al. (eds.), *Stochastic Transport in Upper Ocean Dynamics III*,
Mathematics of Planet Earth 13, https://doi.org/10.1007/978-3-031-70660-8_4

$$u_t - u_0 + \int_0^t \mathcal{L}_{u_s} u_s \, ds - \nu \int_0^t \Delta u_s \, ds - \int_0^t B(u_s) \circ d\mathcal{W}_s + \nabla \rho_t = 0 \qquad (1)$$

where u represents the fluid velocity, $\nu > 0$ the viscosity, ρ the pressure,[1] $\mathcal{W}$ is a Cylindrical Brownian Motion, $\mathcal{L}$ represents the nonlinear term and B is a first order differential operator (the SALT Operator) properly introduced in Sect. 5. Intrinsic to this stochastic methodology is that B is defined relative to a collection of functions (ξ_i) which physically represent spatial correlations. These (ξ_i) can be determined at coarse-grain resolutions from finely resolved numerical simulations, and mathematically are derived as eigenvectors of a velocity-velocity correlation matrix (see [13–15]).

The newfound attention on this class of equations arrives at a time where the analytical literature in stochastic partial differential equations (SPDEs) has established itself as one of mathematics' most exciting prospects. Indeed, Martin Hairer's Fields Medal winning work on *regularity structures*, [31], gave rigorous meaning to a drastically extended class of SPDEs which are ill-defined in standard function spaces due to spatial irregularities. At a similar time, Gubinelli, Imkeller and Perkowski developed a theory of *paracontrolled distributions*, [30], which again provided a toolkit for the seemingly intractable theory of distribution-valued SPDEs. These techniques are of a very different flavour to the more classical *variational approach* pioneered by Étienne Pardoux, [47], whereby the equation is formatted in a Gelfand Triple for a noise valued in the Hilbert Space.

Our Eq. (1), however, finds itself in a structural limbo between these methodologies. The first-order noise operator introduces a singularity into the Hilbert Space of the variational framework, so it fails to fit into this classical theory. The noise, though, is still well-defined in traditional function spaces, suggesting that the heavy machinery and distribution-tailored approaches of Hairer, Gubinelli et al. are not particularly appropriate. Of course (1) is not alone in this, coming as part of a substantial class of equations which continue to expand. Figure 2 of [16] gives a brief overview of just some of the deterministic fluid models, each of which can be stochastically perturbed through a similarly widening array of variational principles (beyond the seminal works [32, 43], see more recently [33, 50]). Therefore, two strong arguments for considering a *general solution theory* for this class of equations present themselves:

1. To complete a highlighted gap in the SPDE literature;
2. To most efficiently deduce their well-posedness for the purpose of modelling and applications.

The goal of this chapter is to discuss two recent results of this kind, one for local-in-time solutions and the other global, demonstrating how these results can be applied to obtain the existence and uniqueness of solutions to the Eq. (1) in different

---

[1] The pressure term is a semimartingale, and an explicit form for the SALT Euler Equation is given in [50] Subsection 3.3.

dimensions, under different boundary conditions, and with differing regularity on the initial condition. It is our hope to inspire a similar treatment for related equations. The structure of the chapter is detailed now:

- Section 2 is devoted to the setup of the problem in terms of notation, along with the basic stochastic framework.
- Section 3 covers the local-in-time theory. Section 3.1 details the functional framework of the abstract SPDE, and is followed by two sets of assumptions in Sects. 3.2 and 3.3. Corresponding solution types and their well-posedness results are given in the remaining two subsections. A stronger solution, which we call H-valued, is given in Sect. 3.4. A weaker notion follows in Sect. 3.5, named U-valued solutions. The role of each solution in terms of the Stochastic Navier-Stokes Equations is given in Sect. 5. The results of this section are proven in [29]. We note that a corresponding theory for *inviscid* fluid equations is given in [2].
- Section 4 is comprised of the global-in-time theory. Following the same structure as Sect. 3, the functional framework is detailed in Sect. 4.1 whilst three sets of assumptions follow in Sects. 4.2, 4.3 and 4.4. These assumptions build upon each other for the corresponding notions of solution: 'martingale weak solutions' of Sect. 4.5, 'weak solutions' of Sect. 4.6 and 'strong solutions' of Sect. 4.7. The results of this section are proven in [26]. We note that a corresponding theory for *inviscid* fluid equations is given in [51].
- Section 5 addresses applications of the above theory to the Eq. (1). The section is split into subsections based upon the domain and boundary conditions imposed: Sect. 5.1 considers (1) posed on the torus, Sect. 5.2 on a smooth bounded domain with no-slip boundary condition and Sect. 5.3 for the Navier boundary conditions. In all cases, the equation is considered in both 2 and 3 dimensions, with varied regularity on the initial condition. The spirit of this chapter is to demonstrate how the Stochastic Navier-Stokes Equations can be set-up in the abstract frameworks, and it is again not our intention to give complete proofs of the results here. We remark that a thorough introduction to (1) is given in [27], and much of the detail of the proofs is provided across [25, 27, 28].

2 Preliminaries

2.1 Elementary Notation

In the following $\mathcal{O}$ will represent either the N-dimensional torus $\mathbb{T}^N$ or a smooth bounded domain $\mathscr{O} \subset \mathbb{R}^N$. Both are equipped with Euclidean norm and Lebesgue measure λ. We consider Banach Spaces as measure spaces equipped with their corresponding Borel σ-algebra. Let $(\mathcal{X}, \mu)$ denote a general topological measure space, $(\mathcal{Y}, \|\cdot\|_{\mathcal{Y}})$ and $(\mathcal{Z}, \|\cdot\|_{\mathcal{Z}})$ be separable Banach Spaces, and $(\mathcal{U}, \langle \cdot, \cdot \rangle_{\mathcal{U}})$, $(\mathcal{H}, \langle \cdot, \cdot \rangle_{\mathcal{H}})$ be general separable Hilbert spaces. We introduce the following spaces of functions.

- $L^p(\mathcal{X}; \mathcal{Y})$ is the class of measurable p-integrable functions from $\mathcal{X}$ into $\mathcal{Y}$, $1 \leq p < \infty$, which is a Banach space with norm

$$\|\phi\|^p_{L^p(\mathcal{X};\mathcal{Y})} := \int_{\mathcal{X}} \|\phi(x)\|^p_{\mathcal{Y}} \, \mu(dx).$$

In particular $L^2(\mathcal{X}; \mathcal{Y})$ is a Hilbert Space when $\mathcal{Y}$ itself is Hilbert, with the standard inner product

$$\langle \phi, \psi \rangle_{L^2(\mathcal{X};\mathcal{Y})} = \int_{\mathcal{X}} \langle \phi(x), \psi(x) \rangle_{\mathcal{Y}} \, \mu(dx).$$

In the case $\mathcal{X} = \mathcal{O}$ and $\mathcal{Y} = \mathbb{R}^N$ note that

$$\|\phi\|^2_{L^2(\mathcal{O};\mathbb{R}^N)} = \sum_{l=1}^{N} \left\|\phi^l\right\|^2_{L^2(\mathcal{O};\mathbb{R})}, \qquad \phi = \left(\phi^1, \ldots, \phi^2\right), \quad \phi^l : \mathcal{O} \to \mathbb{R}.$$

We denote $\|\cdot\|_{L^p(\mathcal{O};\mathbb{R}^N)}$ by $\|\cdot\|_{L^p}$ and $\|\cdot\|_{L^2(\mathcal{O};\mathbb{R}^N)}$ by $\|\cdot\|$.
- $L^\infty(\mathcal{X}; \mathcal{Y})$ is the class of measurable functions from $\mathcal{X}$ into $\mathcal{Y}$ which are essentially bounded. $L^\infty(\mathcal{X}; \mathcal{Y})$ is a Banach Space when equipped with the norm

$$\|\phi\|_{L^\infty(\mathcal{X};\mathcal{Y})} := \inf\{C \geq 0 : \|\phi(x)\|_{\mathcal{Y}} \leq C \text{ for } \mu - a.e.x \in \mathcal{X}\}.$$

- $C(\mathcal{X}; \mathcal{Y})$ is the space of continuous functions from $\mathcal{X}$ into $\mathcal{Y}$.
- $C_w(\mathcal{X}; \mathcal{Y})$ is the space of 'weakly continuous' functions from $\mathcal{X}$ into $\mathcal{Y}$, by which we mean continuous with respect to the given topology on $\mathcal{X}$ and the weak topology on $\mathcal{Y}$.
- $C^m(\mathcal{O}; \mathbb{R})$ is the space of $m \in \mathbb{N}$ times continuously differentiable functions from $\mathcal{O}$ to $\mathbb{R}$, that is $\phi \in C^m(\mathcal{O}; \mathbb{R})$ if and only if for every 2 dimensional multi index $\alpha = \alpha_1, \alpha_2$ with $|\alpha| \leq m$, $D^\alpha \phi \in C(\mathcal{O}; \mathbb{R})$ where D^α is the corresponding classical derivative operator $\partial_{x_1}^{\alpha_1} \partial_{x_2}^{\alpha_2}$.
- $C^\infty(\mathcal{O}; \mathbb{R})$ is the intersection over all $m \in \mathbb{N}$ of the spaces $C^m(\mathcal{O}; \mathbb{R})$.
- $C_0^m(\mathcal{O}; \mathbb{R})$ for $m \in \mathbb{N}$ or $m = \infty$ is the subspace of $C^m(\mathcal{O}; \mathbb{R})$ of functions which have compact support.
- $C^m(\mathcal{O}; \mathbb{R}^N)$, $C_0^m(\mathcal{O}; \mathbb{R}^N)$ for $m \in \mathbb{N}$ or $m = \infty$ is the space of functions from $\mathcal{O}$ to $\mathbb{R}^N$ whose component mappings each belong to $C^m(\mathcal{O}; \mathbb{R})$, $C_0^m(\mathcal{O}; \mathbb{R})$.
- $W^{m,p}(\mathcal{O}; \mathbb{R})$ for $1 \leq p < \infty$ is the sub-class of $L^p(\mathcal{O}, \mathbb{R})$ which has all weak derivatives up to order $m \in \mathbb{N}$ also of class $L^p(\mathcal{O}, \mathbb{R})$. This is a Banach space with norm

$$\|\phi\|^p_{W^{m,p}(\mathcal{O},\mathbb{R})} := \sum_{|\alpha| \leq m} \left\|D^\alpha \phi\right\|^p_{L^p(\mathcal{O};\mathbb{R})},$$

where D^α is the corresponding weak derivative operator. In the case $p = 2$ the space $W^{m,2}(\mathcal{O}, \mathbb{R})$ is Hilbert with inner product

$$\langle \phi, \psi \rangle_{W^{m,2}(\mathcal{O};\mathbb{R})} := \sum_{|\alpha| \leq m} \langle D^\alpha \phi, D^\alpha \psi \rangle_{L^2(\mathcal{O};\mathbb{R})}.$$

- $W^{m,\infty}(\mathcal{O};\mathbb{R})$ for $m \in \mathbb{N}$ is the sub-class of $L^\infty(\mathcal{O},\mathbb{R})$ which has all weak derivatives up to order $m \in \mathbb{N}$ also of class $L^\infty(\mathcal{O},\mathbb{R})$. This is a Banach space with norm

$$\|\phi\|_{W^{m,\infty}(\mathcal{O},\mathbb{R})} := \sup_{|\alpha| \leq m} \left\| D^\alpha \phi \right\|_{L^\infty(\mathcal{O};\mathbb{R}^N)}.$$

- $W^{m,\infty}(\mathcal{O};\mathbb{R}^N)$ is the sub-class of $L^\infty(\mathcal{O},\mathbb{R}^N)$ which has all weak derivatives up to order $m \in \mathbb{N}$ also of class $L^\infty(\mathcal{O},\mathbb{R}^N)$. This is a Banach space with norm

$$\|\phi\|_{W^{m,\infty}} := \sup_{l \leq N} \left\| \phi^l \right\|_{W^{m,\infty}(\mathcal{O};\mathbb{R})}.$$

- $\dot{L}^2(\mathbb{T}^N;\mathbb{R}^N)$ is the sub-class of $L^2(\mathbb{T}^N;\mathbb{R}^N)$ of functions ϕ such that

$$\int_{\mathbb{T}^N} \phi \, d\lambda = 0.$$

- $\dot{W}^{m,2}(\mathbb{T}^N;\mathbb{R}^N)$ is simply the intersection $W^{m,2}(\mathbb{T}^N;\mathbb{R}^3) \cap \dot{L}^2(\mathbb{T}^N;\mathbb{R}^3)$.
- $W_0^{m,p}(\mathcal{O};\mathbb{R})$, $W_0^{m,p}(\mathcal{O};\mathbb{R}^N)$ for $m \in \mathbb{N}$ and $1 \leq p \leq \infty$ is the closure of $C_0^\infty(\mathcal{O};\mathbb{R})$, $C_0^\infty(\mathcal{O};\mathbb{R}^N)$ in $W^{m,p}(\mathcal{O};\mathbb{R})$, $W^{m,p}(\mathcal{O};\mathbb{R}^N)$.
- $\mathcal{L}^2(\mathcal{U};\mathcal{H})$ is the space of Hilbert-Schmidt operators from $\mathcal{U}$ to $\mathcal{H}$, defined as the elements $F \in \mathcal{L}(\mathcal{U};\mathcal{H})$ such that for some basis (e_i) of $\mathcal{U}$,

$$\sum_{i=1}^{\infty} \|Fe_i\|_{\mathcal{H}}^2 < \infty.$$

This is a Hilbert space with inner product

$$\langle F, G \rangle_{\mathcal{L}^2(\mathcal{U};\mathcal{H})} = \sum_{i=1}^{\infty} \langle Fe_i, Ge_i \rangle_{\mathcal{H}}$$

which is independent of the choice of basis (see e.g. [12]).

2.2 Stochastic Framework

Let $(\Omega, \mathcal{F}, (\mathcal{F}_t), \mathbb{P})$ be a fixed filtered probability space satisfying the usual conditions of completeness and right continuity. We take $\mathcal{W}$ to be a cylindrical Brownian motion over some Hilbert Space $\mathfrak{U}$ with orthonormal basis (e_i). Recall (e.g. [40], Definition 3.2.36) that $\mathcal{W}$ admits the representation $\mathcal{W}_t = \sum_{i=1}^{\infty} e_i W_t^i$ as a limit in $L^2(\Omega; \mathfrak{U}')$ whereby the (W^i) are a collection of i.i.d. standard real valued Brownian Motions and $\mathfrak{U}'$ is an enlargement of the Hilbert Space $\mathfrak{U}$ such that the embedding $J : \mathfrak{U} \to \mathfrak{U}'$ is Hilbert-Schmidt and $\mathcal{W}$ is a JJ^*-Cylindrical Brownian Motion over $\mathfrak{U}'$. Given a process $F : [0, T] \times \Omega \to \mathcal{L}^2(\mathfrak{U}; \mathcal{H})$ progressively measurable and such that $F \in L^2(\Omega \times [0, T]; \mathcal{L}^2(\mathfrak{U}; \mathcal{H}))$, for any $0 \le t \le T$ we define the stochastic integral

$$\int_0^t F_s d\mathcal{W}_s := \sum_{i=1}^{\infty} \int_0^t F_s(e_i) dW_s^i,$$

where the infinite sum is taken in $L^2(\Omega; \mathcal{H})$. We can extend this notion to processes F which are such that $F(\omega) \in L^2([0, T]; \mathcal{L}^2(\mathfrak{U}; \mathcal{H}))$ for $\mathbb{P} - a.e.$ ω via the traditional localisation procedure. In this case the stochastic integral is a local martingale in $\mathcal{H}$.[2]

3 Local Theory for SPDEs

3.1 Functional Framework

Our object of study is the Itô SPDE

$$\Psi_t = \Psi_0 + \int_0^t \mathcal{A}(s, \Psi_s) ds + \int_0^t \mathcal{G}(s, \Psi_s) d\mathcal{W}_s \tag{2}$$

which we pose for a triplet of embedded, separable Hilbert Spaces

$$V \hookrightarrow H \hookrightarrow U$$

whereby the embeddings are continuous linear injections. We ask that there is a continuous bilinear form $\langle \cdot, \cdot \rangle_{U \times V} : U \times V \to \mathbb{R}$ such that for $f \in H$ and $\psi \in V$,

$$\langle f, \psi \rangle_{U \times V} = \langle f, \psi \rangle_H. \tag{3}$$

The Eq. (2) is posed on a time interval $[0, T]$ for arbitrary but henceforth fixed $T \ge 0$. The mappings $\mathcal{A}, \mathcal{G}$ are such that $\mathcal{A} : [0, T] \times V \to U, \mathcal{G} : [0, T] \times V \to \mathcal{L}^2(\mathfrak{U}; H)$ are measurable. Understanding $\mathcal{G}$ as a mapping $\mathcal{G} : [0, T] \times V \times \mathfrak{U} \to H$, we introduce the notation $\mathcal{G}_i(\cdot, \cdot) := \mathcal{G}(\cdot, \cdot, e_i)$. We further impose the existence of

---

[2] A complete, direct construction of this integral, a treatment of its properties and the fundamentals of stochastic calculus in infinite dimensions can be found in [24] Section 1.

a system of elements (a_k) of V with the following properties. Let us define the spaces $V_n := \text{span}\{a_1, \ldots, a_n\}$ and $\mathcal{P}_n$ as the orthogonal projection to V_n in U. It is required that the $(\mathcal{P}_n)$ are uniformly bounded in H, which is to say that there exists a constant c independent of n such that for all $\phi \in H$,

$$\|\mathcal{P}_n f\|_H \leq c \|f\|_H. \tag{4}$$

We also suppose that there exists a real valued sequence (μ_n) with $\mu_n \to \infty$ such that for any $f \in H$,

$$\|(I - \mathcal{P}_n)f\|_U \leq \frac{1}{\mu_n} \|f\|_H \tag{5}$$

where I represents the identity operator in U. Specific bounds on the mappings $\mathcal{A}$ and $\mathcal{G}$ will be imposed in Assumption Sets A and B. In order to make the assumptions we introduce some more notation here: we shall let $c. : [0, T] \to \mathbb{R}$ denote any bounded function, and for any constant $p \in \mathbb{R}$ we define the functions $K_U : U \to \mathbb{R}$, $K_H : H \to \mathbb{R}$, $K_V : V \to \mathbb{R}$ by

$$K_U(\phi) = 1 + \|\phi\|_U^p, \quad K_H(\phi) = 1 + \|\phi\|_H^p, \quad K_V(\phi) = 1 + \|\phi\|_V^p.$$

We may also consider these mappings as functions of two variables, e.g. $K_U : U \times U \to \mathbb{R}$ by

$$K_U(\phi, \psi) = 1 + \|\phi\|_U^p + \|\psi\|_U^p.$$

Our assumptions will be stated for 'the existence of a K such that...' where we really mean 'the existence of a p such that, for the corresponding $K, \ldots$'.

3.2 Assumption Set A

Recall the setup and notation of Sect. 3.1. We assume that there exists a $c.$, K and $\gamma > 0$ such that for all $\phi, \psi \in V$, $\phi^n \in V_n$, $f \in H$ and $t \in [0, T]$:

Assumption 3.1

$$\|\mathcal{A}(t, \phi)\|_U^2 + \sum_{i=1}^{\infty} \|\mathcal{G}_i(t, \phi)\|_H^2 \leq c_t K_U(\phi) \left[1 + \|\phi\|_V^2\right], \tag{6}$$

$$\|\mathcal{A}(t, \phi) - \mathcal{A}(t, \psi)\|_U^2 \leq c_t K_V(\phi, \psi) \|\phi - \psi\|_V^2, \tag{7}$$

$$\sum_{i=1}^{\infty} \|\mathcal{G}_i(t, \phi) - \mathcal{G}_i(t, \psi)\|_U^2 \leq c_t K_U(\phi, \psi) \|\phi - \psi\|_H^2. \tag{8}$$

Assumption 3.2

$$2\langle \mathcal{P}_n \mathcal{A}(t, \phi^n), \phi^n \rangle_H + \sum_{i=1}^{\infty} \|\mathcal{P}_n \mathcal{G}_i(t, \phi^n)\|_H^2$$

$$\leq c_t K_U(\phi^n) \left[1 + \|\phi^n\|_H^4 \right] - \gamma \|\phi^n\|_V^2, \tag{9}$$

$$\sum_{i=1}^{\infty} \langle \mathcal{P}_n \mathcal{G}_i(t, \phi^n), \phi^n \rangle_H^2 \leq c_t K_U(\phi^n) \left[1 + \|\phi^n\|_H^6 \right]. \tag{10}$$

Assumption 3.3

$$2\langle \mathcal{A}(t, \phi) - \mathcal{A}(t, \psi), \phi - \psi \rangle_U + \sum_{i=1}^{\infty} \|\mathcal{G}_i(t, \phi) - \mathcal{G}_i(t, \psi)\|_U^2$$

$$\leq c_t K_U(\phi, \psi) \left[1 + \|\phi\|_H^2 + \|\psi\|_H^2 \right] \|\phi - \psi\|_U^2 - \gamma \|\phi - \psi\|_H^2, \tag{11}$$

$$\sum_{i=1}^{\infty} \langle \mathcal{G}_i(t, \phi) - \mathcal{G}_i(t, \psi), \phi - \psi \rangle_U^2$$

$$\leq c_t K_U(\phi, \psi) \left[1 + \|\phi\|_H^2 + \|\psi\|_H^2 \right] \|\phi - \psi\|_U^4. \tag{12}$$

Assumption 3.4

$$2\langle \mathcal{A}(t, \phi), \phi \rangle_U + \sum_{i=1}^{\infty} \|\mathcal{G}_i(t, \phi)\|_U^2 \leq c_t K_U(\phi) \left[1 + \|\phi\|_H^2 \right], \tag{13}$$

$$\sum_{i=1}^{\infty} \langle \mathcal{G}_i(t, \phi), \phi \rangle_U^2 \leq c_t K_U(\phi) \left[1 + \|\phi\|_H^4 \right]. \tag{14}$$

Assumption 3.5

$$\langle \mathcal{A}(t, \phi) - \mathcal{A}(t, \psi), f \rangle_U \leq c_t K_U(\phi, \psi)(1 + \|f\|_H)$$

$$\times \left[1 + \|\phi\|_V + \|\psi\|_V \right] \|\phi - \psi\|_H. \tag{15}$$

3.3   Assumption Set B

Recall the setup and notation of Sect. 3.1. Suppose now that X is a separable Hilbert Space with continuous embedding $U \hookrightarrow X$. We ask that there is a continuous bilinear form $\langle \cdot, \cdot \rangle_{X \times H} : X \times H \to \mathbb{R}$ such that for $\phi \in U$ and $f \in H$,

$$\langle \phi, f \rangle_{X \times H} = \langle \phi, f \rangle_U . \tag{16}$$

Moreover it is now necessary that the system (a_k) forms an orthogonal basis of U.[3] The operators $\mathcal{A}$ and $\mathcal{G}$ must now be extended to the larger spaces, and are such that for any $T > 0$, $\mathcal{A} : [0, T] \times H \to X$ and $\mathcal{G} : [0, T] \times H \to \mathscr{L}^2(\mathfrak{U}; U)$ are measurable. We assume that there exists a $c.$, K and $\gamma > 0$ such that for all $\phi \in V$, $f, g \in H$ and $t \in [0, T]$:

Assumption 3.6

$$\|\mathcal{A}(t, f)\|_X^2 + \sum_{i=1}^{\infty} \|\mathcal{G}_i(t, f)\|_U^2 \leq c_t K_U(f) \left[1 + \|f\|_H^2 \right], \tag{17}$$

$$\|\mathcal{A}(t, f) - \mathcal{A}(t, g)\|_X^2 \leq c_t K_U(f, g) \left[1 + \|f\|_H^2 + \|g\|_H^2 \right] \|f - g\|_H^2 \tag{18}$$

Assumption 3.7

$$2 \langle \mathcal{A}(t, f) - \mathcal{A}(t, g), f - g \rangle_X + \sum_{i=1}^{\infty} \|\mathcal{G}_i(t, f) - \mathcal{G}_i(t, g)\|_X^2 \leq c_t K_U(f, g)$$

$$\times \left[1 + \|f\|_H^2 + \|g\|_H^2 \right] \|f - g\|_X^2 , \tag{19}$$

$$\sum_{i=1}^{\infty} \langle \mathcal{G}_i(t, f) - \mathcal{G}_i(t, g), f - g \rangle_X^2 \leq c_t K_U(f, g)$$

$$\times \left[1 + \|f\|_H^2 + \|g\|_H^2 \right] \|f - g\|_X^4 \tag{20}$$

Assumption 3.8

$$2 \langle \mathcal{A}(t, \phi), \phi \rangle_U + \sum_{i=1}^{\infty} \|\mathcal{G}_i(t, \phi)\|_U^2 \leq c_t K_U(\phi) - \gamma \|\phi\|_H^2 , \tag{21}$$

$$\sum_{i=1}^{\infty} \langle \mathcal{G}_i(t, \phi), \phi \rangle_U^2 \leq c_t K_U(\phi). \tag{22}$$

---

[3] Therefore, the spaces V, H, must be dense in U.

3.4   H-Valued Solutions

We state the definitions and main result for H-valued solutions.

Definition 3.9 Let $\Psi_0 : \Omega \rightarrow H$ be $\mathcal{F}_0$- measurable. A pair (Ψ, τ) where τ is a $\mathbb{P} - a.s.$ positive stopping time and Ψ is a process such that for $\mathbb{P} - a.e.\ \omega$, $\Psi.(\omega) \in C([0, T]; H)$ and $\Psi.(\omega)\mathbb{1}._{\leq \tau(\omega)} \in L^2([0, T]; V)$ with $\Psi.\mathbb{1}._{\leq \tau}$ progressively measurable in V, is said to be an H-valued local strong solution of the Eq. (2) if the identity

$$\Psi_t = \Psi_0 + \int_0^{t \wedge \tau} \mathcal{A}(s, \Psi_s)ds + \int_0^{t \wedge \tau} \mathcal{G}(s, \Psi_s)dW_s \tag{23}$$

holds $\mathbb{P} - a.s.$ in U for all $t \in [0, T]$.

Remark 3.10 If (Ψ, τ) is an H-valued local strong solution of the Eq. (2), then $\Psi. = \Psi._{\wedge \tau}$ due to the identity (23).

Definition 3.11 A pair (Ψ, Θ) such that there exists a sequence of stopping times (θ_j) which are $\mathbb{P} - a.s.$ monotone increasing and convergent to Θ, whereby $(\Psi._{\wedge \theta_j}, \theta_j)$ is an H-valued local strong solution of the Eq. (2) for each j, is said to be an H-valued maximal strong solution of the Eq. (2) if for any other pair (Φ, Γ) with this property then $\Theta \leq \Gamma \ \mathbb{P} - a.s.$ implies $\Theta = \Gamma \ \mathbb{P} - a.s..$

Remark 3.12 We do not require Θ to be finite in this definition, in which case we mean that the sequence (θ_j) is monotone increasing and unbounded for such ω.

Definition 3.13 An H-valued maximal strong solution (Ψ, Θ) of the Eq. (2) is said to be unique if for any other such solution (Φ, Γ), then $\Theta = \Gamma \ \mathbb{P} - a.s.$ and

$$\mathbb{P}\left(\{\omega \in \Omega : \Psi_t(\omega) = \Phi_t(\omega) \quad \forall t \in [0, \Theta)\}\right) = 1.$$

Theorem 3.14 *Let Assumption Set A hold. For any given $\mathcal{F}_0$- measurable $\Psi_0 : \Omega \rightarrow H$, there exists a unique H-valued maximal strong solution (Ψ, Θ) of the Eq. (2). Moreover at $\mathbb{P} - a.e.\ \omega$ for which $\Theta(\omega) < \infty$, we have that*

$$\sup_{r \in [0, \Theta(\omega))} \|\Psi_r(\omega)\|_H^2 + \int_0^{\Theta(\omega)} \|\Psi_r(\omega)\|_V^2 \, dr = \infty. \tag{24}$$

Proof See [29] Theorem 3.15.   □

3.5   U-Valued Solutions

We state the definitions and main result for U-Valued Solutions.

Definition 3.15 Let $\Psi_0 : \Omega \to U$ be $\mathcal{F}_0$-measurable. A pair (Ψ, τ) where τ is a $\mathbb{P} - a.s.$ positive stopping time and Ψ is a process such that for $\mathbb{P} - a.e.\ \omega$, $\Psi.(\omega) \in C([0, T]; U)$ and $\Psi.(\omega)\mathbb{1}_{\leq \tau(\omega)} \in L^2([0, T]; H)$ with $\Psi.\mathbb{1}_{\leq \tau}$ progressively measurable in H, is said to be a U-valued local strong solution of the Eq. (2) if the identity

$$\Psi_t = \Psi_0 + \int_0^{t \wedge \tau} \mathcal{A}(s, \Psi_s)ds + \int_0^{t \wedge \tau} \mathcal{G}(s, \Psi_s)dW_s \qquad (25)$$

holds $\mathbb{P} - a.s.$ in X for all $t \in [0, T]$.

Definition 3.16 A pair (Ψ, Θ) such that there exists a sequence of stopping times (θ_j) which are $\mathbb{P} - a.s.$ monotone increasing and convergent to Θ, whereby $(\Psi_{\cdot \wedge \theta_j}, \theta_j)$ is a U-valued local strong solution of the Eq. (2) for each j, is said to be a U-valued maximal strong solution of the Eq. (2) if for any other pair (Φ, Γ) with this property then $\Theta \leq \Gamma\ \mathbb{P} - a.s.$ implies $\Theta = \Gamma\ \mathbb{P} - a.s..$

Definition 3.17 A U-valued maximal strong solution (Ψ, Θ) of the Eq. (2) is said to be unique if for any other such solution (Φ, Γ), then $\Theta = \Gamma\ \mathbb{P} - a.s.$ and

$$\mathbb{P}(\{\omega \in \Omega : \Psi_t(\omega) = \Phi_t(\omega) \quad \forall t \in [0, \Theta)\}) = 1.$$

Theorem 3.18 *Let Assumption Sets A and B hold. For any given $\mathcal{F}_0$- measurable $\Psi_0 : \Omega \to U$, there exists a unique U-valued maximal strong solution (Ψ, Θ) of the Eq. (2). Moreover at $\mathbb{P} - a.e.\ \omega$ for which $\Theta(\omega) < \infty$, we have that*

$$\sup_{r \in [0, \Theta(\omega))} \|\Psi_r(\omega)\|_U^2 + \int_0^{\Theta(\omega)} \|\Psi_r(\omega)\|_H^2\, dr = \infty. \qquad (26)$$

Proof See [29] Theorem 4.9. □

4 Global Theory for SPDEs

4.1 Functional Framework

Recall that our object of study is the Itô SPDE (2),

$$\Psi_t = \Psi_0 + \int_0^t \mathcal{A}(s, \Psi_s)ds + \int_0^t \mathcal{G}(s, \Psi_s)dW_s$$

which we pose for a triplet of embedded separable Hilbert Spaces

$$V \hookrightarrow H \hookrightarrow U$$

whereby the embeddings are continuous linear injections. The Eq. (2) is posed on a time interval $[0, T]$ for arbitrary but henceforth fixed $T \geq 0$. The mappings $\mathcal{A}, \mathcal{G}$ are such that $\mathcal{A} : [0, T] \times V \to U, \mathcal{G} : [0, T] \times H \to \mathscr{L}^2(\mathfrak{U}; U)$ are measurable. Understanding $\mathcal{G}$ as a mapping $\mathcal{G} : [0, T] \times H \times \mathfrak{U} \to U$, we introduce the notation $\mathcal{G}_i(\cdot, \cdot) := \mathcal{G}(\cdot, \cdot, e_i)$. We further impose the existence of a system of elements (a_k) of V which form an orthogonal basis of U and a basis of H. Let us define the spaces $V_n := \mathrm{span}\{a_1, \ldots, a_n\}$ and $\mathcal{P}_n$ as the orthogonal projection to V_n in U, that is

$$\mathcal{P}_n : f \mapsto \sum_{k=1}^{n} \langle f, a_k \rangle_U \, a_k.$$

It is required that the $(\mathcal{P}_n)$ are uniformly bounded in H, which is to say that there exists a constant c independent of n such that for all $f \in H$,

$$\|\mathcal{P}_n f\|_H \leq c \|f\|_H . \tag{27}$$

Moreover, our setup can be expanded by considering the induced Gelfand Triple

$$H \hookrightarrow U \hookrightarrow H^*$$

defined relative to the inclusion mapping $i : H \to U$; indeed, the embedding of U into H^* is given by the composition of the isomorphism mapping U into U^* with the adjoint $i^* : U^* \to H^*$. In particular, the duality pairing between H and H^*, $\langle \cdot, \cdot \rangle_{H^* \times H}$, is compatible with $\langle \cdot, \cdot \rangle_U$ in the sense that for any $f \in U, g \in H$,

$$\langle f, g \rangle_{H^* \times H} = \langle f, g \rangle_U .$$

We assume that $\mathcal{A} : [0, T] \times H \to H^*$ is measurable. Specific bounds on the mappings $\mathcal{A}$ and $\mathcal{G}$ will be imposed in Assumption Sets 1, 2 and 3. We shall again use the notation of K from Sect. 3.1.

4.2 Assumption Set 1

Recall the setup and notation of Sect. 4.1. We assume that there exists a $c., K$ and $\gamma > 0$ such that for all $\phi, \psi \in V, f \in H$ and $t \in [0, T]$:

Assumption 4.1

$$\|\mathcal{A}(t, f)\|_{H^*} + \sum_{i=1}^{\infty} \|\mathcal{G}_i(t, f)\|_U^2 \leq c_t K_U(f) \left[1 + \|f\|_H^2\right], \tag{28}$$

$$\|\mathcal{A}(t, \phi) - \mathcal{A}(t, \psi)\|_U^2 \leq c_t K_V \|\phi - \psi\|_V^2 , \tag{29}$$

$$\sum_{i=1}^{\infty} \|\mathcal{G}_i(t, \phi) - \mathcal{G}_i(t, \psi)\|_U^2 \leq c_t K_V(\phi, \psi) \|\phi - \psi\|_H^2 . \tag{30}$$

Assumption 4.2

$$2 \langle \mathcal{A}(t, \phi), \phi \rangle_U + \sum_{i=1}^{\infty} \|\mathcal{G}_i(t, \phi)\|_U^2 \leq c_t \left[1 + \|\phi\|_U^2 \right] - \gamma \|\phi\|_H^2 , \tag{31}$$

$$\sum_{i=1}^{\infty} \langle \mathcal{G}_i(t, \phi), \phi \rangle_U^2 \leq c_t \left[1 + \|\phi\|_U^4 \right] . \tag{32}$$

Assumption 4.3[4]

$$\langle \mathcal{A}(t, \phi), f \rangle_U \leq c_t \left[K_U(\phi) + \|\phi\|_H^{\frac{3}{2}} \right] \left[K_U(f) + \|f\|_H^{\frac{3}{2}} \right] , \tag{33}$$

$$\sum_{i=1}^{\infty} \langle \mathcal{G}_i(t, \phi), f \rangle_U^2 \leq c_t K_U(\phi) K_H(f) . \tag{34}$$

Assumption 4.4

$$\langle \mathcal{A}(t, \phi) - A(t, f), \psi \rangle_{H^* \times H} \leq c_t K_V(\psi) \left[1 + \|\phi\|_H + \|f\|_H \right] \|\phi - f\|_U ,$$

$$\tag{35}$$

$$\sum_{i=1}^{\infty} \langle \mathcal{G}_i(t, \phi) - \mathcal{G}_i(t, f), \psi \rangle_U^2 \leq c_t K_V(\psi) \|\phi - f\|_U^2 . \tag{36}$$

4.3 Assumption Set 2

Recall the setup and notation of Sect. 4.1. We assume that there exists a $c.$, K and $\gamma > 0$ such that for all $f, g \in H$ and $t \in [0, T]$:

Assumption 4.5

$$\|\mathcal{A}(t, f)\|_{H^*}^2 \leq c_t K_U(f) \left[1 + \|f\|_H^2 \right] . \tag{37}$$

---

[4] In fact in (33), the exponent 3/2 could be replaced by any $q < 2$.

Assumption 4.6

$$2 \langle \mathcal{A}(t, f) - \mathcal{A}(t, g), f - g \rangle_{H^* \times H} + \sum_{i=1}^{\infty} \|\mathcal{G}_i(t, f) - \mathcal{G}_i(t, g)\|_U^2$$

$$\leq c_t K_U(f, g) \left[1 + \|f\|_H^2 + \|g\|_H^2 \right] \|f - g\|_U^2 - \gamma \|f - g\|_H^2, \qquad (38)$$

$$\sum_{i=1}^{\infty} \langle \mathcal{G}_i(t, f) - \mathcal{G}_i(t, g), f - g \rangle_U^2$$

$$\leq c_t K_U(f, g) \left[1 + \|f\|_H^2 + \|g\|_H^2 \right] \|f - g\|_U^4. \qquad (39)$$

4.4 Assumption Set 3

Recall the setup and notation of Sect. 4.1. We now impose the existence of a new Banach Space $\bar{H}$ which is an extension of H, or precisely, $H \subseteq \bar{H} \subseteq U$ and for every $f \in \bar{H}$, $\|f\|_{\bar{H}} = \|f\|_H$. In addition, $\mathcal{G} : [0, T] \times V \to \mathcal{L}^2(\mathfrak{U}; \bar{H})$ is assumed measurable. We also suppose that there exists a real valued sequence (μ_n) with $\mu_n \to \infty$ such that for any $f \in \bar{H}$,

$$\|(I - \mathcal{P}_n) f\|_U \leq \frac{1}{\mu_n} \|f\|_{\bar{H}} \qquad (40)$$

where I represents the identity operator in U. Furthermore we assume that there exists a $\gamma > 0$ such that for any $\varepsilon > 0$, there exists a $c., K$ (dependent on ε) such that for any $\phi \in V$, $\phi^n \in V_n$ and $t \in [0, T]$:

Assumption 4.7

$$\|\mathcal{A}(t, \phi)\|_U^2 + \sum_{i=1}^{\infty} \|\mathcal{G}_i(t, \phi)\|_{\bar{H}}^2 \leq c_t K_U(\phi) \left[1 + \|\phi\|_H^4 + \|\phi\|_V^2 \right] \qquad (41)$$

Assumption 4.8

$$2 \langle \mathcal{P}_n \mathcal{A}(t, \phi^n), \phi^n \rangle_H + \sum_{i=1}^{\infty} \|\mathcal{P}_n \mathcal{G}_i(t, \phi^n)\|_H^2$$

$$\leq c_t K_U(\phi^n) \left[1 + \|\phi^n\|_H^4 \right] - \gamma \|\phi^n\|_V^2, \qquad (42)$$

$$\sum_{i=1}^{\infty} \langle \mathcal{P}_n \mathcal{G}_i(t, \phi^n), \phi^n \rangle_H^2 \leq c_t K_U(\phi^n) \left[1 + \|\phi^n\|_H^6 \right] + \varepsilon \|\phi^n\|_V^2. \qquad (43)$$

4.5 Martingale Weak Solutions

We now state the definition and main result for martingale weak solutions.

Definition 4.9 Let $\boldsymbol{\Psi}_0 : \Omega \to U$ be $\mathcal{F}_0$-measurable. If there exists a filtered probability space $\left(\tilde{\Omega}, \tilde{\mathcal{F}}, (\tilde{\mathcal{F}}_t), \tilde{\mathbb{P}}\right)$, a Cylindrical Brownian Motion $\tilde{\mathcal{W}}$ over $\mathfrak{U}$ with respect to $\left(\tilde{\Omega}, \tilde{\mathcal{F}}, (\tilde{\mathcal{F}}_t), \tilde{\mathbb{P}}\right)$, an $\mathcal{F}_0$-measurable $\tilde{\boldsymbol{\Psi}}_0 : \tilde{\Omega} \to U$ with the same law as $\boldsymbol{\Psi}_0$, and a progressively measurable process $\tilde{\boldsymbol{\Psi}}$ in H such that for $\tilde{\mathbb{P}} - a.e.$ $\tilde{\omega}$, $\tilde{\boldsymbol{\Psi}}.(\omega) \in C_w\left([0, T]; U\right) \cap L^2\left([0, T]; H\right)^5$ and

$$\tilde{\boldsymbol{\Psi}}_t = \tilde{\boldsymbol{\Psi}}_0 + \int_0^t \mathcal{A}(s, \tilde{\boldsymbol{\Psi}}_s)ds + \int_0^t \mathcal{G}(s, \tilde{\boldsymbol{\Psi}}_s)d\mathcal{W}_s \tag{44}$$

holds $\tilde{\mathbb{P}} - a.s.$ in H^* for all $t \in [0, T]$, then $\tilde{\boldsymbol{\Psi}}$ is said to be a martingale weak solution of the Eq. (2).

Theorem 4.10 *Let Assumption Set 1 hold. For any given $\mathcal{F}_0$-measurable $\boldsymbol{\Psi}_0 \in L^\infty(\Omega; U)$, there exists a martingale weak solution of the Eq. (2).*

Proof See [26] Theorem 2.7. □

4.6 Weak Solutions

We now state the definitions and main result for weak solutions.

Definition 4.11 Let $\boldsymbol{\Psi}_0 : \Omega \to U$ be $\mathcal{F}_0$-measurable. A process $\boldsymbol{\Psi}$ which is progressively measurable in H and such that for $\mathbb{P}-a.e.$ ω, $\boldsymbol{\Psi}.(\omega) \in C\left([0, T]; U\right) \cap L^2\left([0, T]; H\right)$, is said to be a weak solution of the Eq. (2) if the identity (2) holds $\mathbb{P} - a.s.$ in H^* for all $t \in [0, T]$.

Definition 4.12 A weak solution $\boldsymbol{\Psi}$ of the Eq. (2) is said to be the unique solution if for any other such solution $\boldsymbol{\Phi}$,

$$\mathbb{P}\left(\{\omega \in \Omega : \boldsymbol{\Psi}_t(\omega) = \boldsymbol{\Phi}_t(\omega) \quad \forall t \geq 0\}\right) = 1.$$

Theorem 4.13 *Let Assumption Sets 1 and 2 hold. For any given $\mathcal{F}_0$-measurable $\boldsymbol{\Psi}_0 : \Omega \to U$, there exists a unique weak solution of the Eq. (2).*

Proof See [26] Theorem 3.5. □

---

[5] Note that $C_w\left([0, T]; U\right) \subseteq L^\infty\left([0, T]; U\right)$.

4.7 Strong Solutions

We now state the definitions and main result for strong solutions.

Definition 4.14 Let $\Psi_0 : \Omega \to H$ be $\mathcal{F}_0$-measurable. A process Ψ which is progressively measurable in V and such that for $\mathbb{P} - a.e.\ \omega$, $\Psi_\cdot(\omega) \in L^\infty([0, T]; H) \cap L^2([0, T]; V)$, is said to be a strong solution of the Eq. (2) if the identity (2) holds $\mathbb{P} - a.s.$ in U for all $t \in [0, T]$.

Note that a strong solution necessarily has continuous paths in U, from the evolution equation satisfied in this space.

Definition 4.15 A strong solution Ψ of the Eq. (2) is said to be unique if for any other such solution Φ,

$$\mathbb{P}(\{\omega \in \Omega : \Psi_t(\omega) = \Phi_t(\omega) \quad \forall t \geq 0\}) = 1.$$

Theorem 4.16 *Let Assumption Sets 1, 2 and 3 hold. For any given $\mathcal{F}_0$-measurable $\Psi_0 : \Omega \to H$, there exists a unique strong solution of the Eq. (2).*

Proof See [26] Theorem 4.5.   □

5 Stochastic Navier-Stokes Equations

We recall the SALT Navier-Stokes Equation (1) stated in the introduction, given by

$$u_t = u_0 - \int_0^t \mathcal{L}_{u_s} u_s \, ds + \nu \int_0^t \Delta u_s \, ds + \int_0^t B(u_s) \circ d\mathcal{W}_s - \nabla \rho_t.$$

A complete introduction to this equation is given in [27], with full technical details that are glossed over here, although we note that our results can be applied for a variety of additive, multiplicative and transport noise structures. The choice of SALT noise is particularly challenging and demonstrates the efficacy of the frameworks from Sects. 3 and 4. The SALT operator B is given by the actions of its components B_i on a vector field ϕ, relative to the collection of vector fields (ξ_i), by

$$B_i : \phi \mapsto \sum_{j=1}^{N} \left(\xi_i^j \partial_j \phi + \phi^j \nabla \xi_i^j \right)$$

in N-dimensions, where the superscript denotes the jth component mapping. We note this is the sum of a classical transport term and a zeroth-order term. Here and throughout this section, we work with $N = 2$ or 3. The nonlinear term is formally defined by

$$\mathcal{L}_f g = \sum_{j=1}^{N} f^j \partial_j g$$

with Laplacian $\Delta f = \sum_{j=1}^{N} \partial_j^2 f$. The Eq. (1) is to be posed on either the N-dimensional torus $\mathbb{T}^N$ or a smooth bounded domain $\mathcal{O}$ with the no-slip or Navier boundary conditions, on a fixed time interval $[0, T]$. In all cases we require the divergence-free property of solutions, which is to say that $\sum_{j=1}^{N} \partial_j u^j = 0$. To facilitate the analysis, we introduce some additional function spaces. Recall that any function $f \in L^2(\mathbb{T}^N; \mathbb{R}^N)$ admits the representation

$$f(x) = \sum_{k \in \mathbb{Z}^N} f_k e^{ik \cdot x} \tag{45}$$

where by each $f_k \in \mathbb{C}^N$ is such that $f_k = \overline{f_{-k}}$ and the infinite sum is defined as a limit in $L^2(\mathbb{T}^N; \mathbb{R}^N)$, see e.g. [49] Subsection 1.5 for details.

Definition 5.1 We define $L^2_\sigma(\mathbb{T}^N; \mathbb{R}^N)$ as the subset of $\dot{L}^2(\mathbb{T}^N; \mathbb{R}^N)$ of functions f whereby for all $k \in \mathbb{Z}^N$, $k \cdot f_k = 0$ with f_k as in (45). For general $m \in \mathbb{N}$ we introduce $W^{m,2}_\sigma(\mathbb{T}^N; \mathbb{R}^N)$ as the intersection of $W^{m,2}(\mathbb{T}^N; \mathbb{R}^N)$ respectively with $L^2_\sigma(\mathbb{T}^N; \mathbb{R}^N)$.

Definition 5.2 We define $C^\infty_{0,\sigma}(\mathcal{O}; \mathbb{R}^N)$ as the subset of $C^\infty_0(\mathcal{O}; \mathbb{R}^N)$ of functions which are divergence-free. $L^2_\sigma(\mathcal{O}; \mathbb{R}^N)$ is defined as the completion of $C^\infty_{0,\sigma}(\mathcal{O}; \mathbb{R}^N)$ in $L^2(\mathcal{O}; \mathbb{R}^N)$, whilst we introduce $W^{1,2}_\sigma(\mathcal{O}; \mathbb{R}^N)$ as the intersection of $W^{1,2}_0(\mathcal{O}; \mathbb{R}^N)$ with $L^2_\sigma(\mathcal{O}; \mathbb{R}^N)$ and $W^{2,2}_\sigma(\mathcal{O}; \mathbb{R}^N)$ as the intersection of $W^{2,2}(\mathcal{O}; \mathbb{R}^N)$ with $W^{1,2}_\sigma(\mathcal{O}; \mathbb{R}^N)$.

Henceforth, we shall use the notation L^2_σ, $W^{m,2}_\sigma$ to represent the above spaces where the domain and dimensionality are clear from the context: we do the same for the general L^2, $W^{m,2}$ spaces. We define the Leray Projector $\mathcal{P}$ as the orthogonal projection in L^2 onto L^2_σ. In any context, for $m = 1, 2$, the inner product

$$\langle f, g \rangle_m := \left\langle (-\mathcal{P}\Delta)^{m/2} f, (-\mathcal{P}\Delta)^{m/2} g \right\rangle \tag{46}$$

is equivalent to the usual $W^{m,2}$ inner product on $W^{m,2}_\sigma$ and we consider $W^{m,2}_\sigma$ as a Hilbert Space equipped with this inner product. Further details can be found in [11] Proposition 4.12, [49] Exercises 2.12, 2.13 and the discussion in Subsection 2.3. To study (1) more freely, we commit two manipulations of it; the first is to project via $\mathcal{P}$, and the second is to convert to Itô Form. With this, we arrive at

$$u_t = u_0 - \int_0^t \mathcal{P}\mathcal{L}_{u_s} u_s \, ds + \nu \int_0^t \mathcal{P}\Delta u_s \, ds$$

$$+ \frac{1}{2} \int_0^t \sum_{i=1}^\infty \mathcal{P}B_i^2 u_s ds + \int_0^t \mathcal{P}B(u_s) d\mathcal{W}_s \tag{47}$$

which is now in the form of (2). We emphasise again that a thorough overview of this process is given in [27]. Various applications are given below.

5.1 The Torus

The torus represents, mathematically, the simplest domain on which to pose (47). The zero-average and divergence-free constraints are imposed, both of which are included in $W_\sigma^{1,2}$. Thus, with the right function spaces, a proper formulation of the problem becomes simple. We first consider global solutions, provided by the theory of Sect. 4. The functional framework of Sect. 4.1 is satisfied for the spaces

$$V := W_\sigma^{2,2}, \qquad H := W_\sigma^{1,2}, \qquad U := L_\sigma^2$$

and the system (a_k) of eigenfunctions of the Stokes Operator $-\mathcal{P}\Delta$ (see [49] Theorem 2.24). The mappings $\mathcal{P}\mathcal{L}$, $\mathcal{P}\Delta$ are understood from $W_\sigma^{1,2}$ into $\left(W_\sigma^{1,2}\right)^*$ by the duality pairings for $f, g \in W_\sigma^{1,2}$ of

$$\left\langle \mathcal{P}\mathcal{L}_f f, g \right\rangle_{\left(W_\sigma^{1,2}\right)^* \times W_\sigma^{1,2}} = \left\langle \mathcal{L}_f f, g \right\rangle_{L^{6/5} \times L^6}$$

$$\left\langle \mathcal{P}\Delta f, g \right\rangle_{\left(W_\sigma^{1,2}\right)^* \times W_\sigma^{1,2}} = -\left\langle f, g \right\rangle_1 .$$

The first expression arises from the Sobolev Embedding of $W^{1,2}(\mathcal{O}; \mathbb{R}^3)$ into $L^6(\mathcal{O}; \mathbb{R}^3)$, and the Hölder conjugation between $L^6(\mathcal{O}; \mathbb{R}^3)$ and $L^{6/5}(\mathcal{O}; \mathbb{R}^3)$; see e.g. [28] equation (25). As a direct application of Theorem 4.10, we can obtain martingale weak solutions.

Theorem 5.3 *Let* $N = 2$ *or* 3, $u_0 \in L^\infty\left(\Omega; L_\sigma^2\right)$ *be* $\mathcal{F}_0$-*measurable,* $(\xi_i) \in W_\sigma^{1,2} \cap W^{2,\infty}$ *with* $\sum_{i=1}^\infty \|\xi_i\|_{W^{2,\infty}}^2 < \infty$. *Then there exists a filtered probability space* $\left(\tilde{\Omega}, \tilde{\mathcal{F}}, (\tilde{\mathcal{F}}_t), \tilde{\mathbb{P}}\right)$, *a Cylindrical Brownian Motion* $\tilde{\mathcal{W}}$ *over* $\mathfrak{U}$ *with respect to* $\left(\tilde{\Omega}, \tilde{\mathcal{F}}, (\tilde{\mathcal{F}}_t), \tilde{\mathbb{P}}\right)$, *an* $\mathcal{F}_0$-*measurable* $\tilde{u}_0 : \tilde{\Omega} \to U$ *with the same law as* u_0, *and a progressively measurable process* $\tilde{u}$ *in* $W_\sigma^{1,2}$ *such that for* $\tilde{\mathbb{P}} - a.e.$ $\tilde{\omega}$, $\tilde{u}.(\omega) \in L^\infty\left([0, T]; L_\sigma^2\right) \cap C_w\left([0, T]; L_\sigma^2\right) \cap L^2\left([0, T]; W_\sigma^{1,2}\right)$ *and*

$$\tilde{u}_t = \tilde{u}_0 - \int_0^t \mathcal{P}\mathcal{L}_{\tilde{u}_s} \tilde{u}_s \, ds + \nu \int_0^t \mathcal{P}\Delta\tilde{u}_s \, ds + \frac{1}{2} \int_0^t \sum_{i=1}^\infty \mathcal{P}B_i^2 \tilde{u}_s ds$$

$$+ \int_0^t \mathcal{P}B(\tilde{u}_s) d\tilde{\mathcal{W}}_s$$

holds $\tilde{\mathbb{P}} - a.s.$ *in* $\left(W_\sigma^{1,2}\right)^*$ *for all* $t \in [0, T]$.

In fact we can do better in 2D, as an application of Theorem 4.13.

Theorem 5.4 *Let* $N = 2$, $u_0 : \Omega \to L_\sigma^2$ *be* $\mathcal{F}_0$-*measurable,* $(\xi_i) \in W_\sigma^{1,2} \cap W^{2,\infty}$ *with* $\sum_{i=1}^\infty \|\xi_i\|_{W^{2,\infty}}^2 < \infty$. *Then there exists a unique progressively measurable process* u *in* $W_\sigma^{1,2}$ *such that for* $\mathbb{P} - a.e.$ $\tilde{\omega}$, $u.(\omega) \in C\left([0, T]; L_\sigma^2\right) \cap L^2\left([0, T]; W_\sigma^{1,2}\right)$ *and (47) holds* $\mathbb{P} - a.s.$ *in* $\left(W_\sigma^{1,2}\right)^*$ *for all* $t \in [0, T]$.

These results are proven in [28] Theorems 1.9 and 1.10, stated slightly differently although Lemma 3.9 connects the definitions. The results there are proven in the case of the no-slip boundary condition, but there is no difference for the torus. More than that, we can invoke Theorem 4.16 to obtain the existence of strong solutions of the Eq. (47).

Theorem 5.5 *Let* $N = 2$, $u_0 : \Omega \to W_\sigma^{1,2}$ *be* $\mathcal{F}_0$-*measurable,* $(\xi_i) \in L_\sigma^2 \cap W^{3,\infty}$ *with* $\sum_{i=1}^\infty \|\xi_i\|_{W^{3,\infty}}^2 < \infty$. *Then there exists a unique progressively measurable process* u *in* $W_\sigma^{2,2}$ *such that for* $\mathbb{P} - a.e.$ ω, $u.(\omega) \in C\left([0, T]; W_\sigma^{1,2}\right) \cap L^2\left([0, T]; W_\sigma^{2,2}\right)$ *and (47) holds* $\mathbb{P} - a.s.$ *in* L_σ^2 *for all* $t \in [0, T]$.

Here, use of [26] Lemma 4.14 is required to obtain the continuity. The extension $\bar{H}$ of H in Assumption Set 3, Sect. 4.4, can simply be taken as H itself. With some concessions, the corresponding existence result in the case of a bounded domain is also obtainable as seen in the following subsections. However, use of the torus allows us to obtain a strong existence result in 3D; the analogous result for the no-slip boundary condition is still open, due to a problematic boundary integral arising in controlling the noise. This comes as an application of the local theory of Sect. 3. The functional framework of Sect. 3.1 and Assumption Set B is satisfied for the spaces

$$V := W_\sigma^{3,2}, \qquad H := W_\sigma^{2,2}, \qquad U := W_\sigma^{1,2}, \qquad X := L_\sigma^2$$

where $W_\sigma^{3,2}$ is equipped with the $\langle \cdot, \cdot \rangle_3$ inner product as defined in (46).[6] As a direct application of Theorem 3.18, we obtain the following.

Theorem 5.6 *Let* $N = 3$, $u_0 : \Omega \to W_\sigma^{1,2}$ *be* $\mathcal{F}_0$-*measurable,* $(\xi_i) \in L_\sigma^2 \cap W^{3,\infty}$ *with* $\sum_{i=1}^\infty \|\xi_i\|_{W^{3,\infty}}^2 < \infty$. *Then there exists a unique* U-*valued maximal strong solution* (u, Θ) *of the Eq. (47) in the sense of Definitions 3.15, 3.16, 3.17. Moreover at* $\mathbb{P} - a.e.$ ω *for which* $\Theta(\omega) < \infty$, *we have that*

$$\sup_{r \in [0, \Theta(\omega))} \|u_r(\omega)\|_1^2 + \int_0^{\Theta(\omega)} \|u_r(\omega)\|_2^2 \, dr = \infty. \tag{48}$$

The result in the case of the H-valued solution, coming from Theorem 3.14, is particularly interesting. The additional degree of regularity obtained in these solu-

---

[6] For the bounded domain, this does not define an equivalent inner product to the usual $W^{3,2}$ one on $W_\sigma^{3,2}$.

tions is pertinent for the Itô-Stratonovich conversion, which has been little-discussed here but was greatly emphasised in [24] Subsection 2.3, [27] Subsection 3.1. We first state the following proposition, which was [27] Proposition 3.2.

Proposition 5.7 *Suppose that* (u, τ) *are such that:* τ *is a* $\mathbb{P} - a.s.$ *positive stopping time and* u *is a process whereby for* $\mathbb{P} - a.e.$ ω, $u.(\omega) \in C\left([0, T]; W_\sigma^{2,2}\right)$ *and* $u.(\omega)\mathbb{1}._{\leq \tau(\omega)} \in L^2\left([0, T]; W_\sigma^{3,2}\right)$ *with* $u.\mathbb{1}._{\leq \tau}$ *progressively measurable in* $W_\sigma^{3,2}$, *and moreover satisfying the identity*

$$u_t = u_0 - \int_0^{t \wedge \tau} \mathcal{P}\mathcal{L}_{u_s} u_s \, ds + v \int_0^{t \wedge \tau} \mathcal{P}\Delta u_s \, ds$$
$$+ \frac{1}{2} \int_0^{t \wedge \tau} \sum_{i=1}^\infty \mathcal{P} B_i^2 u_s ds + \int_0^{t \wedge \tau} \mathcal{P} B u_s dW_s$$

$\mathbb{P} - a.s.$ *in* $W_\sigma^{1,2}$ *for all* $t \in [0, T]$. *Then the pair* (u, τ) *satisfies the identity*

$$u_t = u_0 - \int_0^{t \wedge \tau} \mathcal{P}\mathcal{L}_{u_s} u_s \, ds + v \int_0^{t \wedge \tau} \mathcal{P}\Delta u_s \, ds + \int_0^{t \wedge \tau} \mathcal{P} B u_s \circ dW_s$$

$\mathbb{P} - a.s.$ *in* L_σ^2 *for all* $t \in [0, T]$.

As a consequence of this, by applying Theorem 3.14, we obtain:

Theorem 5.8 *Let* $N = 3$, $u_0 : \Omega \to W_\sigma^{1,2}$ *be* $\mathcal{F}_0$-*measurable,* $(\xi_i) \in L_\sigma^2 \cap W^{3,\infty}$ *with* $\sum_{i=1}^\infty \|\xi_i\|_{W^{3,\infty}}^2 < \infty$. *Then there exists a pair* (u, τ) *such that:* τ *is a* $\mathbb{P} - a.s.$ *positive stopping time and* u *is a process whereby for* $\mathbb{P} - a.e.$ ω, $u.(\omega) \in C\left([0, T]; W_\sigma^{2,2}\right)$ *and* $u.(\omega)\mathbb{1}._{\leq \tau(\omega)} \in L^2\left([0, T]; W_\sigma^{3,2}\right)$ *with* $u.\mathbb{1}._{\leq \tau}$ *progressively measurable in* $W_\sigma^{3,2}$, *and moreover satisfying the identity*

$$u_t = u_0 - \int_0^{t \wedge \tau} \mathcal{P}\mathcal{L}_{u_s} u_s \, ds + v \int_0^{t \wedge \tau} \mathcal{P}\Delta u_s \, ds + \int_0^{t \wedge \tau} \mathcal{P} B u_s \circ dW_s$$

$\mathbb{P} - a.s.$ *in* L_σ^2 *for all* $t \in [0, T]$.

This result is proven in [27] Theorem 3.1.

5.2 No-Slip Boundary Condition

Let us impose the boundary condition $u = 0$ on $\partial\mathcal{O}$; this is the so-called no-slip boundary condition. Considering the global theory, the functional framework of Sect. 4.1 is satisfied for the spaces

$$V := W_\sigma^{2,2}, \qquad H := W_\sigma^{1,2}, \qquad U := L_\sigma^2$$

and the system (a_k) of eigenfunctions of the Stokes Operator $-\mathcal{P}\Delta$ (see [49] Theorem 2.24). In this setting, the identical results of Theorems 5.3 and 5.4 are achieved. The existence of strong solutions in 2D is far more challenging, however, and remains unsolved. Towards a success, we note the necessity of considering an extension $\bar{H}$ of H in Assumption Set 3, Sect. 4.4. The Leray Projector $\mathcal{P}$ does not preserve the zero-trace property, so $\mathcal{P}B_i$ does not map from $W_\sigma^{2,2}$ into $W_\sigma^{1,2}$, but instead an extended space $\bar{W}_\sigma^{1,2}$ defined as the intersection of $W^{1,2}$ with L_σ^2. This is again a Hilbert Space with $\langle \cdot, \cdot \rangle_1$ inner product; see [25] Subsection 1.2.

Nevertheless, we are still unable to verify (40) and Assumption 4.8. This owes to the fact that $\mathcal{P}_n$ is self-adjoint only on $W_\sigma^{1,2}$ and not $\bar{W}_\sigma^{1,2}$, leaving us stuck with the finite dimensional projection in a way which offers no clear solution. The situation is different for the Navier boundary conditions.

5.3 Navier Boundary Conditions

In two spatial dimensions we can impose different boundary conditions for (47), namely the Navier boundary conditions. These are defined on $\partial\mathscr{O}$ by

$$u \cdot \mathbf{n} = 0, \qquad 2(Du)\mathbf{n} \cdot \iota + \alpha u \cdot \iota = 0 \tag{49}$$

where $\mathbf{n}$ is the unit outwards normal vector, ι the unit tangent vector, Du is the rate of strain tensor $(Du)^{k,l} := \frac{1}{2}\left(\partial_k u^l + \partial_l u^k\right)$ and $\alpha \in C^2(\partial\mathscr{O}; \mathbb{R})$ represents a friction coefficient which determines the extent to which the fluid slips on the boundary relative to the tangential stress. These conditions were first proposed by Navier in [45, 46], and have been derived in [[42]] from the kinetic theory of gases and in [41] as a hydrodynamic limit. Furthermore these conditions have proven viable for modelling rough boundaries as seen in [3, 22, 48]. To fit the framework of this chapter we again have to embed the boundary conditions into useful function spaces. We shall use the space $\bar{W}_\sigma^{1,2}$, which was the intersection of $W^{1,2}$ with L_σ^2, and contains the divergence-free and impermeable boundary condition (that $u \cdot \mathbf{n} = 0$). The remaining component of (49) has to be included at the $W^{2,2}$ level, as we are concerned with the trace of a derivative which needs more than $W^{1,2}$ regularity to be understood in the usual sense. Thus, we define

$$\bar{W}_\alpha^{2,2} := \left\{ f \in W^{2,2}(\mathscr{O}; \mathbb{R}^2) \cap \bar{W}_\sigma^{1,2} : 2(Df)\mathbf{n} \cdot \iota + \alpha f \cdot \iota = 0 \text{ on } \partial\mathscr{O} \right\}.$$

Of course this space does not appear in the definition of a weak solution, so it is perhaps unclear how the boundary conditions (49) inform the weak solution. The answer comes from how to extend the Stokes Operator $-\mathcal{P}\Delta$ to $\bar{W}_\sigma^{1,2}$. In [35] equation (5.1), c.f. [25] Lemma 1.4, it is verified that for $\phi \in \bar{W}_\alpha^{2,2}$, $f \in \bar{W}_\sigma^{1,2}$,

$$\langle \mathcal{P}\Delta\phi, f \rangle_{L^2} = -\langle \phi, f \rangle_1 + \langle (\kappa - \alpha)\phi, f \rangle_{L^2(\partial\mathscr{O}; \mathbb{R}^2)}$$

where $\kappa : \partial\mathcal{O} \to \mathbb{R}$ represents the curvature of the boundary. Therefore, for each α as in (49), we extend the Stokes Operator from $\bar{W}_\alpha^{2,2}$ to $\bar{W}_\sigma^{1,2}$ as a mapping into $\left(\bar{W}_\sigma^{1,2}\right)^*$ by the duality pairing for g, $f \in \bar{W}_\sigma^{1,2}$ of

$$\langle \mathcal{P}\Delta g, f \rangle_{\left(\bar{W}_\sigma^{1,2}\right)^* \times \bar{W}_\sigma^{1,2}} = -\langle g, f \rangle_1 + \langle (\kappa - \alpha)g, f \rangle_{L^2(\partial\mathcal{O};\mathbb{R}^2)}.$$

The nonlinear term requires no special attention to be understood in the weak sense, similarly to the no-slip boundary condition. We equip $\bar{W}_\alpha^{2,2}$ with the inner product $\langle f, g \rangle_2 := \langle \mathcal{P}\Delta f, \mathcal{P}\Delta g \rangle_{L^2}$ which is equivalent to the standard $W^{2,2}$ inner product (see [25] Lemma 1.2), and $\bar{W}_\sigma^{1,2}$ with the $\langle \cdot, \cdot \rangle_1$ inner product. Then as a direct application of Theorem 4.13 we obtain the following.

Theorem 5.9 Let $\alpha \in C^2(\partial\mathcal{O}; \mathbb{R})$, $u_0 : \Omega \to L_\sigma^2$ be $\mathcal{F}_0$-measurable, $(\xi_i) \in W_\sigma^{1,2} \cap W^{2,\infty}$ with $\sum_{i=1}^\infty \|\xi_i\|_{W^{2,\infty}}^2 < \infty$. Then there exists a progressively measurable process u in $\bar{W}_\sigma^{1,2}$ such that for $\mathbb{P} - a.e.$ $\tilde{\omega}$, $u.(\omega) \in C\left([0, T]; L_\sigma^2\right) \cap L^2\left([0, T]; \bar{W}_\sigma^{1,2}\right)$ and (47) holds $\mathbb{P} - a.s.$ in $\left(\bar{W}_\sigma^{1,2}\right)^*$ for all $t \in [0, T]$.

This result is given in [25] Theorem 1.14, and a strong existence result is proven as Theorem 1.15 in the same chapter. For this, we need to make some adjustments; to verify Assumption 4.8 we cannot use the $\langle \cdot, \cdot \rangle_1$ inner product for H as this leads to an uncontrollable boundary integral. Instead, we have to manufacture a more reasonable boundary integral into our inner product. Thus, we instead equip $\bar{W}_\sigma^{1,2}$ with

$$\langle f, g \rangle_H := \langle f, g \rangle_1 + \langle (\kappa - \alpha)f, g \rangle_{L^2(\partial\mathcal{O};\mathbb{R}^2)}$$

which is an inner product equivalent to the usual $W^{1,2}$ form when $\alpha \geq \kappa$ everywhere on $\partial\mathcal{O}$. This requirement appears in the result, obtainable through Theorem 4.16.

Theorem 5.10 Let $\alpha \in C^2(\partial\mathcal{O}; \mathbb{R})$ be such that $\alpha \geq \kappa$, $u_0 : \Omega \to \bar{W}_\sigma^{1,2}$ be $\mathcal{F}_0$-measurable, $(\xi_i) \in L_\sigma^2 \cap W_0^{3,2} \cap W^{3,\infty}$ with $\sum_{i=1}^\infty \|\xi_i\|_{W^{3,\infty}}^2 < \infty$. Then there exists a progressively measurable process u in $\bar{W}_\alpha^{2,2}$ such that for $\mathbb{P}-a.e.$ $\tilde{\omega}$, $u.(\omega) \in C\left([0, T]; \bar{W}_\sigma^{1,2}\right) \cap L^2\left([0, T]; \bar{W}_\alpha^{2,2}\right)$ and (47) holds $\mathbb{P}-a.s.$ in L_σ^2 for all $t \in [0, T]$.

We note that continuity is obtainable in this instance as the extension $\bar{H}$ of H in Assumption Set 3, Sect. 4.4, can simply be $\bar{W}_\sigma^{1,2}$ itself (such that [26] Lemma 4.14 can be applied). In addition, it should be noted that the requirement $\alpha \geq \kappa$ is rather reasonable; at least heuristically, as α grows large then $u \cdot \iota = 0$ dominates the second identity of (49), which would then result in the traditional no-slip condition. Given the wide acceptance of the no-slip condition, deviation from it with Navier boundary conditions is only expected for large α. A rigorous result regarding the convergence of solutions to the deterministic Navier-Stokes equation with Navier boundary conditions to the no-slip solution for $\alpha \to \infty$ is available in [35] Section 9.

Of course this invites the question as to how the Navier boundary conditions solve the issue of $\mathcal{P}_n$ present for the no-slip case. More detail is given in the conclusion of [25], but in essence, this owes to the fact that the basis of eigenfunctions of the Stokes Operator satisfying the Navier boundary conditions are dense in the range of the Leray Projector in $W^{1,2}$. That is, these eigenfunctions form a basis of $\bar{W}_\sigma^{1,2}$ instead of $W_\sigma^{1,2}$, so the Leray Projector mapping only into $\bar{W}_\sigma^{1,2}$ is now non-problematic.

Thanks I would like to give my sincerest thanks to Dan Crisan for the regular and extended discussions around the chapter, his feedback on it, and overall guidance during this process. I would also like to thank the anonymous reviewers for their time, effort, and useful comments.

Acknowledgments The author was supported by the Engineering and Physical Sciences Research Council (EPSCR) Project 2478902.

References

1. Alonso-Orán, D., Bethencourt de León, A., Holm, D.D., Takao, S.: Modelling the climate and weather of a 2D Lagrangian-averaged Euler–Boussinesq equation with transport noise. Journal of Statistical Physics **179**(5), 1267–1303 (2020)
2. Alonso-Orán, D., Rohde, C., Tang, H.: A local-in-time theory for singular SDEs with applications to fluid models with transport noise. Journal of Nonlinear Science **31**, 1–55 (2021)
3. Basson, A., Gérard-Varet, D.: Wall laws for fluid flows at a boundary with random roughness. Communications on Pure and Applied Mathematics: A Journal Issued by the Courant Institute of Mathematical Sciences **61**(7), 941–987 (2008)
4. Bensoussan, A.: Stochastic navier-stokes equations. Acta Applicandae Mathematica **38**, 267–304 (1995)
5. Bensoussan, A., Temam, R.: Équations stochastiques du type Navier–Stokes. J. Funct. Anal. **13**, 195–222 (1973)
6. Breit, D., Hofmanova, M.: Stochastic Navier-Stokes equations for compressible fluids. Indiana University Mathematics Journal pp. 1183–1250 (2016)
7. Brzeźniak, Z., Capiński, M., Flandoli, F.: Stochastic Navier-Stokes equations with multiplicative noise. Stochastic Analysis and Applications **10**(5), 523–532 (1992)
8. Brzezniak, Z., Peszat, S.: Strong local and global solutions for stochastic Navier-Stokes equations. Infinite dimensional stochastic analysis pp. 85–98 (1999)
9. Brzeźniak, Z.a., Motyl, E.: Existence of a martingale solution of the stochastic Navier-Stokes equations in unbounded 2D and 3D domains. J. Differential Equations **254**(4), 1627–1685 (2013). DOI 10.1016/j.jde.2012.10.009. URL http://dx.doi.org/10.1016/j.jde.2012.10.009
10. Capiński, M., Cutland, N.: Stochastic Navier-Stokes equations. Acta Applicandae Mathematica **25**, 59–85 (1991)
11. Constantin, P., Foias, C.: Navier-Stokes Equations. University of Chicago Press (1988)
12. Conway, J.B.: A course in operator theory. American Mathematical Soc. (2000)
13. Cotter, C., Crisan, D., Holm, D., Pan, W., Shevchenko, I.: Data assimilation for a quasi-geostrophic model with circulation-preserving stochastic transport noise. Journal of Statistical Physics **179**(5), 1186–1221 (2020)
14. Cotter, C., Crisan, D., Holm, D.D., Pan, W., Shevchenko, I.: Modelling uncertainty using stochastic transport noise in a 2-layer quasi-geostrophic model. arXiv preprint arXiv:1802.05711 (2018)

15. Cotter, C., Crisan, D., Holm, D.D., Pan, W., Shevchenko, I.: Numerically modeling stochastic Lie transport in fluid dynamics. Multiscale Modeling & Simulation **17**(1), 192–232 (2019)
16. Crisan, D., Holm, D.D., Luesink, E., Mensah, P.R., Pan, W.: Theoretical and computational analysis of the thermal quasi-geostrophic model. arXiv preprint arXiv:2106.14850 (2021)
17. Dufée, B., Mémin, E., Crisan, D.: Stochastic parametrization: an alternative to inflation in Ensemble Kalman filters. Quarterly Journal of the Royal Meteorological Society **148**(744), 1075–1091 (2022)
18. Flandoli, F., Gatarek, D.: Martingale and stationary solutions for stochastic Navier–Stokes equations. Probab. Theory Relat. Fields **102**(3), 367–391 (1995)
19. Flandoli, F., Luongo, E.: Stochastic partial differential equations in fluid mechanics, vol. 2328. Springer Nature (2023)
20. Flandoli, F., Pappalettera, U.: 2D Euler equations with Stratonovich transport noise as a large-scale stochastic model reduction. Journal of Nonlinear Science **31**(1), 1–38 (2021)
21. Flandoli, F., Pappalettera, U.: From additive to transport noise in 2D fluid dynamics. Stochastics and Partial Differential Equations: Analysis and Computations pp. 1–41 (2022)
22. Gérard-Varet, D., Masmoudi, N.: Relevance of the slip condition for fluid flows near an irregular boundary. Communications in Mathematical Physics **295**(1), 99–137 (2010)
23. Glatt-Holtz, N., Ziane, M., et al.: Strong pathwise solutions of the stochastic Navier-Stokes system. Advances in Differential Equations **14**(5/6), 567–600 (2009)
24. Goodair, D.: Stochastic Calculus in Infinite Dimensions and SPDEs. arXiv preprint arXiv:2203.17206 (2022)
25. Goodair, D.: Navier-Stokes Equations with Navier Boundary Conditions and Stochastic Lie Transport: Well-Posedness and Inviscid Limit. arXiv preprint arXiv:2308.04290 (2023)
26. Goodair, D.: Weak and Strong Solutions to Nonlinear SPDEs with Unbounded Noise. arXiv preprint arXiv:2401.10076 (2024)
27. Goodair, D., Crisan, D.: On the 3D Navier-Stokes Equations with Stochastic Lie Transport. in: Stochastic Transport in Upper Ocean Dynamics II: STUOD 2022 Workshop, London, UK, September 26–29, vol. 11, p. 53. Springer Nature (2023)
28. Goodair, D., Crisan, D.: The Zero Viscosity Limit of Stochastic Navier-Stokes Flows. arXiv preprint arXiv:2305.18836 (2023)
29. Goodair, D., Crisan, D., Lang, O.: Existence and uniqueness of maximal solutions to SPDEs with applications to viscous fluid equations. Stochastics and Partial Differential Equations: Analysis and Computations pp. 1–64 (2023)
30. Gubinelli, M., Imkeller, P., Perkowski, N.: Paracontrolled distributions and singular PDEs. in: Forum of Mathematics, Pi, vol. 3, p. e6. Cambridge University Press (2015)
31. Hairer, M.: A theory of regularity structures. Inventiones mathematicae **198**(2), 269–504 (2014)
32. Holm, D.D.: Variational principles for stochastic fluid dynamics. Proceedings of the Royal Society A: Mathematical, Physical and Engineering Sciences **471**(2176), 20140,963 (2015)
33. Holm, D.D., Luesink, E.: Stochastic wave–current interaction in thermal shallow water dynamics. Journal of Nonlinear Science **31**(2), 1–56 (2021)
34. Holm, D.D., Luesink, E., Pan, W.: Stochastic circulation dynamics in the ocean mixed layer. arXiv preprint arXiv:2006.05707 (2020)
35. Kelliher, J.P.: Navier–Stokes equations with Navier boundary conditions for a bounded domain in the plane. SIAM journal on mathematical analysis **38**(1), 210–232 (2006)
36. Lang, O., Pan, W.: A pathwise parameterisation for stochastic transport. arXiv preprint arXiv:2202.10852 (2022)
37. Langa, J.A., Real, J., Simon, J.: Existence and regularity of the pressure for the stochastic Navier–Stokes equations. Applied Mathematics and Optimization **48**, 195–210 (2003)
38. van Leeuwen, P.J., Crisan, D., Lang, O., Potthast, R.: Bayesian Inference for Fluid Dynamics: A Case Study for the Stochastic Rotating Shallow Water Model. arXiv preprint arXiv:2112.15216 (2021)
39. Liu, W., Röckner, M.: Stochastic partial differential equations: an introduction. Springer (2015)
40. Lototsky, S.V., Rozovsky, B.L., et al.: Stochastic partial differential equations. Springer (2017)

41. Masmoudi, N., Saint-Raymond, L.: From the Boltzmann equation to the Stokes-Fourier system in a bounded domain. Communications on Pure and Applied Mathematics: A Journal Issued by the Courant Institute of Mathematical Sciences 56(9), 1263–1293 (2003)
42. Maxwell, J.C.: VII. On stresses in rarified gases arising from inequalities of temperature. Philosophical Transactions of the royal society of London 7(170), 231–256 (1879)
43. Mémin, E.: Fluid flow dynamics under location uncertainty. Geophysical & Astrophysical Fluid Dynamics 108(2), 119–146 (2014)
44. Menaldi, J.L., Sritharan, S.S.: Stochastic 2-D Navier–Stokes equation. Appl. Math. Optim. 46(1), 31–53 (2002). DOI 10.1007/s00245-002-0734-6. URL http://dx.doi.org/10.1007/s00245-002-0734-6
45. Navier, C.: Mémoire sur les lois du mouvement des fluides. éditeur inconnu (1822)
46. Navier, C.: Sur les lois de l'équilibre et du mouvement des corps élastiques. Mem. Acad. R. Sci. Inst. France 6(369), 1827 (1827)
47. Pardoux, E.: Equations aux dérivées partielles stochastiques monotones, These, Univ (1975)
48. Paré; s, C.: Existence, uniqueness and regularity of solution of the equations of a turbulence model for incompressible fluids. Applicable Analysis 43(3–4), 245–296 (1992)
49. Robinson, J.C., Rodrigo, J.L., Sadowski, W.: The three-dimensional Navier–Stokes equations: Classical theory, vol. 157. Cambridge university press (2016)
50. Street, O.D., Crisan, D.: Semi-martingale driven variational principles. Proceedings of the Royal Society A 477(2247), 20200,957 (2021)
51. Tang, H., Wang, F.Y.: A general framework for solving singular SPDEs with applications to fluid models driven by pseudo-differential noise. arXiv preprint arXiv:2208.08312 (2022)

Geometric Theory of Perturbation Dynamics Around Non-equilibrium Fluid Flows

Darryl D. Holm, Ruiao Hu, and Oliver D. Street

1 Introduction

We are dealing with the stability analysis of ideal fluid dynamics when the perturbations are in the form of displacement vector fields. In our approach, the displacement vector fields are shown to possess their own dynamics which depend functionally on the unperturbed fluid flow. The methodology we employ to derive the dynamics of the unperturbed flow and its perturbations is the Euler-Poincaré variational principle [20] with higher-order variations. In the Euler-Poincaré variational principle, the $1st$ order variations yield essentially all of the well known models of ideal fluid dynamics [21]. When the stationary condition of the $1st$ order variations are satisfied, the $2nd$ order variations yield the dynamics of the perturbations of the steady flows arising from the $1st$ order variations. The advantage of this variational approach is that the dynamics of the perturbations arising from the $2nd$ order variations are linear perturbation equations for arbitrary time-dependent flows. Thus, it is a generalisation to the fluid equilibrium flows that are typically used in stability analysis.

The emergence of fluid models from the $1st$ order variations in Hamilton's Principle was first revealed long before VI Arnold's observation [2] that the solutions of Euler's fluid equations represent time (t) dependent geodesic paths g_t on the manifold of smooth invertible maps (diffeomorphisms). That is, $g_t \in \mathrm{Diff}(M)$ acts on the fluid reference configuration ($\mathrm{Diff} \times M \to M$) in the flow domain M, whose paths $\mathbf{x}_t(\mathbf{x}_0) = g_t\mathbf{x}_0$ with $g_0\mathbf{x}_0 = \mathbf{x}_0$ are the trajectories of Lagrangian fluid parcels. Arnold's observation opened the flood gates of new mathematical research in fluid dynamics. For a review, see e.g., [3]. In Arnold [2], the variational principle

D. D. Holm (✉) · R. Hu · O. D. Street
Imperial College London, Mathematics, London, UK
e-mail: d.holm@imperial.ac.uk; ruiao.hu15@imperial.ac.uk; o.street18@imperial.ac.uk

© The Author(s) 2025
B. Chapron et al. (eds.), *Stochastic Transport in Upper Ocean Dynamics III*,
Mathematics of Planet Earth 13, https://doi.org/10.1007/978-3-031-70660-8_5

which governs the Euler fluid motion was found to be the Hamilton principle $\delta S = 0$ with $S = \int_0^T \ell(u)\,dt$ whose Lagrangian $\ell(u)$ is the fluid kinetic energy $\frac{1}{2}\|u\|_{L^2}^2$ for fluid velocity, u. The fluid kinetic energy serves as the metric on the tangent space of the diffeomorphisms, expressed in terms of the *Eulerian* velocity vector fields. The variations are taken with respect to the infinitesimal action of the diffeomorphisms on volume-preserving (spatial) vector fields. Since this observation, the framework was generalised to semidirect product spaces where the symmetry is broken by the inclusion of advected quantities [21]. As a result, the characterisation of a broad class of models through the variational approach became available. These developments related *symmetry reduction* to fluid dynamics.

The *2nd* order variations of the Euler-Poincaré variational principle have already been effective in deriving mean-flow equations [17–19] as small-amplitude gener-alised Lagrangian mean (called glm) equations leading to turbulence models such as the Navier-Stokes-alpha model and its ideal version the Euler-alpha model. The alpha turbulence models introduced in [9, 21] were derived by applying the Lagrangian mean to the *2nd* order variations for fluid dynamics in [17–19]. They were then analysed mathematically in [11, 12] and applied computationally to primitive-equation global ocean circulation models in [14–16]. Furthermore, a rela-tionship exists between the *2nd* order variations of the Euler-Poincaré variational principle, Jacobi fields and the Jacobi equations. The Jacobi field equations for the evolution of an initial Lagrangian displacement away from a geodesic flow are usually treated in terms of covariant derivatives. However, here we will discuss Jacobi fields in the Euler-Poincaré framework. For examples and references to the classical approach to the treatment of Jacobi fields, see [5, 10, 23–25, 27, 28, 30, 31].

Stability analysis is a vast field with a long history. Below, we will briefly review the Jacobi approach to stability analysis to better illuminate how this chapter contributes to the literature.

Brief Review of the Jacobi Approach to Stability Analysis Stationary variational principles deal with balance and closure in dynamical systems. The dynamics near balance and the discovery of imbalance is the province of perturbation theory. Higher order variational principles which govern the perturbations can tell us about instability and the initial phases of imbalance. Of course, for nonlinear systems tipping points also may exist and those can take the system to states far away from balance. For viscous fluids, the primary tipping point is the onset of turbulence, whose true nature remains fascinating but elusive and beyond the treatment of linear imbalance and instability of ideal fluid flow considered here.

The classical studies of imbalance and instability refer to the dynamical behaviour of solutions near equilibria. One of the most beautiful mathematical theories of imbalance was introduced by Jacobi to describe solutions near geodesic flows. Besides Jacobi, this topic also stimulated research by the likes of Dirichlet, Dedekind, Riemann, Poincaré, and Lyapunov [29]. Jacobi's theory and Riemann's results inspired Chandrasekhar's focus on ellipsoidal figures of equilibrium of rotating self-gravitating fluids [6]. That focus on steady ellipsoidal fluid configurations in turn led Chandrasekhar to his work on the emission of gravity waves by rotating ellipsoidal fluid masses in [7], and eventually to Chandrasekhar's mathematical theory of black holes in [8]. The angular momentum of the rotating fluid body is the source of the emission of gravity waves. However, in the current chapter we shall be concerned with the dual of angular momentum introduced by Dedekind: namely, we shall focus on the fluid circulation in a fixed frame.

In Arnold [1], a stationary flow of an ideal fluid is shown to be Lyapunov stable if the quadratic form given by the second variation of the kinetic energy restricted to coadjoint orbits in the algebra of smooth divergence-free vector fields is either positive, or sufficiently negative. The Hamiltonian version of Arnold's stability result for the Euler fluid equations was obtained by applying the Legendre transformation to their application of Hamilton's principle. This step led to new methods for determining sufficient conditions for nonlinear stability of fluid and plasma equilibria which include potential energy such as thermodynamics and magnetic energy, as well as kinetic energy; see, e.g., [22]. The present work investigates the quadratic form given by the second variation of the Hamilton principle whose fluid Lagrangian contains both the kinetic and potential energy. In addition, this work investigates the second-variation Hamilton's principle for time-dependent fluid flows, and is not limited to time-independent fluid equilibria.

Plan of the Chapter Section 2 sets the stage for the remainder of the chapter, by defining symmetry-reduced variational principles at $1st$ and $2nd$ order. By considering geodesics of a right-invariant metric on a Lie group, the equations resulting from the $2nd$ variation are shown to be an extension of the literature on Jacobi fields. Section 3 presents several examples of linearisation of well-known Euler-Poincaré fluid equations based on second-order symmetry-reduced variational principles. The examples in Sect. 3 demonstrate that the current approach is not limited to geodesic motions, nor is it limited to steady flows. Section 4 presents numerical simulations for the Euler-Bousinessq equations and their perturbation equations resulting from the $2nd$ order variations in a vertical slice domain. In Sect. 5, we summarise the present results and discuss future developments.

2   Euler-Poincaré Variational Principles and Their Linearisation

Following Arnold's identification of fluid flows as paths g_t on the manifold of smooth invertible maps, it is natural to consider the dynamics of imbalance induced by perturbations of nonequilibrium fluid flows in the light of the Jacobi equations for geodesic flows [24, 25]. Here, we are concerned with time-parameterised curves g_t in the diffeomorphism group, and a family of variations of these curves, parameterised by $\epsilon \in \mathbb{R}$, $g_{\epsilon,t}$ such that $g_{0,t} = g_t$. We seek to extend the Jacobi equations of displacement dynamics near a geodesic flow to apply in semidirect product spaces which admit the dynamics of an Eulerian vector field $\xi(\mathbf{x}, t) \in \mathfrak{X}(M)$ representing the displacement defined by

$$\xi(\mathbf{x}_t, t) := \left.\frac{\partial g_{t,\epsilon}}{\partial \epsilon}\right|_{\epsilon=0} g_t^{-1} \mathbf{x}_t = \delta g_t \mathbf{x}_0 =: \delta g_t g_t^{-1} \mathbf{x}_t \,, \tag{2.1}$$

for each fluid element in the flow $\mathbf{x}_t = g_t \mathbf{x}_0$ initially at the reference position $g_0 \mathbf{x}_0 = \mathbf{x}_0$, where $g_{t,\epsilon}$ are arbitrary disturbances of g_t near the identity of the group of diffeomorphisms. This infinitesimal displacement vector field appears naturally in the Euler-Poincaré variational principle which we will discuss next before considering the correspondence to Jacobi fields and the Jacobi equation.

2.1 The Euler-Poincaré and Lie-Poisson Equations

In the Euler-Poincaré theory of ideal fluid dynamics [20], the fluid is formally described by the elements in the semidirect product space comprising the vector fields $\mathfrak{X}(M)$, and the space of advected quantities, V^*, on which there exists a right representation of $\mathrm{Diff}(M)$ by pullback. The space of vector fields $\mathfrak{X}(M)$ contains the fluid's Eulerian velocity vector field u, defined in terms of curves, g_t, in the diffeomorphism group as $u := \dot{g}_t g_t^{-1}$. The vector space $V^*(M)$, defined following convention as the dual to a vector space $V(M)$, is the space of advected quantities. Let $a \in V^*(M)$, we say that a is an advected quantity if it satisfies the pushforward relation $a_t = a_0 g_t^{-1}$. This a_t is the global solution of the advection relation,

$$\partial_t a_t = -\mathcal{L}_{u_t} a_t \,, \tag{2.2}$$

in which $\mathcal{L}_{u_t}$ denotes Lie derivative with respect to the time-dependent fluid velocity vector field $u_t \in \mathfrak{X}(M)$. Advected quantities are tensor fields, examples of which are mass density (a volume form) or potential temperature (a 0-form). Arising from the variations of the paths g_t on the manifold of diffeomorphisms, the corresponding variations of the fluid velocity $u = \dot{g}_t g_t^{-1}$ and the advected quantity $a_t = a_0 g_t^{-1}$ are given, respectively, by [20]

$$\delta u = \partial_t \xi - \mathrm{ad}_u \xi := \partial_t \xi + [u, \xi] \quad \text{and} \quad \delta a = -\mathcal{L}_\xi a \,. \tag{2.3}$$

Here, the displacement vector field $\xi \in \mathfrak{X}(M)$ is defined in Eq. (2.1) and $-\mathrm{ad}_u \xi := [u, \xi] \in \mathfrak{X}(M)$ for right adjoint Lie algebra action is the Jacobi-Lie bracket of the vector fields u and ξ.

Let $\ell(u, a)$ denote the fluid Lagrangian. By applying Hamilton's principle, $\delta S = 0$, to the action integral $S = \int_0^T \ell(u, a)\, dt$ with the constrained variations of fluid variables (u, a) as given in (2.3), we find

$$0 = \delta S = \delta \int_0^T \ell(u, a)\, dt$$

$$= \int_0^T \left\langle \frac{\delta \ell}{\delta u}, \delta u \right\rangle + \left\langle \frac{\delta \ell}{\delta a}, \delta a \right\rangle dt$$

$$= \int_0^T \left\langle \frac{\delta \ell}{\delta u} , \, \partial_t \xi - \mathrm{ad}_u \xi \right\rangle + \left\langle \frac{\delta \ell}{\delta a} , \, -\mathcal{L}_\xi a \right\rangle dt$$

$$= \int_0^T \left\langle -(\partial_t + \mathrm{ad}_u^*) \frac{\delta \ell}{\delta u} + \frac{\delta \ell}{\delta a} \diamond a , \, \xi \right\rangle dt + \left\langle \frac{\delta \ell}{\delta u} , \, \xi \right\rangle \Big|_0^T , \tag{2.4}$$

where the brackets $\langle \cdot , \, \cdot \rangle$ denote the L^2 pairing on the flow manifold M and the diamond ($\diamond$) operator is defined as

$$\left\langle \frac{\delta \ell}{\delta a} , \, -\mathcal{L}_\xi a \right\rangle_{V \times V^*} =: \left\langle \frac{\delta \ell}{\delta a} \diamond a , \, \xi \right\rangle_{\mathfrak{X}^*(M) \times \mathfrak{X}(M)} \tag{2.5}$$

in which we have applied natural boundary conditions when integrating by parts in space. The stationarity condition in Hamilton's principle $\delta S = 0$ with vanishing endpoint conditions in time on $\xi = \delta g \cdot g^{-1}$ then yields the Euler-Poincaré equation of fluid motion

$$\left(\partial_t + \mathcal{L}_u\right) \frac{\delta \ell}{\delta u} = \frac{\delta \ell}{\delta a} \diamond a , \tag{2.6}$$

where the advected quantities a satisfy the advection relation,

$$\left(\partial_t + \mathcal{L}_u\right) a = 0 . \tag{2.7}$$

Note that the fact that a is an advected quantity is encoded within the variational procedure (2.4) within the form of the variation δa. This calculation leads us to the Kelvin-Noether theorem, written below.

Theorem 2.1 (Kelvin-Noether Theorem [20]) *Given the local advection of mass by fluid transport,*

$$\left(\partial_t + \mathcal{L}_u\right)(D d^n x) = 0 , \tag{2.8}$$

implied by the push-forward relation $D_t d^n x_t = g_{t*}(D_0 d^n x_0)$ for each fluid element in the flow $\mathbf{x}_t = g_t \mathbf{x}_0$ initially at the reference position $g_0 \mathbf{x}_0 = \mathbf{x}_0$ and volume element $d^n x$ in n dimensions, then the Euler-Poincaré equation of fluid motion (2.6) implies the Kelvin-Noether relation,

$$\frac{d}{dt} \oint_{C(u)} \frac{1}{D} \frac{\delta \ell}{\delta u} = \oint_{C(u)} \frac{1}{D} \frac{\delta \ell}{\delta a} \diamond a , \tag{2.9}$$

for any material loop $C(u)$ moving with the flow velocity $u = \dot{g}_t g_t^{-1}$.

The Euler-Poincaré equation (2.6), together with the advection relation for a, can be shown to be equivalent to semidirect-product Lie-Poisson equations on $\mathfrak{X}^* \ltimes V^*$, where $\ltimes$ denotes the semidirect product. Indeed, we make the following Legendre transformation from $\mathfrak{X}$ to $\mathfrak{X}^*$

$$m = \frac{\delta\ell}{\delta u}, \quad \text{and} \quad h(m, a) = \langle m, u \rangle - \ell(u, a).$$

Under the assumption that the map $u \to m$ is a diffeomorphism from $\mathfrak{X}$ to $\mathfrak{X}^*$, we have that $u = \delta h/\delta m$ and the Lie-Poisson equations arise from the following application of Hamilton's principle.

$$
\begin{aligned}
0 = \delta S &= \delta \int_0^T \ell(u, a)\, dt = \delta \int_0^T \langle m, u \rangle - h(m, a)\, dt \\
&= \int_0^T \langle m, \delta u \rangle + \left\langle \delta m, u - \frac{\delta h}{\delta m} \right\rangle - \left\langle \frac{\delta h}{\delta a}, \delta a \right\rangle dt \\
&= \int_0^T \langle m, \partial_t \xi - \mathrm{ad}_u \xi \rangle + \left\langle \delta m, u - \frac{\delta h}{\delta m} \right\rangle + \left\langle \frac{\delta h}{\delta a}, \mathcal{L}_\xi a \right\rangle dt \\
&= \int_0^T \left\langle -(\partial_t + \mathrm{ad}_u^*)m - \frac{\delta h}{\delta a} \diamond a, \xi \right\rangle + \left\langle \delta m, u - \frac{\delta h}{\delta m} \right\rangle dt + \left\langle \frac{\delta\ell}{\delta u}, \xi \right\rangle \Big|_0^T.
\end{aligned}
$$

This calculation yields the implicit form of the following Lie-Poisson equation for fluid motion

$$\left(\partial_t + \mathcal{L}_{\delta h/\delta m} \right)m = -\frac{\delta h}{\delta a} \diamond a, \quad \text{and} \quad \left(\partial_t + \mathcal{L}_{\delta h/\delta m} \right)a = 0. \tag{2.10}$$

Note that these equations are equivalent to the Lie-Poisson equation

$$(\partial_t + \mathrm{ad}_{\delta h/\delta \mu}^*)\mu = 0, \quad \text{for} \quad \mu = (m, a) \in \mathfrak{X}^* \times V^*, \tag{2.11}$$

where ad^* is the coadjoint representation of $\mathfrak{X} \ltimes V$ acting on its dual $\mathfrak{X}^* \times V^*$.

2.2 The Second Variation

The first and second variations are defined as

$$\delta f(\mu; \delta\mu) := \frac{d}{d\epsilon}\Big|_{\epsilon=0} f(\mu + \epsilon\delta\mu), \quad \text{and} \quad \delta^2 f(\mu; \delta\mu) := \frac{d^2}{d\epsilon^2}\Big|_{\epsilon=0} f(\mu + \epsilon\delta\mu), \tag{2.12}$$

respectively. In each example we consider in this article, the second variation produces a symmetric bilinear form, which we will denote also by $\delta^2 f(\delta\mu, \delta\mu)$. Recall the definition of a functional derivative,

$$\left\langle \frac{\delta f}{\delta \mu}, \delta \mu \right\rangle := \delta f(\mu; \delta \mu).$$

That is, $\langle \delta f/\delta \mu, \delta \mu \rangle$ is the first term of the expansion around $\epsilon = 0$ of the first derivative of $f(\mu + \epsilon \delta \mu)$ in ϵ. When taking second variations, we will often wish to go further in this expansion. Indeed, we see that

$$\frac{d}{d\epsilon} f(\mu + \epsilon \delta \mu) = \frac{d}{d\epsilon}\bigg|_{\epsilon=0} f(\mu + \epsilon \delta \mu) + \epsilon \frac{d^2}{d\epsilon^2}\bigg|_{\epsilon=0} f(\mu + \epsilon \delta \mu) + \mathcal{O}(\epsilon^2)$$

$$= \left\langle \frac{\delta f}{\delta \mu}, \delta \mu \right\rangle + \epsilon \delta^2 f(\delta \mu, \delta \mu) + \mathcal{O}(\epsilon^2)$$

$$= \left\langle \frac{\delta f}{\delta \mu}, \delta \mu \right\rangle + \frac{\epsilon}{2} \left\langle \frac{\delta(\delta^2 f)}{\delta(\delta \mu)}, \delta \mu \right\rangle + \mathcal{O}(\epsilon^2),$$

(2.13)

where the final equality holds since $\delta^2 f(\delta \mu, \delta \mu)$ is a symmetric bilinear form. This calculation allows us to take functional derivatives to the next order of the expansion, since the term of order ϵ involves a pairing of an element of $\mathfrak{g}$ against $\delta \mu$.

2.3 Linearised Euler-Poincaré and Lie-Poisson Equations

We may consider an expansion of the Lie-Poisson equation

$$\partial_t \mu + \mathrm{ad}^*_{\frac{\delta h}{\delta \mu}} \mu = 0,$$

(2.14)

by exploiting the calculation in Eq. (2.13) to notice that

$$\left(\partial_t + \mathrm{ad}^*_{\frac{\delta h}{\delta \mu} + \frac{\epsilon}{2} \frac{\delta(\delta^2 h)}{\delta(\delta \mu)}} \right) (\mu + \epsilon \delta \mu) = \mathcal{O}(\epsilon^3).$$

(2.15)

Notice that the first order terms in this equation correspond exactly to the Lie-Poisson equation (2.14), and the next order terms give the linearised equation for the perturbation $\delta \mu$ around the solution μ to the Lie-Poisson system

$$\partial_t \delta \mu + \mathrm{ad}^*_{\frac{\delta h}{\delta \mu}} \delta \mu = -\frac{1}{2} \mathrm{ad}^*_{\frac{\delta(\delta^2 h)}{\delta(\delta \mu)}} \mu.$$

(2.16)

This can be written in terms of Poisson bracket-like objects as

$$\partial_t f(\delta \mu) = \left\langle \delta \mu, \left[\frac{\delta f}{\delta(\delta \mu)}, \frac{\delta h}{\delta \mu} \right] \right\rangle + \frac{1}{2} \left\langle \mu, \left[\frac{\delta f}{\delta(\delta \mu)}, \frac{\delta(\delta^2 h)}{\delta(\delta \mu)} \right] \right\rangle.$$

(2.17)

For a fixed solution, μ, of the Lie-Poisson equation, the second bracket here is a *frozen* Lie-Poisson bracket. Notice that there is a direct connection here to the motion considered in a previous study [22], where it was observed that for an equilibrium solution, μ_e, corresponding to a critical point of $h+C$ for some Casimir, C, this equation for $\delta\mu$ is Hamiltonian with respect to the frozen Lie-Poisson bracket with Hamiltonian $\delta^2 h_C$. The full Eq. (2.16) considered here is *not* itself a Lie-Poisson system.

Hamiltonian Systems on Semidirect Product Spaces and Continuum Dynamics
When the configuration space is a semidirect product Lie group, $G \ltimes V$, where G acts on V through a right representation, the equations of motion on the Lie co-algebra, $\mathfrak{g}^* \ltimes V^*$, are

$$\partial_t(\delta\mu, \delta a) + \text{ad}^*_{\frac{\delta h}{\delta(\mu,a)}}(\delta\mu, \delta a) = -\frac{1}{2}\text{ad}^*_{\frac{\delta(\delta^2 h)}{\delta(\delta\mu,\delta a)}}(\mu, a). \tag{2.18}$$

These equations can be manipulated into a more convenient form by determining ad^* for the semidirect product space in the usual way. That is, an alternative form of Eq. (2.18) can be directly deduced from Eq. (2.17) by inserting the standard Lie bracket for semidirect product spaces as

$$\begin{aligned}
\partial_t f(\delta\mu, \delta a) &= \left\langle (\delta\mu, \delta a), \left[\frac{\delta f}{\delta(\delta\mu, \delta a)}, \frac{\delta h}{\delta(\mu, a)}\right]\right\rangle \\
&\quad + \frac{1}{2}\left\langle (\mu, a), \left[\frac{\delta f}{\delta(\delta(\mu, a))}, \frac{\delta(\delta^2 h)}{\delta(\delta\mu, \delta a)}\right]\right\rangle \\
&= \left\langle \delta\mu, \text{ad}_{\frac{\delta f}{\delta(\delta\mu)}}\frac{\delta h}{\delta\mu}\right\rangle + \left\langle \delta a, \mathcal{L}^T_{\frac{\delta f}{\delta(\delta\mu)}}\frac{\delta h}{\delta a} - \mathcal{L}^T_{\frac{\delta h}{\delta\mu}}\frac{\delta f}{\delta(\delta a)}\right\rangle \\
&\quad + \frac{1}{2}\left\langle \mu, \text{ad}_{\frac{\delta f}{\delta(\delta\mu)}}\frac{\delta(\delta^2 h)}{\delta(\delta\mu)}\right\rangle \\
&\quad + \frac{1}{2}\left\langle a, \mathcal{L}^T_{\frac{\delta f}{\delta(\delta\mu}}\frac{\delta(\delta^2 h)}{\delta(\delta a)} - \mathcal{L}^T_{\frac{\delta(\delta^2 h)}{\delta(\delta\mu)}}\frac{\delta f}{\delta(\delta a)}\right\rangle,
\end{aligned} \tag{2.19}$$

where $\mathcal{L}^T$ is the transpose of the Lie derivative. Integrating by parts gives the following equations for the dynamics of the perturbations,

$$\left(\partial_t + \text{ad}^*_{\frac{\delta h}{\delta\mu}}\right)\delta\mu = -\frac{1}{2}\text{ad}^*_{\frac{\delta(\delta^2 h)}{\delta(\delta\mu)}}\mu - \frac{\delta h}{\delta a}\diamond\delta a - \frac{1}{2}\frac{\delta(\delta^2 h)}{\delta(\delta a)}\diamond a, \tag{2.20}$$

$$\left(\partial_t + \mathcal{L}_{\frac{\delta h}{\delta\mu}}\right)\delta a = -\frac{1}{2}\mathcal{L}_{\frac{\delta(\delta^2 h)}{\delta(\delta\mu)}}a. \tag{2.21}$$

These comprise a useful form in which to write the equations, since the operator $\partial_t + \text{ad}^*_{\delta h/\delta\mu}$ is the standard geometric form of the advective derivative and mirrors the left hand side of the regular Lie-Poisson and Euler-Poincaré equations. It then remains only to determine the right hand sides of the equation in an analogous fashion to how fluid models are derived from the Euler-Poincaré equation (2.6), as discussed in [20]. It is well known that the equations with advected quantities which break the relabelling symmetry under the entire diffeomorphism group, whilst not Euler-Poincaré equations on the semidirect product space $\mathfrak{g} \ltimes V^*$, are the standard Lie-Poisson equations on the dual space $\mathfrak{g}^* \ltimes V^*$. Here, we have demonstrated that the same is true for the linearised equation on the semidirect product space. That is, the linearised system of Eqs. (2.20) and (2.21) has the same geometric form as the linearised Lie-Poisson equation (2.16).

Euler-Poincaré Equations for Semidirect Product Spaces When the Legendre transform is well defined, the Eqs. (2.20) and (2.21) have equivalent forms in terms of the Lagrangian, $\ell : \mathfrak{g} \times V^* \to \mathbb{R}$. As was achieved for the Lie-Poisson equations, one may deduce these equations directly from the following Euler-Poincaré equations

$$\left(\partial_t + \mathrm{ad}_u^*\right) \frac{\delta \ell}{\delta u} = \frac{\delta \ell}{\delta a} \diamond a \,, \tag{2.22}$$

$$\left(\partial_t + \mathcal{L}_u\right) a = 0 \,. \tag{2.23}$$

Again utilising the calculation performed in Eq. (2.13), we find the expansion of these equations as

$$\left(\partial_t + \mathrm{ad}_{u+\epsilon\delta u}^*\right)\left(\frac{\delta \ell}{\delta u} + \frac{\epsilon}{2}\frac{\delta(\delta^2 \ell)}{\delta(\delta u)}\right) = \left(\frac{\delta \ell}{\delta a} + \frac{\epsilon}{2}\frac{\delta(\delta^2 \ell)}{\delta(\delta a)}\right) \diamond (a + \epsilon\delta a) + \mathcal{O}(\epsilon^3) \,,$$

$$\left(\partial_t + \mathcal{L}_{u+\epsilon\delta u}\right)(a + \epsilon\delta a) = \mathcal{O}(\epsilon^3) \,.$$

As for the Hamiltonian case, the first order terms are the Euler-Poincaré equations and the linearised equations are given by the order ϵ terms as

$$\left(\partial_t + \mathrm{ad}_u^*\right)\left(\frac{1}{2}\frac{\delta(\delta^2 \ell)}{\delta(\delta u)}\right) = -\,\mathrm{ad}_{\delta u}^* \frac{\delta \ell}{\delta u} + \frac{\delta \ell}{\delta a} \diamond \delta a + \frac{1}{2}\frac{\delta(\delta^2 \ell)}{\delta(\delta a)} \diamond a \,, \tag{2.24}$$

$$\left(\partial_t + \mathcal{L}_u\right)\delta a + \mathcal{L}_{\delta u}a = 0 \,. \tag{2.25}$$

2.4 Second Order Variations of the Euler-Poincaré Variational Principle

In the previous linearised equations, the perturbations $(\delta u, \delta a)$ have been arbitrary. When understanding such equations *within* the variational principle itself, these perturbations become constrained.

The Euler-Poincaré Variational Principle The Eqs. (2.24) and (2.25) can also be deduced by considering the next variation in Hamilton's principle. When deducing symmetry reduced equations from the variational principle, arbitrary variations in the group are not arbitrary in the algebra. In particular, for an Eulerian velocity vector field $u = \partial_t g \cdot g^{-1} \in \mathfrak{g}$, where concatenation denotes the lifted right translation of $\partial_t g \in T_g G$ by g^{-1}, an arbitrary variation δg of the group element allows the variation of u to be expressed in terms of an arbitrary vector field $\xi = \delta g \cdot g^{-1} \in \mathfrak{g}$. We may deduce the forms of the first and second variation by expanding the vector field $u_\epsilon = \partial_t g_{t,\epsilon} \cdot g_{t,\epsilon}^{-1}$ in a Taylor series in powers of a small parameter $\epsilon \ll 1$ around the identity $\epsilon = 0$ as

$$u_\epsilon = u + \epsilon(\partial_t \xi - \mathrm{ad}_u \xi) + \frac{\epsilon^2}{2}(\partial_t \delta \xi - \mathrm{ad}_{\delta u}\xi - \mathrm{ad}_u \delta \xi) + \mathcal{O}(\epsilon^2)$$

$$=: u + \epsilon \delta u + \frac{\epsilon}{2}\delta^2 u + \mathcal{O}(\epsilon^2). \tag{2.26}$$

Likewise, for an advected quantity which evolves by push-forward as $a_t = a_0 g_t^{-1}$ one defines the Taylor series for the variation as

$$a_\epsilon = u|_{\epsilon=0} - \epsilon \mathcal{L}_\xi a - \frac{\epsilon^2}{2}\left(\mathcal{L}_{\delta \xi} a + \mathcal{L}_\xi \delta a\right) + \mathcal{O}(\epsilon^2) =: a + \epsilon \delta a + \frac{\epsilon}{2}\delta^2 a + \mathcal{O}(\epsilon^2). \tag{2.27}$$

Equations (2.26) and (2.27) comprise a second order extension of the *Lin constraints* used to derive the original Euler-Poincaré equations (see e.g. [4]).

As in Sect. 2.1, the Euler-Poincaré equations (2.22) and (2.23) can be deduced from Hamilton's principle by making use of these variations

$$0 = \delta S = \delta \int \ell(u, a)\, dt = \int \left\langle \frac{\delta \ell}{\delta u}, \delta u \right\rangle + \left\langle \frac{\delta \ell}{\delta a}, \delta a \right\rangle dt$$

$$= \int \left\langle \frac{\delta \ell}{\delta u}, \partial_t \xi - \mathrm{ad}_u \xi \right\rangle - \left\langle \frac{\delta \ell}{\delta a}, \mathcal{L}_\xi a \right\rangle dt \tag{2.28}$$

$$= -\int \left\langle (\partial_t + \mathrm{ad}_u^*)\frac{\delta \ell}{\delta u} - \frac{\delta \ell}{\delta a} \diamond a, \xi \right\rangle dt.$$

At the second order, we have

$$0 = \delta^2 S = \int \left\langle \frac{1}{2} \frac{\delta(\delta^2\ell)}{\delta(\delta u)}, \delta u \right\rangle + \left\langle \frac{1}{2} \frac{\delta(\delta^2\ell)}{\delta(\delta a)}, \delta a \right\rangle + \left\langle \frac{\delta\ell}{\delta u}, \delta^2 u \right\rangle + \left\langle \frac{\delta\ell}{\delta a}, \delta^2 a \right\rangle dt$$

$$= \int \left\langle \frac{1}{2} \frac{\delta(\delta^2\ell)}{\delta(\delta u)}, \partial_t \xi - \mathrm{ad}_u \xi \right\rangle - \left\langle \frac{1}{2} \frac{\delta(\delta^2\ell)}{\delta(\delta a)}, \mathcal{L}_\xi a \right\rangle$$

$$+ \left\langle \frac{\delta\ell}{\delta u}, \partial_t \delta\xi - \mathrm{ad}_{\delta u}\xi - \mathrm{ad}_u \delta\xi \right\rangle - \left\langle \frac{\delta\ell}{\delta a}, \mathcal{L}_{\delta\xi} a + \mathcal{L}_\xi \delta a \right\rangle dt$$

$$= \int \left\langle -(\partial_t + \mathrm{ad}_u^*) \frac{\delta\ell}{\delta u} + \frac{\delta\ell}{\delta a} \diamond a, \delta\xi \right\rangle$$

$$+ \left\langle -(\partial_t + \mathrm{ad}_u^*) \frac{1}{2} \frac{\delta(\delta^2\ell)}{\delta(\delta u)} + \frac{1}{2} \frac{\delta(\delta^2\ell)}{\delta(\delta a)} \diamond a + \frac{\delta\ell}{\delta a} \diamond \delta a - \mathrm{ad}_{\delta u}^* \frac{\delta\ell}{\delta u}, \xi \right\rangle dt.$$

$$(2.29)$$

The arbitrary nature of the vector field ξ and, by extension, $\delta\xi$ gives the Euler-Poincaré equation (2.22) and the Eq. (2.24) for the linear perturbation. This is the symmetry-reduced version of taking second order variations in first order variational principle used in e.g. [5] to derive dynamics of Jacobi fields.

In practice, this process will give us equations for δu and δa. Since $(\delta u, \delta a)$ can be expressed in terms of the arbitrary variable $\xi = \delta g \cdot g^{-1}$, these imply an equation for ξ. However, as the following Proposition demonstrates, deriving an equation for ξ only requires the equation for δu, since the equation for δa is satisfied trivially.

Proposition 2.1 *Given the constrained form of the first variations*

$$\delta u = \partial_t \xi - \mathrm{ad}_u \xi, \quad and \quad \delta a = -\mathcal{L}_u a, \tag{2.30}$$

and the fact that $a \in V^*$ is advected by $u \in \mathfrak{X}(M)$, we find that the Eq. (2.25) is satisfied.

Proof We may verify this claim by direct computation. Indeed, notice that

$$\partial_t \delta a + \mathcal{L}_u \delta a + \mathcal{L}_{\delta u} a = -\partial_t \mathcal{L}_\xi a - \mathcal{L}_a \mathcal{L}_\xi a + \mathcal{L}_{\partial_t \xi - \mathrm{ad}_u \xi} a$$

$$= -\mathcal{L}_{\partial_t \xi} a - \mathcal{L}_\xi \partial_t a - \mathcal{L}_a \mathcal{L}_\xi a + \mathcal{L}_{\partial_t \xi} a - \mathcal{L}_{\mathrm{ad}_u \xi} a$$

$$= \mathcal{L}_\xi \mathcal{L}_u a - \mathcal{L}_u \mathcal{L}_\xi a - \mathcal{L}_{\mathrm{ad}_u \xi} a = 0,$$

where the final line is a consequence of the standard relationship between the adjoint representation and the Lie bracket for right-invariant systems. □

The Legendre Transform and Lie-Poisson Equations As was described in Sect. 2.1, the Lie-Poisson equations on semidirect product Lie co-algebras can be deduced by Legendre transforming within the application of Hamilton's Principle illustrated by Eq. (2.28). That is, we consider the following variational problem

$$0 = \delta S(\mu, a; u) = \delta \int \langle \mu, u \rangle - h(\mu, a)\, dt$$

$$= \int \left\langle \delta\mu, u - \frac{\delta h}{\delta\mu} \right\rangle + \langle \mu, \delta u \rangle - \left\langle \frac{\delta h}{\delta a}, \delta a \right\rangle dt$$

$$= \int \left\langle \delta\mu, u - \frac{\delta h}{\delta\mu} \right\rangle + \langle \mu, \partial_t \xi - \mathrm{ad}_u \xi \rangle + \left\langle \frac{\delta h}{\delta a}, \mathcal{L}_\xi a \right\rangle dt$$

$$= \int \left\langle \delta\mu, u - \frac{\delta h}{\delta\mu} \right\rangle - \left\langle (\partial_t + \mathrm{ad}_u^*)\mu + \frac{\delta h}{\delta a} \diamond a, \xi \right\rangle dt ,$$

(2.31)

which yields Lie-Poisson equation (2.10). Again, considering the second variation of this we have

$$0 = \delta^2 S(\mu, a; u) = \int \left\langle \delta\mu, \delta u - \frac{1}{2}\frac{\delta(\delta^2 h)}{\delta(\delta\mu)} \right\rangle + \left\langle \delta^2\mu, u - \frac{\delta h}{\delta\mu} \right\rangle$$

$$+ \langle \delta\mu, \delta u \rangle - \left\langle \frac{1}{2}\frac{\delta(\delta^2 h)}{\delta(\delta a)}, \delta a \right\rangle + \langle \mu, \delta^2 u \rangle - \left\langle \frac{\delta h}{\delta a}, \delta^2 a \right\rangle dt$$

$$= \int \left\langle \delta\mu, \delta u - \frac{1}{2}\frac{\delta(\delta^2 h)}{\delta(\delta\mu)} \right\rangle + \left\langle \delta^2\mu, u - \frac{\delta h}{\delta\mu} \right\rangle$$

$$+ \langle \delta\mu, \partial_t \xi - \mathrm{ad}_u \xi \rangle + \left\langle \frac{1}{2}\frac{\delta(\delta^2 h)}{\delta(\delta a)}, \mathcal{L}_\xi a \right\rangle$$

$$+ \langle \mu, \partial_t \delta\xi - \mathrm{ad}_{\delta u}\xi - \mathrm{ad}_u \delta\xi \rangle + \left\langle \frac{\delta h}{\delta a}, \mathcal{L}_\xi \delta a + \mathcal{L}_{\delta\xi} a \right\rangle dt$$

(2.32)

$$= \int \left\langle \delta\mu, \delta u - \frac{1}{2}\frac{\delta(\delta^2 h)}{\delta(\delta\mu)} \right\rangle + \left\langle \delta^2\mu, u - \frac{\delta h}{\delta\mu} \right\rangle$$

$$+ \left\langle -(\partial_t + \mathrm{ad}_u^*)\mu - \frac{\delta h}{\delta a} \diamond a, \delta\xi \right\rangle$$

$$+ \left\langle -(\partial_t + \mathrm{ad}_u^*)\delta\mu - \frac{1}{2}\frac{\delta(\delta^2 h)}{\delta(\delta a)} \diamond a - \frac{\delta h}{\delta a} \diamond \delta a - \mathrm{ad}_{\delta u}^* \mu, \xi \right\rangle dt .$$

The first two terms in the final line of this calculation give us the identities

$$u = \frac{\delta h}{\delta u}, \quad \text{and} \quad \delta u = \frac{1}{2}\frac{\delta(\delta^2 h)}{\delta(\delta\mu)},$$

the arbitrary nature of $\delta\xi$ gives us the Lie-Poisson equation, and since ξ is arbitrary we have the Eq. (2.20) for the linear perturbation expressed in terms of the Hamiltonian.

2.5 Jacobi Fields and Geodesics of a Right-Invariant Metric on a Lie Group

In this section, we will specialise to the case in which the Euler-Poincaré equation and its linearisation arise from a Lagrangian which is defined by a right-invariant inner product. Since we are in the Euler-Poincaré setting [20] and are motivated by fluid dynamics, it is natural to derive these equations in terms of pull-backs by diffeomorphisms and Lie derivatives with respect to the smooth vector fields which generate the diffeomorphisms. However, the literature on Jacobi fields in the context of Lie groups is predominantly concerned with expressions on the group, G, or algebra, $\mathfrak{g}$, rather than its dual. As such, the Jacobi equation is most familiarly expressed in terms of geometric objects such as the covariant derivative and Riemannian curvature, as opposed to the adjoint representation, Lie derivative, and diamond operation we have thus far employed. Of course, the transformation from Lie derivatives to covariant derivatives can be performed using standard methods for Riemannian spaces, [24, 25, 31].

In this section, we will relate these fields by showing that, for the Lagrangian defined by our right-invariant inner product, the resulting Euler-Poincaré equation on $\mathfrak{g}^*$ is the geodesic equation. This will be illustrated by writing the equation in its equivalent form on $\mathfrak{g}$. Furthermore, since the equation is then expressed on $\mathfrak{g}$, we will demonstrate that the linearised Euler-Poincaré equation is equivalent to the Jacobi equation in which the Jacobi field is defined by the arbitrary variations in the Euler-Poincaré variational principle. This will involve a shift in notation in this section relative to the others in this chapter.

For a Lie group, G, there exists a natural duality pairing between its algebra $\mathfrak{g}$, and its co-algebra $\mathfrak{g}^*$, which we will denote by $\langle \cdot, \cdot \rangle_{\mathfrak{g}^* \times \mathfrak{g}} : \mathfrak{g}^* \times \mathfrak{g} \to \mathbb{R}$. If the Lie group is augmented further with a right-invariant Riemannian metric, then the algebra possesses a (weak) right-invariant inner product, denoted by $\langle \cdot, \cdot \rangle_{\mathfrak{g}} : \mathfrak{g} \times \mathfrak{g} \to \mathbb{R}$. This metric permits the discussion of *geodesics* in this setting. Indeed, consider the following application of Hamilton's Principle

$$0 = \delta \int \frac{1}{2} \langle u^\flat, u \rangle_{\mathfrak{g}^* \times \mathfrak{g}} \, dt, \qquad (2.33)$$

where $\flat : \mathfrak{g} \to \mathfrak{g}^*$ is the musical mapping defined by $\langle u^\flat, v \rangle_{\mathfrak{g}^* \times \mathfrak{g}} = \langle u, v \rangle_{\mathfrak{g}}$ for $u, v \in \mathfrak{g}$. We note that when the inner product is weak, the musical mapping $\flat$ is not surjective [26]. Thus, the inverse musical mapping $\sharp : \mathfrak{g}^* \to \mathfrak{g}$ is defined, on the image of $\flat$, by $\langle \alpha, v \rangle_{\mathfrak{g}^* \times \mathfrak{g}} = \langle \alpha^\sharp, v \rangle_{\mathfrak{g}}$ for $\alpha \in \mathfrak{g}^*$ and $v \in \mathfrak{g}$. When the inner product is strong, the musical mappings become isomorphisms. Following the calculations in Sect. 2.4, we see that the variational problem (2.33) implies the following Euler-Poincaré equation

$$\partial_t u^\flat + \mathrm{ad}_u^* u^\flat = 0. \qquad (2.34)$$

The equation for δu, given by going to the second order in Hamilton's Principle, is

$$\partial_t \delta u^\flat + \mathrm{ad}_u^* \, \delta u^\flat = -\, \mathrm{ad}_{\delta u}^* \, u^\flat \,, \tag{2.35}$$

and, when considered together with the constrained variations (2.30), implies the following equation,

$$(\partial_t + \mathrm{ad}_u^*)\big((\partial_t \xi - \mathrm{ad}_u \xi)^\flat \big) = -\, \mathrm{ad}_{\partial_t \xi - \mathrm{ad}_u \xi}^* \, u^\flat \,, \tag{2.36}$$

written in terms of the variable $\xi = \delta g \cdot g^{-1} \in \mathfrak{g}$, which is a *Jacobi field*. We will now seek to formalise this notion.

In what follows, will use definitions and notation following Michor [25] and readers should consult this text for additional details of this construction. Consider a family of time-parameterised geodesics, $\{g_{t,s}\}_{s\in\mathbb{R}}$. Then we define $u, \xi \in \mathfrak{g}$ by $u = \big[(\partial_t g_{t,s}) \cdot g_{t,s} \big]_{s=0}$ and $\xi = \big[(\partial_s g_{t,s}) \cdot g_{t,s} \big]_{s=0}$, and these associations can be understood as smooth maps $C^\infty(G, \mathfrak{g})$. Furthermore, the lifted right action by $g_{t,s}$ provides a map from $C^\infty(G, \mathfrak{g})$ to the space of vector fields on G, and as such we have an isomorphism, $C^\infty(G, \mathfrak{g}) \cong \mathfrak{X}(G)$. Notice that $\partial_t g_{t,s}, \partial_s g_{t,s} \in T_{g_{t,s}} G$ and the pushforward of ∂_t and ∂_s by $g_{t,\epsilon}$ can be understood as vector fields in G. This permits us to introduce the notation $\nabla_{\partial_t} \xi$ by which we mean the Levi-Civita covariant derivative of ξ along the curve g_t, which can be interpreted as an element of $C^\infty(G, \mathfrak{g})$. Since we have constructed a covariant derivative in this manner, we may similarly define the Riemannian curvature as a function of u and ξ, $\mathcal{R}(\xi, u)$: $C^\infty(G, \mathfrak{g}) \to C^\infty(G, \mathfrak{g})$. In particular, this is understood in the usual sense in terms of $\mathfrak{X}(G)$, where the isomorphism is applied at each stage to identify the vector fields on G with elements of Lie algebra.

Theorem 2.2 *Suppose u is a solution of the geodesic Eq. (2.34), then we have the following infinite dimensional analogue of the Jacobi equation*

$$\nabla_{\partial_t} \nabla_{\partial_t} \xi + \mathcal{R}(\xi, u) u = 0 \,. \tag{2.37}$$

Proof Following on from the equations derived in Sect. 2.4, we have Eqs. (2.33) and (2.35). In order to prevent overuse of the musical isomorphisms, and to better connect with the existing literature in this direction, we will define $\mathrm{ad}_\square^\dagger \square : \mathfrak{g} \times \mathfrak{g} \to \mathfrak{g}$ in terms of $\mathrm{ad}_\square^* \square : \mathfrak{g} \times \mathfrak{g}^* \to \mathfrak{g}^*$ as

$$\mathrm{ad}_u^\dagger v = \big(\mathrm{ad}_u^* v^\flat \big)^\sharp \,. \tag{2.38}$$

Note that this is simply the consequence of ad^* being the dual operator to ad with respect to the natural pairing $\langle \cdot, \cdot \rangle_\mathfrak{g}$ and $\mathrm{ad}^\dagger$ being defined in the same manner with respect to the inner product $(\cdot, \cdot)_\mathfrak{g}$ on $\mathfrak{g}$. This allows us to write the equations entirely on the Lie algebra, without needing the dual space. Firstly, notice that applying Hamilton's Principle instead to the action $\frac{1}{2} \int \langle u, u \rangle_\mathfrak{g} \, dt$ yields the following equations

$$\partial_t u + \mathrm{ad}_u^\dagger u = 0 \,, \tag{2.39}$$

$$(\partial_t + \mathrm{ad}_u^\dagger)(\partial_t \xi - \mathrm{ad}_u \xi) = - \mathrm{ad}_{\partial_t \xi - \mathrm{ad}_u \xi}^\dagger u \,, \tag{2.40}$$

which correspond to rewriting Eqs. (2.33) and (2.35) in terms of the operator $\mathrm{ad}^\dagger$. By using the linearity of $\mathrm{ad}^\dagger$ and moving terms in Eq. (2.40) to the right hand side, we have

$$\partial_{tt}\xi = - \mathrm{ad}_{\partial_t \xi}^\dagger u + \mathrm{ad}_u \partial_t \xi - \mathrm{ad}_u^\dagger \partial_t \xi + \mathrm{ad}_{\mathrm{ad}_u \xi}^\dagger u$$

$$+ \mathrm{ad}_u^\dagger \mathrm{ad}_u \xi + \mathrm{ad}_{\partial_t u} \xi$$

using Eq. (2.39) $\quad = - \mathrm{ad}_{\partial_t \xi}^\dagger u + \mathrm{ad}_u \partial_t \xi - \mathrm{ad}_u^\dagger \partial_t \xi + \mathrm{ad}_{\mathrm{ad}_u \xi}^\dagger u$

$$+ \mathrm{ad}_u^\dagger \mathrm{ad}_u \xi - \mathrm{ad}_{\mathrm{ad}_u^\dagger u} \xi$$

antisymmetry of the Lie bracket $\quad = - \mathrm{ad}_{\partial_t \xi}^\dagger u + \mathrm{ad}_u \partial_t \xi - \mathrm{ad}_u^\dagger \partial_t \xi - \mathrm{ad}_{\mathrm{ad}_\xi u}^\dagger u$

$$- \mathrm{ad}_u^\dagger \mathrm{ad}_\xi u + \mathrm{ad}_\xi \mathrm{ad}_u^\dagger u$$

$$= - \mathrm{ad}_{\partial_t \xi}^\dagger u + \mathrm{ad}_u \partial_t \xi - \mathrm{ad}_u^\dagger \partial_t \xi$$

$$+ [\mathrm{ad}_\xi^\dagger + \mathrm{ad}_\xi, \mathrm{ad}_u^\dagger]u \,,$$

where, in the final line, we have used the identity $- \mathrm{ad}_{\mathrm{ad}_\xi u}^\dagger u = [\mathrm{ad}_\xi^\dagger, \mathrm{ad}_u^\dagger]u$. As was shown by Michor [25], $\nabla_{\partial_t} \nabla_{\partial_t} \xi + \mathcal{R}(\xi, u)u$ is zero when the above equation is true. $\quad\square$

3   Examples

In each example, we will first derive the linearised equations in terms of arbitrary perturbations of our variables $(\delta u, \delta a)$. Following this, we will illustrate the equation in terms of the arbitrary vector field $\xi = \delta g \cdot g^{-1}$.

3.1   The Incompressible Euler Equations

The n-dimensional Euler equations on a manifold M correspond to a Lagrangian, $\ell_E : \mathfrak{X}(M) \times \mathrm{Den}(M) \rightarrow \mathbb{R}$, defined by

$$\ell_E(u, Dd^n x; \pi) = \int_M \frac{D}{2} |\boldsymbol{u}|^2 - \pi(D-1)\, d^n x \,, \tag{3.1}$$

and thus
$$\frac{1}{2}\delta^2\ell_E = \int_M \frac{D}{2}|\delta u|^2 + \delta Du \cdot \delta u - \delta\pi\delta D \, d^n x \,, \tag{3.2}$$

where $u = \boldsymbol{u} \cdot \nabla$ and $\delta u = \boldsymbol{\delta u} \cdot \nabla$ denote the vector fields expressed in terms of a basis. From Eqs. (3.1) and (3.2), we may compute the following variational derivatives

$$\frac{\delta\ell_E}{\delta u} = Du^\flat \otimes d^n x \,, \quad \frac{\delta\ell_E}{\delta D} = \frac{1}{2}|u|^2 - \pi \,, \quad \frac{\delta\ell}{\delta\pi} = (D-1) \, d^n x \,,$$

$$\frac{1}{2}\frac{\delta(\delta^2\ell_E)}{\delta(\delta u)} = D\delta u^\flat \otimes d^n x + \delta Du^\flat \otimes d^n x \,,$$

$$\frac{1}{2}\frac{\delta(\delta^2\ell_E)}{\delta(\delta D)} = u \cdot \delta u - \delta\pi \,, \quad \frac{1}{2}\frac{\delta(\delta^2\ell_E)}{\delta(\delta\pi)} = -\delta D \, d^n x \,.$$

We may assemble the variational derivatives of the Lagrangian into the Euler-Poincaré equations (2.22) and (2.23) gives

$$(\partial_t + \mathcal{L}_u)(Du^\flat \otimes d^n x) = Dd\left(\frac{1}{2}|u|^2 - \pi\right) \otimes d^n x \,, \quad \text{and} \quad (\partial_t + \mathcal{L}_u)(D \, d^n x) = 0 \,.$$

The second of these equations, together with the variation in π giving $D = 1$, implies that u is a divergence free vector field. The first of these equations is Euler's momentum equation in its geometric form. We may similarly assemble the variational derivatives of $\delta^2\ell/2$ into the Eqs. (2.24) and (2.25) as follows

$$(\partial_t + \mathcal{L}_u)\left(D\delta u^\flat \otimes d^n x + \delta Du^\flat \otimes d^n x\right) = -\mathcal{L}_{\delta u}\left(Du^\flat \otimes d^n x\right)$$

$$+ \delta Dd\left(\frac{1}{2}|u|^2 - \pi\right) \otimes d^n x + Dd\left(u \cdot \delta u - \delta\pi\right) \otimes d^n x \,, \tag{3.3}$$

$$(\partial_t + \mathcal{L}_u)(\delta D \, d^n x) = -\mathcal{L}_{\delta u}(D \, d^n x) \,. \tag{3.4}$$

The variations in π and $\delta\pi$ imply that $D = 1$ and $\delta D = 0$. Hence δu is a divergence free vector field and we have

$$(\partial_t + \mathcal{L}_u)(\delta u^\flat) = -\mathcal{L}_{\delta u}u^\flat + d(u \cdot \delta u - \delta\pi) \,, \quad \text{and} \quad (*d*)u^\flat = (*d*)\delta u^\flat = 0 \,, \tag{3.5}$$

where $*$ denotes the Hodge star. In three dimensions, this can be represented in vector calculus notation as

$$\partial_t \boldsymbol{\delta u} - \boldsymbol{u} \times \operatorname{curl} \boldsymbol{\delta u} = \boldsymbol{\delta u} \times \operatorname{curl} \boldsymbol{u} - \nabla(\boldsymbol{u} \cdot \boldsymbol{\delta u} + \delta\pi) \,, \quad \text{and} \quad \operatorname{div} \boldsymbol{u} = \operatorname{div} \boldsymbol{\delta u} = 0 \,, \tag{3.6}$$

or, equivalently,

$$\partial_t \delta u + u \cdot \nabla \delta u + \delta u \cdot \nabla u = -\nabla(\delta \pi). \tag{3.7}$$

Taking the approach described in Sect. 2.4, these equations, when written in terms of $\xi = (\delta g) \cdot g^{-1}$ give

$$(\partial_t + \mathrm{ad}_u^*)((\partial_t \xi - \mathrm{ad}_u \xi)^\flat) = -\mathrm{ad}_{\partial_t \xi - \mathrm{ad}_u \xi}^* u^\flat + d((\partial_t \xi - \mathrm{ad}_u \xi) \lrcorner u^\flat - \delta \pi). \tag{3.8}$$

This equation results from simply substituting in the relationship $\delta u = \partial_t \xi - \mathrm{ad}_u \xi$ into the Eq. (3.5). To obtain (3.8) in vector calculus notation, the equation $\delta u = \partial_t \xi + \xi \cdot \nabla u - u \cdot \nabla \xi$ can be considered alongside the Eq. (3.7).

3.2   The Stratified Thermal Rotating Euler Equations

We take a constant gravitational force to act in the direction of one of our coordinates, z, and introduce an advected parameter, ρ, which models thermal effects. Furthermore, we introduce a variable representing the effects of rotation, $\mathbf{R}$, which is a given function of $\mathbf{x}$ satisfying $\mathrm{curl}\mathbf{R}(\mathbf{x}) = 2\mathbf{\Omega}(\mathbf{x}) = f(\mathbf{x})\hat{\mathbf{z}}$, where f is the Coriolis parameter. The thermal rotating Euler equations then correspond to the following Lagrangian, $\ell_{tE} : \mathfrak{X}(M) \times \mathrm{Den}(M) \times \Lambda^0(M) \to \mathbb{R}$

$$\ell_{tE}(u, Dd^n x, \rho; \pi) = \int_M D\rho \left(\frac{1}{2}|u|^2 + u \cdot \mathbf{R} - gz \right) - \pi(D-1) d^n x, \tag{3.9}$$

$$\frac{1}{2}\delta^2 \ell_{tE} = \int_M \frac{D\rho}{2}|\delta u|^2 + (\delta D\rho + D\delta\rho)(u \cdot \delta u + \delta u \cdot \mathbf{R})$$

$$+ \delta D\delta\rho \left(\frac{|u|^2}{2} + u \cdot \mathbf{R} - gz \right) - \delta\pi\delta D\, d^n x. \tag{3.10}$$

The variational derivatives can be computed in a manner analogous to those found in Sect. 3.1. The Euler-Poincaré equations are

$$(\partial_t + \mathrm{ad}_u^*)\left(D\rho(u^\flat + \mathbf{R} \cdot d\mathbf{x}) \otimes d^n x \right) = Dd(\rho\varpi - \pi) \otimes d^n x - D\varpi d\rho \otimes d^n x, \tag{3.11}$$

$$(\partial_t + \mathcal{L}_u)(D\, d^n x) = 0, \tag{3.12}$$

$$(\partial_t + \mathcal{L}_u)\rho = 0, \tag{3.13}$$

where

$$\varpi := \frac{1}{2}|u|^2 + u \cdot \mathbf{R} - gz. \tag{3.14}$$

Making use of the advection equations and the pressure constraint $D = 1$, the first of these is

$$\rho(\partial_t + \mathrm{ad}_u^*)(u^\flat + \mathbf{R} \cdot d\mathbf{x}) = \rho d\varpi - d\pi \ . \tag{3.15}$$

In three dimensions, these equations can be expressed in vector calculus form as

$$\partial_t u + u \cdot \nabla u - u \cdot \mathrm{curl}\,\mathbf{R} = -g\hat{z} - \frac{1}{\rho}\nabla\pi \ ,$$

$$\partial_t \rho + u \cdot \nabla\rho = 0 \ ,$$

$$\nabla \cdot u = 0 \ ,$$

where $\hat{z}$ is the unit vector in the z direction. Similarly, the variational derivatives of the functional defined in Eq. (3.10) can be substituted into the Eqs. (2.24) and (2.25) to give

$$(\partial_t + \mathrm{ad}_u^*)\left(D\rho\delta u^\flat \otimes d^n x + (\delta D\rho + D\delta\rho)(u^\flat + \mathbf{R} \cdot d\mathbf{x}) \otimes d^n x\right)$$

$$= -\mathcal{L}_{\delta u}\left(D\rho(u^\flat + \mathbf{R} \cdot d\mathbf{x}) \otimes d^n x\right) + \delta Dd\,(\rho\varpi - \pi) \otimes d^n x$$

$$+ Dd\,(\rho(u \cdot \delta u + \delta u \cdot \mathbf{R}) + \delta\rho\varpi - \delta\pi) \otimes d^n x$$

$$- D\varpi d(\delta\rho) \otimes d^n x - (D(u \cdot \delta u + \delta u \cdot \mathbf{R}) + \delta D\varpi)\,d\rho \otimes d^n x \ , \tag{3.16}$$

$$(\partial_t + \mathcal{L}_u)(\delta D\,d^n x) = -\mathcal{L}_{\delta u}(D\,d^n x) \ , \tag{3.17}$$

$$(\partial_t + \mathcal{L}_u)\delta\rho = -\mathcal{L}_{\delta u}\rho \ . \tag{3.18}$$

The first of these equations is simplified when using the constraints $D = 1$ and $\delta D = 0$ and, after applying the product rule, we have

$$\rho(\partial_t + \mathrm{ad}_u^*)\delta u^\flat + \delta\rho(\partial_t + \mathrm{ad}_u^*)\left(u^\flat + \mathbf{R} \cdot d\mathbf{x}\right) = \delta\rho\,d\varpi + \rho d\,(u \cdot \delta u + \delta u \cdot \mathbf{R})$$

$$- \rho\mathcal{L}_{\delta u}\left(u^\flat + \mathbf{R} \cdot d\mathbf{x}\right) - d\,(\delta\pi) \ . \tag{3.19}$$

Making use of the Euler-Poincaré equation (3.15), this equation is

$$(\partial_t + \mathrm{ad}_u^*)\delta u^\flat = \frac{\delta\rho}{\rho^2}d\pi + d\,(u \cdot \delta u + \delta u \cdot \mathbf{R}) - \mathcal{L}_{\delta u}\left(u^\flat + \mathbf{R} \cdot d\mathbf{x}\right) - \frac{1}{\rho}d\,(\delta\pi) \ . \tag{3.20}$$

In vector calculus notation, the equations are given in three dimensions by

$$\partial_t \delta u + u \cdot \nabla\delta u + \delta u \cdot \nabla u - \delta u \times \mathrm{curl}\,\mathbf{R} = \frac{\delta\rho}{\rho^2}\nabla\pi - \frac{1}{\rho}\nabla(\delta\pi) \ , \tag{3.21}$$

$$\partial_t \delta\rho + \boldsymbol{u} \cdot \nabla \delta\rho + \delta\boldsymbol{u} \cdot \nabla\rho = 0\,, \tag{3.22}$$

$$\nabla \cdot \boldsymbol{u} = \nabla \cdot \delta\boldsymbol{u} = 0\,. \tag{3.23}$$

The equation for $\xi = \delta g \cdot g^{-1} = \boldsymbol{\xi} \cdot \nabla$ is given by substituting the equations

$$\delta\boldsymbol{u} = \partial_t \xi - \mathrm{ad}_u\, \xi\,, \quad \text{and} \quad \delta\rho = -\mathcal{L}_\xi \rho\,, \tag{3.24}$$

or

$$\delta\boldsymbol{u} = \partial_t \boldsymbol{\xi} + \boldsymbol{\xi} \cdot \nabla\boldsymbol{u} - \boldsymbol{u} \cdot \nabla\boldsymbol{\xi}\,, \quad \text{and} \quad \delta\rho = -\boldsymbol{\xi} \cdot \nabla\rho\,, \tag{3.25}$$

into Eq. (3.20) or (3.21) respectively.

3.3   The Euler-Boussinesq Equations

Notice that the introduction of thermal effects in the previous example results in the pressure, π, from the fluid remains in the equation governing the linear perturbation. This is not the case if one makes the Boussinesq approximation as follows. The Lagrangian is now defined to be

$$\ell_{EB}(u, Dd^n x, \rho; \pi) = \int_M D\left(\frac{1}{2}|\boldsymbol{u}|^2 + \boldsymbol{u} \cdot \boldsymbol{R} - g\rho z\right) - \pi(D-1)\, d^n x\,, \tag{3.26}$$

$$\frac{1}{2}\delta^2 \ell_{EB} = \int_M \frac{D}{2}|\delta\boldsymbol{u}|^2 + \delta D\, (\boldsymbol{u} \cdot \delta\boldsymbol{u} + \delta\boldsymbol{u} \cdot \boldsymbol{R})$$
$$- gz\delta D\delta\rho - \delta\pi\delta D\, d^n x\,. \tag{3.27}$$

Proceeding as in the previous examples, the Euler-Poincaré equation is

$$(\partial_t + \mathrm{ad}_u^*)(u^\flat + \boldsymbol{R} \cdot d\mathbf{x}) = d\left(\frac{1}{2}|\boldsymbol{u}|^2 + \boldsymbol{u} \cdot \boldsymbol{R} - \pi\right) - g\rho dz\,, \tag{3.28}$$

where the incompressibility constraint and the Eq. (2.22) has been rearranged after substituting in the variational derivatives of the Lagrangian, ℓ_{EB}. In three dimensions, this equation is given by

$$\partial_t \boldsymbol{u} + \boldsymbol{u} \cdot \nabla\boldsymbol{u} - \boldsymbol{u} \times \mathrm{curl}\,\boldsymbol{R} = -\nabla\pi - g\rho\hat{\boldsymbol{z}}\,,$$

$$\partial_t \rho + \boldsymbol{u} \cdot \nabla\rho = 0\,,$$

$$\nabla \cdot \boldsymbol{u} = 0\,.$$

Computing the variational derivatives with respect to δu, δD, and $\delta \pi$ and substituting the results into Eqs. (2.24) and (2.25), we have

$$
\begin{aligned}
(\partial_t + \mathrm{ad}_u^*)&(D\delta u^\flat \otimes d^n x + \delta D(u^\flat + \mathbf{R} \cdot d\mathbf{x}) \otimes d^n x) \\
&= -\mathcal{L}_{\delta u}(D(u^\flat + \mathbf{R} \cdot d\mathbf{x}) \otimes d^n x) \\
&\quad + Dgz\, d(\delta\rho) \otimes d^n x + \delta Dgz\, d\rho \otimes d^n x \\
&\quad + \delta Dd\left(\frac{1}{2}|u|^2 + u \cdot \mathbf{R} - g\rho z - \delta\pi\right) \otimes d^n x \\
&\quad + Dd\left(u \cdot \delta u + \delta u \cdot \mathbf{R} - gz\delta\rho - \delta\pi\right) \otimes d^n x \,.
\end{aligned}
\tag{3.29}
$$

Making use of the constraints $D = 1$ and $\delta D = 0$, resulting from the arbitrary variation in π and $\delta\pi$, we have

$$
\begin{aligned}
(\partial_t + \mathrm{ad}_u^*)\delta u^\flat = &- \mathcal{L}_{\delta u}(u^\flat + \mathbf{R} \cdot d\mathbf{x}) + gz\, d(\delta\rho) \\
&+ d\left(u \cdot \delta u + \delta u \cdot \mathbf{R} - gz\delta\rho - \delta\pi\right) \,.
\end{aligned}
\tag{3.30}
$$

The equations, in three dimensions, are therefore

$$
\partial_t \delta u + u \cdot \nabla \delta u + \delta u \cdot \nabla u - \delta u \times \mathrm{curl}\, \mathbf{R} = -\nabla(\delta\pi) - g\,\delta\rho\hat{z}\,,
\tag{3.31}
$$

$$
\partial_t \delta\rho + u \cdot \nabla\delta\rho + \delta u \cdot \nabla\rho = 0\,,
\tag{3.32}
$$

$$
\nabla \cdot u = \nabla \cdot \delta u = 0\,.
\tag{3.33}
$$

As for the previous example, the equation for ξ follows from substituting the Eqs. (3.24) into (3.30) or Eqs. (3.25) into (3.31).

3.4 The 2D Thermal Rotating Shallow Water Equations

We here consider the two dimensional thermal rotating shallow water equations, which can be interpreted as an approximation to three dimensional models using vertical averaging. The Lagrangian is

$$
\ell_{TRSW}(u, \eta\, d^n x, \rho) = \int_{M\subseteq\mathbb{R}^2}\left(\frac{|u|^2}{2} + u \cdot \mathbf{R} - \frac{g\rho}{2}(\eta - 2h)\right)\eta\, d^2 x\,,
\tag{3.34}
$$

$$
\begin{aligned}
\frac{1}{2}\delta^2\ell_{TRSW} = \int_{M\subseteq\mathbb{R}^2}\Big(&\frac{\eta}{2}|\delta u|^2 + \delta\eta\, \delta u \cdot (u + \mathbf{R}) \\
&- g\,\delta\rho\, \delta\eta\,(\eta - h) - \frac{g\rho}{2}\delta\eta^2\Big)\, d^2 x\,.
\end{aligned}
\tag{3.35}
$$

The variational derivatives are computed as follows

$$\frac{\delta\ell}{\delta u} = \eta(u^\flat + \mathbf{R}\cdot d\mathbf{x})\otimes d^2x\,, \qquad \frac{\delta\ell}{\delta\rho} = -\frac{g\eta}{2}(\eta - 2h)\,d^2x\,,$$

$$\frac{\delta\ell}{\delta\eta} = \frac{|\mathbf{u}|^2}{2} + \mathbf{u}\cdot\mathbf{R} - g\rho(\eta - h)\,,$$

$$\frac{1}{2}\frac{\delta(\delta^2\ell)}{\delta(\delta u)} = (\eta\,\delta u^\flat + \delta\eta(u^\flat + \mathbf{R}\cdot d\mathbf{x}))\otimes d^2x\,, \qquad \frac{1}{2}\frac{\delta(\delta^2\ell)}{\delta(\delta\rho)} = -g\,\delta\eta(\eta - h)\,d^2x\,,$$

$$\frac{1}{2}\frac{\delta(\delta^2\ell)}{\delta(\delta\eta)} = \delta\mathbf{u}\cdot(\mathbf{u} + \mathbf{R}) - g\,\delta\rho(\eta - h) - g\rho\,\delta\eta\,.$$

Assembling these into the Euler-Poincaré equation, making use of the fact that η and ρ are advected quantities, we have

$$(\partial_t + \mathrm{ad}_u^*)(u^\flat + \mathbf{R}\cdot d\mathbf{x}) = d\left(\frac{|\mathbf{u}|^2}{2} + \mathbf{u}\cdot\mathbf{R}\right) - g\rho\,d(\eta - h) - \frac{g\eta}{2}\,d\rho\,, \qquad (3.36)$$

which, in vector calculus notation, is

$$\partial_t\mathbf{u} + \mathbf{u}\cdot\nabla\mathbf{u} - (\nabla^\perp\cdot\mathbf{R})\mathbf{u}^\perp = -g\rho\nabla(\eta - h) - \frac{1}{2}g\eta\nabla\rho\,, \qquad (3.37)$$

$$\partial_t\rho + \mathbf{u}\cdot\nabla\rho = 0\,, \qquad (3.38)$$

$$\partial_t\eta + \nabla\cdot(\eta\mathbf{u}) = 0\,. \qquad (3.39)$$

Assembling the variational derivatives into the Eqs. (2.24) and (2.25), we have

$$\begin{aligned}
(\partial_t + \mathrm{ad}_u^*)&\left((\eta\,\delta u^\flat + \delta\eta(u^\flat + \mathbf{R}\cdot d\mathbf{x}))\otimes d^2x\right) \\
&= -\mathcal{L}_{\delta u}(\eta(u^\flat + \mathbf{R}\cdot d\mathbf{x})\otimes d^2x) \\
&\quad + \delta\eta\,d\left(\frac{|\mathbf{u}|^2}{2} + \mathbf{u}\cdot\mathbf{R} - g\rho(\eta - h)\right)\otimes d^2x \\
&\quad + \eta\,d\,(\delta\mathbf{u}\cdot(\mathbf{u} + \mathbf{R}) - g\,\delta\rho(\eta - h) - g\rho\,\delta\eta)\otimes d^2x \\
&\quad + \frac{g\eta}{2}(\eta - 2h)\,d(\delta\rho)\otimes d^2x \\
&\quad + g\,\delta\eta(\eta - h)\,d\rho\otimes d^2x\,.
\end{aligned} \qquad (3.40)$$

These equations can be simplified by applying the product rule and making use of the Euler-Poincaré equation (3.36) to give

$$(\partial_t + \mathrm{ad}_u^*)\delta u^\flat = -\mathcal{L}_{\delta u}(u^\flat + \mathbf{R} \cdot d\mathbf{x}) + d\left(\frac{|\boldsymbol{u}|^2}{2} + \boldsymbol{u} \cdot \mathbf{R}\right)$$

$$- g\,\delta\rho\,d(\eta - h) - g\rho\,d(\delta\eta) - \frac{g\eta}{2}\,d(\delta\rho) - \frac{g}{2}\delta\eta\,d\rho\,. \tag{3.41}$$

The equations, in vector calculus form, are

$$\partial_t \delta\boldsymbol{u} + \boldsymbol{u} \cdot \nabla\delta\boldsymbol{u} + \delta\boldsymbol{u} \cdot \nabla\boldsymbol{u} - (\nabla^\perp \cdot \mathbf{R})\delta\boldsymbol{u}^\perp$$

$$= \frac{1}{2}g\,\delta\eta\nabla\rho - \frac{1}{2}g\eta\nabla(\delta\rho) - g\,\delta\rho\nabla(\eta - h) - g\rho\nabla(\delta\eta)\,, \tag{3.42}$$

$$\partial_t \delta\rho + \boldsymbol{u} \cdot \nabla\delta\rho + \delta\boldsymbol{u} \cdot \nabla\rho = 0\,, \tag{3.43}$$

$$\partial_t \delta\eta + \nabla \cdot (\delta\rho\boldsymbol{u}) + \nabla \cdot (\rho\delta\boldsymbol{u}) = 0\,. \tag{3.44}$$

The equation for $\xi = \delta g \cdot g^{-1}$ is given, in its geometric form, by substituting the equations

$$\delta u = \partial_t \xi - \mathrm{ad}_u \xi\,, \quad \delta\rho = -\mathcal{L}_\xi \rho\,, \quad \text{and} \quad \delta\eta = -\mathcal{L}_\xi \eta\,, \tag{3.45}$$

into Eq. (3.41). Alternatively, in vector calculus notation, this equation corresponds to substituting

$$\delta\boldsymbol{u} = \partial_t \boldsymbol{\xi} + \boldsymbol{\xi} \cdot \nabla\boldsymbol{u} - \boldsymbol{u} \cdot \nabla\boldsymbol{\xi}\,, \quad \delta\rho = -\boldsymbol{\xi} \cdot \nabla\rho\,, \quad \text{and} \quad \delta\eta = -\nabla \cdot (\eta\boldsymbol{\xi})\,, \tag{3.46}$$

into Eq. (3.42).

4 Numerical Simulations

In this section, we consider numerical simulations of the linearised Euler-Poincaré equations and the associated dynamics of the perturbation vector field for the example of the Euler-Boussinesq (EB) equations given in Sect. 3.3. Simplifying to a 2D vertical domain, the incompresibility conditions of both $\boldsymbol{u}$ and $\delta\boldsymbol{u}$ allow us to express the EB equations and their linearised equations in streamfunction and vorticty formulation. Using the Jacobian operator $\mathcal{J} : C^\infty(M) \times C^\infty(M) \to \mathbb{R}$ in the vertical (x, z)-plane defined by

$$\mathcal{J}(f, g) = -\hat{\boldsymbol{y}} \cdot \nabla f \times \nabla g = \partial_x f \partial_z g - \partial_z f \partial_x g\,, \tag{4.1}$$

we have the following equivalent formulation of Eq. (3.31)

$$\partial_t \omega + \mathcal{J}(\psi, \omega) + g\mathcal{J}(z, b) = 0\,,$$

$$\partial_t b + \mathcal{J}(\psi, b) = 0\,, \quad \text{where} \quad \omega := \triangle\psi \quad \text{and} \quad \boldsymbol{u} = \nabla^\perp\psi\,, \tag{4.2}$$

$$\partial_t \delta\omega + \mathcal{J}(\psi, \delta\omega) + \mathcal{J}(\delta\psi, \omega) + g\mathcal{J}(z, \delta b) = 0\,,$$

$$\partial_t \delta b + \mathcal{J}(\psi, \delta b) + \mathcal{J}(\delta\psi, b) = 0\,, \quad \text{where} \quad \delta\omega := \Delta\delta\psi \quad \text{and} \quad \boldsymbol{\delta u} = \nabla^\perp \delta\psi\,.$$
$$(4.3)$$

Substituting the Ansatz $\delta b = -\mathcal{L}_\xi b = -\boldsymbol{\xi} \cdot \nabla b$ where $\boldsymbol{\xi} \in \mathfrak{X}(M)$ is the perturbation vector field, we have the equivalent form of the linearised dynamics as

$$\partial_t \delta\omega + \mathcal{J}(\psi, \delta\omega) + \mathcal{J}(\delta\psi, \omega) - g\mathcal{J}(z, \boldsymbol{\xi} \cdot \nabla b) = 0\,,$$

$$\partial_t \boldsymbol{\xi} + (\boldsymbol{\xi} \cdot \nabla)\boldsymbol{u} - \boldsymbol{u} \cdot \nabla\boldsymbol{\xi} = \boldsymbol{\delta u}\,, \quad \text{where} \quad \delta\omega := \Delta\delta\psi \quad \text{and} \quad \boldsymbol{\delta u} = \nabla^\perp \delta\psi\,.$$
$$(4.4)$$

The configurations of the numerical simulations are as follows. The computational domain is $\Omega = [0, 1] \times [0, 1]$ which is discretized using 256×256 finite element cells. The boundary conditions for $\boldsymbol{u}$ and $\boldsymbol{\delta u}$ are periodic in the x direction and free slip in z. These boundary conditions can be enforced through their definition from the stream functions ψ and $\delta\psi$ respectively, both of which have homogeneous Dirichlet boundary conditions in $y = 0, 1$ and are periodic in x. In the absence of viscosity, no other boundary conditions are required. The fluid vorticity ω, perturbed vorticity $\delta\omega$ and buoyancy b are approximated with the $2nd$ order discontinuous Galerkin finite element space ($DG1$); the stream function ψ and the perturbation stream function $\delta\psi$ are approximated with the $2nd$ order continuous Galerkin finite element space ($CG1$); Lastly, the perturbation vector field $\boldsymbol{\xi}$, fluid velocity $\boldsymbol{u}$, and perturbed fluid velocity $\boldsymbol{\delta u}$ are approximated with the vectorised $2nd$ order continuous Galerkin finite element space. The numerical method is implemented using the firedrake software [13] and we ran the simulation for a total of 16 time units. The snapshots of interest are presented in Figs. 1, 2, and 3.

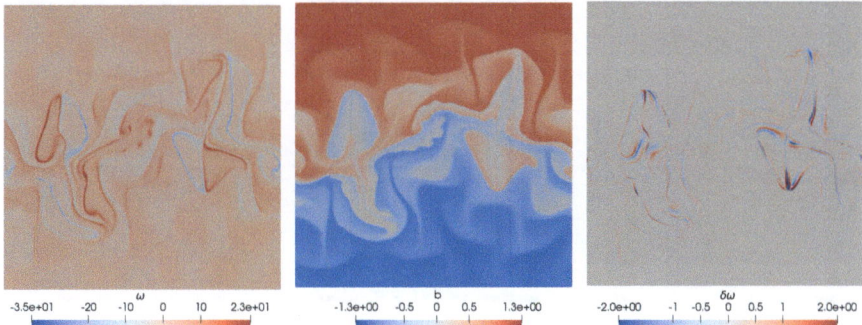

Fig. 1 At $t = 5$, one observes the initial phases of Kelvin Helmholtz instabilities generated by the initial conditions in fluid vorticity and buoyancy gradients in the snapshots of ω (left) and b (middle). The inhomogeneous regions in the δw snapshot (right) closely track the fronts of these instabilities, notably the downwards plume and upwards plume located at the left and right side of domain centre. This feature allows the perturbation vorticity $\delta\omega$ to be used a diagnostic for instabilities

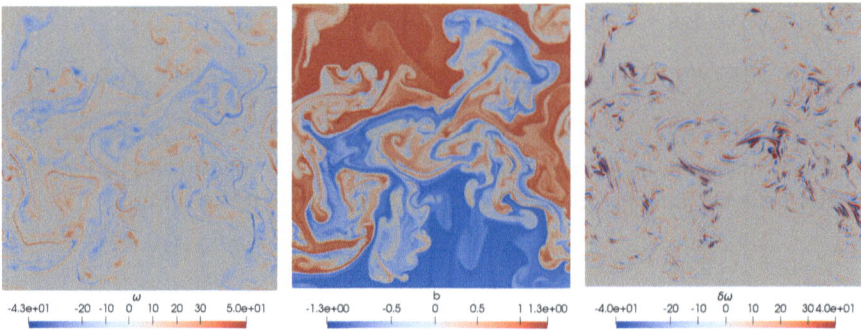

Fig. 2 At $t = 8$, the Kelvin Helmholtz instabilities are fully developed as they are visible in the ω (left) and b (middle) snapshots. The perturbation vorticity $\delta\omega$ (right) shows strong correlations with these instabilities whilst the magnitude are ≈ 10 times larger than the $\delta\omega$ snapshot at $t = 5$. We note that the region of largest $\delta\omega$ originated from the initial Kelvin Helmholtz instability shown in Fig. 1. This is indeed expected as $\delta\omega$ is advected by the fluid velocity, $\boldsymbol{u}$, with an additional forcing term, $\delta\boldsymbol{u} \cdot \nabla\omega$, that generates circulation of $\delta\boldsymbol{u}$

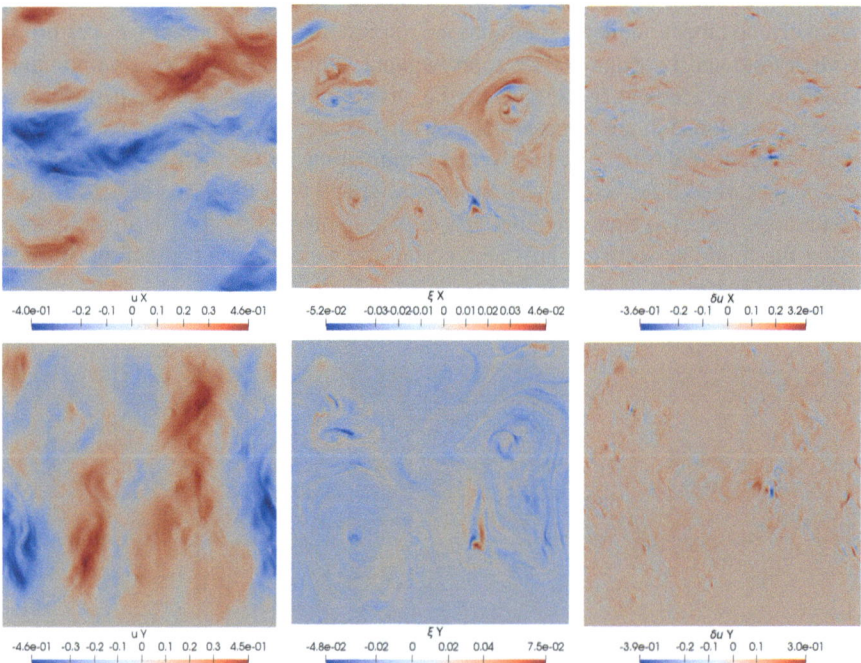

Fig. 3 Snapshots of the fluid velocity field $\boldsymbol{u}$ (left), perturbation vector field ξ (middle), and the perturbed velocity field $\delta\boldsymbol{u}$ (right) at time $t = 8$

5 Summary, Open Problems and Outlook

The present chapter has treated higher-order variations in the Euler-Poincaré variational principles with applications to ideal fluid dynamics. The $1st$ variations yield the well known models of ideal fluid dynamics via the Euler-Poincaré approach [20]. The $2nd$ variations yield equations for linear perturbations propagating in the frame of the fluid motion arising from the $1st$ variation. The advantage of this approach is that the stability equations arising from the $2nd$ variation are perturbation equations for time-dependent flows, not only for equilibrium time-independent flows, although fluid equilibrium solutions are permitted. For the EPDiff equation for geodesics on a Lie group, the linearised equation derived using the methods discussed in this chapter is shown to be equivalent to Jacobi's equation and thus the arbitrary vector field employed in the $1st$ variation is the Jacobi field. The physical examples of fluid models given in this chapter are expressed naturally on semidirect product spaces and are thus not geodesic equations. It remains to understand how the broken symmetry of ideal fluid dynamics with advected quantities influences the Jacobi equation for the vector field $\xi = \delta g g^{-1}$, when expressed in terms of covariant derivatives, and hence the behaviour of nearby trajectories in the Lie group.

Acknowledgments We are grateful to C. Cotter, D. Crisan, J.-M. Leahy, A. Lobbe, J. Woodfield, as well as H. Dumpty for several thoughtful suggestions during the course of this work which have improved or clarified the interpretation of its results. DH and RH were partially supported during the present work by Office of Naval Research (ONR) grant award N00014-22-1-2082, "Stochastic Parameterization of Ocean Turbulence for Observational Networks". DH and OS were partially supported during the present work by European Research Council (ERC) Synergy grant "Stochastic Transport in Upper Ocean Dynamics" (STUOD)—DLV-856408.

References

1. Arnold, V.I., 1965. Conditions for nonlinear stability of the stationary plane curvilinear flows of an ideal fluid. Doklady Mat. Nauk., 162 (5), pp. 773–777
2. Arnold, V.I., 1966. Sur la géométrie différentielle des groupes de Lie de dimension infinie et ses applications à l'hydrodynamique des fluides parfaits. In Annales de l'institut Fourier (Vol. 16, No. 1, pp. 319–361).
3. Arnold, V.I. and Khesin, B., 1998. Topological Methods in Hydrodynamics, Springer, New York.
4. Bloch, A., Krishnaprasad, P.S., Marsden, J.E., Ratiu T.S., 1996. The Euler-Poincaré equations and double bracket dissipation. Commun.Math. Phys. 175, 1–42. https://doi.org/10.1007/BF02101622
5. Casciaro, B. and Francaviglia, M., On the second variation for first order calculus of variations on fibered manifolds. I: Generalized Jacobi equations, Rend. Mat., Serie VII,16 (1996), pp. 233–264.
6. Chandrasekhar, S., 1969. Ellipsoidal Figures of Equilibrium (New Haven. Conn.: Yale University).
7. Chandrasekhar, S., 1970. Solutions of two problems in the theory of gravitational radiation. Phys. Rev. Lett., 24(11), p.611.

8. Chandrasekhar, S., 1983. The Mathematical Theory of Black Holes. Oxford University Press, Oxford, 1983, xxii / 646 pp.

9. Chen, S. Foias, C. Holm, D.D. Olson, E.J., Titi, E.S. and Wynne, S., 1998. The Camassa-Holm equations as a closure model for turbulent channel and pipe flows, *Phys. Rev. Lett.*, **81** (1998) 5338–5341, https://doi.org/10.1103/PhysRevLett.81.5338

10. Chiaffredo, F., Fatibene, L., Ferraris, M., Ricossa, E. and Usseglio, D., 2023. A variational framework for higher order perturbations. arXiv preprint arXiv:2310.12907.

11. Foias, C., Holm, D.D. and Titi, E.S., 2001. The Navier-Stokes-alpha model of fluid turbulence. Physica D (152) 505–519. https://doi.org/10.1016/S0167-2789(01)00191-9

12. Foias, C., Holm, D.D. and Titi, E.S., 2002. The three dimensional viscous Camassa-Holm equations, and their relation to the Navier-Stokes equations and turbulence theory. J. Dyn. and Diff. Eqns. (14) 1–35. https://doi.org/10.1023/A:1012984210582

13. Ham D. A., Kelly P.H.J., Mitchell L., Cotter C.J., et al., Firedrake User Manual. Imperial College London and University of Oxford and Baylor University and University of Washington, 2023. 10.25561/104839

14. Hecht, M.W., Holm, D.D. Petersen, M.R. and Wingate, B.A., The LANS-alpha and Leray turbulence parameterizations in primitive equation ocean modeling. J. Phys. A: Math. Theor. **41** (2008) 344009 https://doi.org/10.1088/1751-8113/41/34/344009

15. Implementation of the LANS-alpha turbulence model in a primitive equation ocean model. MW Hecht, DD Holm, MR Petersen, BA Wingate, J. Comp. Physics (227) (2008) 5691. https://doi.org/10.1016/j.jcp.2008.02.018

16. Efficient form of the LANS-alpha turbulence model in a primitive-equation ocean model. MW Hecht, DD Holm, MR Petersen, BA Wingate, J. Comp. Physics (227) (2008) 5717. https://doi.org/10.1016/j.jcp.2008.02.017

17. Holm, D.D., 1999. Fluctuation effects on 3D Lagrangian mean and Eulerian mean fluid motion. Physica D: Nonlinear Phenomena, 133(1–4), pp.215–269. https://doi.org/10.1016/S0167-2789(02)00552-3

18. Holm, D.D., 2002. Lagrangian averages, averaged Lagrangians, and the mean effects of fluctuations in fluid dynamics. Chaos, 12(2), pp.518–530. https://doi.org/10.1063/1.1460941

19. Holm, D.D., 2002. Averaged Lagrangians and the mean effects of fluctuations in ideal fluid dynamics. Physica D: Nonlinear Phenomena, 170(3–4), pp.253–286. https://doi.org/10.1016/S0167-2789(02)00552-3

20. Holm, D.D., Marsden, J.E. and Ratiu, T.S., 1998. The Euler–Poincaré equations and semidirect products with applications to continuum theories. Advances in Mathematics, 137(1), pp.1–81. https://doi.org/10.1006/aima.1998.1721

21. Holm, D.D., Marsden, J.E., Ratiu, T.S., 1998. Euler–Poincaré models of ideal fluids with nonlinear dispersion. Phys. Rev. Lett., (80) 4173–4177. https://doi.org/10.1103/PhysRevLett.80.4173

22. Holm, D.D., Marsden, J.E., Ratiu, T.S. and Weinstein, A., 1985. Nonlinear stability of fluid and plasma equilibria. Physics reports, 123(1–2), pp.1–116. https://doi.org/10.1016/0370-1573(85)90028-6

23. Joharinad, P. and Jost, J., 2023. Metric Spaces and Manifolds. In Mathematical Principles of Topological and Geometric Data Analysis (pp. 115–164). Cham: Springer International Publishing.

24. Jost, J., 2008. Geodesics and Jacobi Fields. Riemannian Geometry and Geometric Analysis, pp.179–241.

25. Michor, P.W. (2006). Some geometric evolution equations arising as geodesic equations on groups of diffeomorphisms including the Hamiltonian approach. In: Bove, A., Colombini, F., Del Santo, D. (eds) Phase Space Analysis of Partial Differential Equations. Progress in Nonlinear Differential Equations and Their Applications, vol 69. Birkhäuser Boston. https://doi.org/10.1007/978-0-8176-4521-2_11.

26. Michor, P.W., 2015. Manifolds of mappings and shapes. Arxiv. https://arxiv.org/abs/1505.02359

27. Modin, K. and Perrot, M., 2023. Eulerian and Lagrangian stability in Zeitlin's model of hydrodynamics. arXiv preprint arXiv:2305.08479.
28. Preston, S.C., 2004. For ideal fluids, Eulerian and Lagrangian instabilities are equivalent. Geometric and Functional Analysis, 14(5), pp.1044–1062.
29. Sreenivasan, K. R., 2019. Chandrasekhar's Fluid Dynamics. Annual Review of Annual Review of Fluid Mechanics, 51:1–24. https://doi.org/10.1146/annurev-fluid-010518-040537
30. Washabaugh, P. and Preston, S.C., The geometry of axisymmetric ideal fluid flows with swirl - Arnold Mathematical Journal, 2017 - Springer
31. Younes, L., 2007. Jacobi fields in groups of diffeomorphisms and applications. Quarterly of applied mathematics, pp.113–134.

On Forward–Backward SDE Approaches to Conditional Estimation

Jin Won Kim and Sebastian Reich

1 Introduction

In this chapter, we revisit the problem of partially observed diffusion processes and their associated continuous time filtering problems [1, 11]. While the detailed problem formulation appears in Sect. 2, the goal of the continuous time filtering problem is to characterise the distribution of the hidden state X_T at time T, given the observation path $Z_{0:T}$. Let us denote the conditional distribution at time T by π_T. Instead of computing π_T as a probability measure, we focus on constructing an estimator of $f(X_T)$ for given a measurable and bounded function f in the form of

$$\mathcal{S}_T[f] = \mu[Y_0] - \int_0^T \mathcal{U}_t^T dZ_t, \tag{1}$$

where μ is the density of states at initial time $t = 0$ which is assumed to be known and $\mathcal{U}_t$ is a weight on the incoming observation. The function $Y_0(x)$ is chosen such that the corresponding estimator $\mathcal{S}_T[f]$ becomes an unbiased estimator of $f(X_T)$—that is, $\mathbb{E}[\mathcal{S}_T[f]] = \mathbb{E}[f(X_T)]$—for any admissible $\mathcal{U}_t$. Our interest has been triggered by recent progress on the topic [15–17], which has extended the dual optimal control perspective for linear problems, as originally introduced by [12, 14], to nonlinear filtering problems. In the work of [15–17], the weight $\mathcal{U}_t$ is considered as a control input to drive a function-valued stochastic process $Y_t(x)$ for $t \in [0, T]$ to satisfy the backward stochastic partial differential equation (BSPDE)

$$- dY_t = \mathcal{L}^x Y_t dt - V_t^T dZ_t + (\mathcal{U}_t + V_t)^T h \, dt, \quad Y_T = f. \tag{2}$$

J. W. Kim (✉) · S. Reich
Institut für Mathematik, Universität Potsdam, Potsdam, Germany
e-mail: sebastian.reich@uni-potsdam.de

© The Author(s) 2025
B. Chapron et al. (eds.), *Stochastic Transport in Upper Ocean Dynamics III*,
Mathematics of Planet Earth 13, https://doi.org/10.1007/978-3-031-70660-8_6

Here $\mathcal{L}^x$ denotes the generator of the underlying signal process and $h(x)$ is the forward map in the observation process. See Sect. 2 for a more detailed definition. The minimum variance property of the conditional expectation is then expressed in terms of an appropriate cost function which determined the optimal $\mathcal{U}_{0:T}$ uniquely, such that the optimal estimator satisfies

$$\mathcal{S}_T[f] = \pi_T[f] := \mathbb{E}[f(X_T)|Z_{0:T}]. \tag{3}$$

Instead of the BSPDE (2) and the minimum variance properties, we address the definition of (1) from the perspective of forward-backward stochastic differential equations (FBSDEs) [7] instead. In particular, instead of the controlled process in function space, we consider the backward-in-time process to be an uncontrolled representation of the target value. Our approach provides a unified perspective on the following four scenarios:

(i) estimating $\pi_T[f]$ using an estimator of the form (1),
(ii) estimating $\sigma_T[f]$ using an estimator of the form (1), where σ_t denotes the non-normalised filtering distribution satisfying the Zakai equation [1, 11],
(iii) estimating $\pi_T[f]$ using the innovation

$$I_t = Z_t - \int_0^t \pi_s[h]ds \tag{4}$$

in the estimator (1) instead of the observations Z_t,
(iv) estimating $\sigma_T[f]$ using the induced observation error

$$W_t = Z_t - \int_0^t h(X_s)ds \tag{5}$$

instead of the observations Z_t directly.

Our FBSDE approach allows for a transparent definition of estimators that exactly match the corresponding conditional expectations in all four cases. It is found that (ii)–(iv) lead to deterministic backward Kolmogorov equations [7, 19] in the unknown $Y_t(x)$ instead of the BSPDE formulation (2), which arises from (i). Furthermore, while (ii) and (iii) lead to the standard backward Kolmogorov equation, (iv) introduces an additional drift term in the spirit of Feynman-Kac formulations [7]. By strictly following an FBSDE approach, a numerical treatment also becomes feasible following established computational methods for FBSDEs [8]. This aspect will be explored in a future publication.

We emphasise that we assume throughout this chapter that the observable $f(x)$ is a deterministic function independent of the data $Z_{0:T}$. On the contrary, applications of estimators of the form (1) to the problem of filter stability have been discussed in [15, 18]. In this context, the terminal condition in (2) becomes data dependent and the analysis of the resulting FBSDEs becomes more involved. Such an extension is left to future work.

Estimators of the form (1) extend to conditional expectation values

$$V_T := \int_0^T \pi_t[c_t]dt + \pi_T[f], \tag{6}$$

which naturally arise from partially observed stochastic optimal control problems. Here $c_t(x)$, $t \in [0, T]$ denotes the running cost and $f(x)$ the terminal cost at some finite time horizon T [3, 5]. We briefly discuss such an extension and its connection to our FBSDE formulations later in the chapter.

The remainder of this chapter is structured as follows. Sect. 2 provides the necessary background on time-continuous nonlinear filtering and introduces two sets of forward SDEs (FSDEs), which will play an essential role in deriving the optimal estimators for scenarios (i)–(iv). The estimators for the conditional expectation values are discussed for all four scenarios separately in Sect. 3. Emphasis is put on a unified framework using appropriate FBSDEs. It is demonstrated that the FBSDE become equivalent to a backward Kolmogorov equation under scenarios (ii) and (iii). The FBSDE perspective also leads to novel formulations of the stochastic optimal estimation problem (i) while scenario (iv) results in a backward PDE formulation in the spirit of Feynman–Kac. Sect. 4 provides an application of (6) to problems from optimal control of partially observed diffusion processes [3, 5]. Our chapter ends with some conclusions.

2 The Continuous Time Filtering Problem

The joint measure over paths $(X_{0:T}, Z_{0:T})$ in state variable x and observed variable z is determined by the FSDE

$$dX_t = b(X_t)dt + \sigma dB_t, \qquad X_0 \sim \mu, \tag{7a}$$

$$dZ_t = h(X_t)dt + dW_t, \qquad Z_0 = 0, \tag{7b}$$

for time $t \in [0, T]$. It is assumed that $X_t \in \Re^{d_x}$ and $Z_t \in \Re^{d_z}$. $b : \Re^{d_x} \rightarrow \Re^{d_x}$ is the drift function and $h : \Re^{d_x} \rightarrow \Re^{d_z}$ is called the observation function. (B_t, W_t) are assumed to be independent Brownian motions on $\Re^{d_x} \times \Re^{d_z}$, while we assume σ is a scalar for simplicity. We denote the infinitesimal generator of the signal process X_t by $\mathcal{L}^x$ which is given by

$$\mathcal{L}^x g = b^{\mathsf{T}}\nabla_x g + \frac{\sigma^2}{2}\nabla_x^2 g. \tag{8}$$

We denote the induced path measure by $\mathbb{P}$ and the expectation value of a function $g(x, z)$ by $\mathbb{E}[g(X_t, Z_t)]$. We are interested in the conditional path measure $X_t|Z_{0:t}$ and denotes its density at time t, by π_t, $t \in [0, T]$ [1, 11]. Recall that the filtering distribution π_t satisfies the Kushner–Stratonovitch (KS) equation of nonlinear filtering [1, 11].

It is also well-known that the innovation process (4) behaves like Brownian motion independent of B_t under the path measure $\mathbb{P}$ and we introduce the FSDE

$$dX_t = b(X_t)dt + \sigma dB_t, \qquad\qquad X_0 \sim \mu, \qquad (9a)$$

$$dD_t = D_t\,(h(X_t) - \pi_t[h])^\mathsf{T}dI_t, \qquad D_0 = 1, \qquad (9b)$$

which are driven by the independent Brownian processes (B_t, I_t). We denote the path measure induced by the FSDE (9) by $\mathbb{P}^*$ and its expectation operator by $\mathbb{E}^*$. It holds that

$$\pi_t[f] = \mathbb{E}^*[D_t f(X_t)|I_{0:t}]. \qquad (10)$$

We also introduce the shorthand notation

$$\mathbb{E}_I^*[g(X_t, D_t)] = \mathbb{E}^*[g(X_t, D_t)|I_{0:t}], \qquad (11)$$

for any suitable function $g(x, d)$.

The SDE (9b) is of mean-field type, which becomes more transparent when rewritten in the form

$$dD_t = D_t\,(h(X_t) - \mathbb{E}_I^*[D_t h(X_t)])^\mathsf{T}dI_t. \qquad (12)$$

Hence, alternatively, we consider the change of measure

$$\frac{d\tilde{\mathbb{P}}}{d\mathbb{P}}(X_t) = \tilde{D}_t^{-1}, \qquad (13)$$

where

$$d\tilde{D}_t = \tilde{D}_t\,h(X_t)^\mathsf{T}dZ_t, \qquad \tilde{D}_0 = 1. \qquad (14)$$

We recall that this change of measure arises from an application of Girsanov's theorem [19] and that Z_t behaves like Brownian motion independent of B_t under the new path measure $\tilde{\mathbb{P}}$ [1]. We, hence, introduce a second pair of FSDEs

$$dX_t = b(X_t)dt + \sigma dB_t, \qquad\qquad X_0 \sim \mu, \qquad (15a)$$

$$d\tilde{D}_t = \tilde{D}_t\,h(X_t)^\mathsf{T}dZ_t, \qquad \tilde{D}_0 = 1, \qquad (15b)$$

which are now driven by the independent Brownian processes (B_t, Z_t). The induced path measure is denoted by $\tilde{\mathbb{P}}^*$ and its associated expectation operator by $\tilde{\mathbb{E}}^*$.

Furthermore, one can introduce the non-normalised filtering density σ_t, which satisfies

$$\sigma_t[f] = \tilde{\mathbb{E}}^*[\tilde{D}_t f(X_t)|Z_{0:t}] \tag{16}$$

for fixed observation process $Z_{0:t}$. Again we introduce the shorthand notation

$$\tilde{\mathbb{E}}_Z^*[g(X_t, \tilde{D}_t)] = \tilde{\mathbb{E}}^*[g(X_t, \tilde{D}_t)|Z_{0:t}]. \tag{17}$$

Recall that σ_t satisfies the Zakai equation of nonlinear filtering [1, 11] and that

$$\pi_t[f] = \frac{\sigma_t[f]}{\sigma_t[1]}. \tag{18}$$

3   FBSDE Estimators for Conditional Expectation Values

We wish to construct estimators for the conditional expectation values $\pi_T[f]$ and $\sigma_T[f]$, respectively, of the following form:

$$S_T^\sigma[f] := \tilde{Y}_0 - \int_0^T \tilde{U}_t^{\mathrm{T}} \mathrm{d}Z_t, \tag{19a}$$

$$S_T^\pi[f] := \check{Y}_0 - \int_0^T \check{U}_t^{\mathrm{T}} \mathrm{d}I_t, \tag{19b}$$

$$\hat{S}_T^\pi[f] := \hat{Y}_0 - \int_0^T \hat{U}_t^{\mathrm{T}} \mathrm{d}Z_t, \tag{19c}$$

$$\bar{S}_T^\sigma[f] := \bar{Y}_0 - \int_0^T \bar{U}_t^{\mathrm{T}} \mathrm{d}W_t. \tag{19d}$$

The required random variables $\tilde{Y}_0$, $\check{Y}_0$, $\hat{Y}_0$, $\bar{Y}_t$ and controls $\tilde{U}_{0:T}$, $\check{U}_{0:T}$, $\hat{U}_{0:T}$, $\bar{U}_t$, respectively, are chosen such that

$$\tilde{\mathbb{E}}_Z^*[S_T^\sigma[f]] = \sigma_T[f], \tag{20a}$$

$$\mathbb{E}_I^*[S_T^\pi[f]] = \pi_T[f], \tag{20b}$$

$$\mathbb{E}_I^*[\hat{S}_T^\pi[f]] = \pi_T[f], \tag{20c}$$

$$\tilde{\mathbb{E}}_Z^*[\bar{S}_t^\sigma[f]] = \sigma_T[f] \tag{20d}$$

for any suitable observable $f(x)$.

We derive appropriate FBSDE formulations for each of the four estimators in the following subsections. See [4, 7] for an introduction to BSDEs and FBSDEs.

3.1 Observation-Based Estimator I

We start with the estimator (19a) and introduce the BSDE

$$d\tilde{Y}_t = \tilde{Q}_t^\mathrm{T} dB_t + \tilde{V}_t^\mathrm{T} dZ_t, \qquad \tilde{Y}_T = \tilde{D}_T f(X_T), \tag{21}$$

where $(X_T, \tilde{D}_T)$ is defined by the forward Eq. (15). The equation is a BSDE as the terminal condition $\tilde{Y}_T$ is given. While both $B_{0:T}$ and $Z_{0:T}$ are the driving martingales of the BSDE (21), the right-hand side is expressed as a single stochastic integral:

$$\tilde{Q}_t^\mathrm{T} dB_t + \tilde{V}_t^\mathrm{T} dZ_t = \begin{bmatrix} \tilde{Q}_t^\mathrm{T}; & \tilde{V}_t^\mathrm{T} \end{bmatrix} \begin{bmatrix} dB_t \\ dZ_t \end{bmatrix} =: \tilde{P}_t^\mathrm{T} \begin{bmatrix} dB_t \\ dZ_t \end{bmatrix} \tag{22}$$

where $\tilde{P}_t^\mathrm{T} = \begin{bmatrix} \tilde{Q}_t^\mathrm{T}; & \tilde{V}_t^\mathrm{T} \end{bmatrix}$. The solution to the BSDE is given by a pair of processes $(\tilde{Y}_t, \tilde{P}_t)$. However, we keep $\tilde{Q}_t$ and $\tilde{V}_t$ separate throughout this chapter rather than using the concatenated vector notation $\tilde{P}_t$.

We note that

$$\tilde{D}_T f(X_T) - S_T^\sigma[f] = \tilde{D}_T f(X_T) - \tilde{Y}_0 + \int_0^T \tilde{U}_t^\mathrm{T} dZ_t \tag{23a}$$

$$= \int_0^T \tilde{Q}_t^\mathrm{T} dB_t + \int_0^T (\tilde{U}_t + \tilde{V}_t)^\mathrm{T} dZ_t. \tag{23b}$$

Therefore taking conditional expectation $\tilde{\mathbb{E}}_Z^*[\cdot]$ on both sides yields

$$\sigma_T[f] - \tilde{\mathbb{E}}_Z^*[S_T^\sigma[f]] = \tilde{\mathbb{E}}_Z^* \left[\int_0^T (\tilde{U}_t + \tilde{V}_t)^\mathrm{T} dZ_t \right]. \tag{24}$$

We immediately obtain the condition

$$\tilde{U}_t = -\tilde{V}_t. \tag{25}$$

Remark 1 In the work of [17], U is interpreted as a "control" term while the optimisation objective is the error variance of the estimator of the form (19c) from $f(X_T)$. Similar consideration on minimum variance estimation is also possible because

$$\tilde{\mathbb{E}}_Z^* \left[\left(\tilde{D}_T f(X_T) - S_T^\sigma[f] \right)^2 \right] = \tilde{\mathbb{E}}_Z^* \left[\left(\tilde{D}_T f(X_T) - \tilde{Y}_0 + \int_0^T \tilde{U}_t^\mathrm{T} dZ_t \right)^2 \right]$$

$$\tag{26a}$$

$$= \tilde{\mathbb{E}}_Z^* \left[\left(\int_0^T \tilde{Q}_t^{\mathsf{T}} \mathrm{d}B_t + \int_0^T (\tilde{U}_t + \tilde{V}_t)^{\mathsf{T}} \mathrm{d}Z_t \right)^2 \right]$$

(26b)

$$= \tilde{\mathbb{E}}_Z^* \left[\int_0^T \left(\| \tilde{Q}_t \|^2 + \| \tilde{U}_t + \tilde{V}_t \|^2 \right) \mathrm{d}t \right] + \tilde{M}_T,$$

(26c)

where $\tilde{M}_T$ is a martingale term which vanishes under expectations with respect to observations $Z_{0:T}$. Therefore the optimal choice is again $\tilde{U}_t = -\tilde{V}_t$. Note that our FBSDE formulation allows the estimator to be exact, and therefore we directly use the backward-in-time representation of the state process itself.

In order to gain a better insight into the solution structure of the FBSDEs (15) and (21) and to find an explicit expression for the optimal weight $\tilde{U}_{0:T}$, we introduce the notations $\tilde{Y}_s^{t,x,d}$, $\tilde{V}_s^{t,x,d}$, $\tilde{Q}_s^{t,x,d}$, and $\tilde{U}_s^{t,x,d}$ for solutions of the FBSDE considered over the restricted time interval $s \in [t, T]$ with initial condition $X_t = x$ and $\tilde{D}_t = d$. Equation (21) now implies

$$\tilde{Y}_t^{t,x,d} = \tilde{\mathbb{E}}^* \left[\tilde{Y}_{t+\tau}^{t,x,d} \right]$$

(27)

for $\tau \in (0, T - t]$, where $X_s^{t,x}$ denotes solutions of the forward SDE (15a) with $X_t = x$ for $s \geq t$ and similarly for $\tilde{D}_s^{t,x,d}$. Here $\tilde{\mathbb{E}}^*$ denotes now expectation with respect to the Brownian noise $B_{t:t+\tau}$ in (15a) and the observations $Z_{t:t+\tau}$ in (15b) under the path measure $\tilde{\mathbb{P}}^*$ for fixed initial conditions $X_t = x$ and $\tilde{D}_t = d$. In particular, we can set $t + \tau = T$ and use $\tilde{Y}_T^{t,x,d} = \tilde{D}_T^{t,x,d} f(X_T^{t,x})$.

Furthermore, let us denote the generator of the FSDEs (15) by $\tilde{\mathcal{L}}$, which acts on functions $g(x, d)$, that is,

$$\tilde{\mathcal{L}} g(x, d) = b(x)^{\mathsf{T}} \nabla_x g(x, d) + \frac{\sigma^2}{2} \nabla_x^2 g(x, d) + \frac{d^2}{2} h(x)^{\mathsf{T}} h(x) \partial_d^2 g(x, d).$$

(28)

Then, under appropriate smoothness assumptions, the deterministic function

$$\tilde{y}_t(x, d) := \tilde{Y}_t^{t,x,d}$$

(29)

satisfies the backward Kolmogorov equation

$$- \partial_t \tilde{y}_t(x, d) = \tilde{\mathcal{L}} \tilde{y}_t(x, d), \qquad \tilde{y}_T(x, d) = d f(x),$$

(30)

since

$$d\tilde{y}_t(X_t, \tilde{D}_t) = (\partial_t \tilde{y}_t(X_t, \tilde{D}_t) + \tilde{\mathcal{L}} \tilde{y}_t(X_t, \tilde{D}_t))dt \tag{31a}$$

$$+ \sigma \nabla_x \tilde{y}_t(X_t, \tilde{D}_t)dB_t + \partial_d \tilde{y}_t(X_t, \tilde{D}_t)\tilde{D}_t h(X_t)^\mathsf{T} dZ_t \tag{31b}$$

$$= d\tilde{Y}_t = \tilde{Q}_t^\mathsf{T} dB_t + \tilde{V}_t^\mathsf{T} dZ_t. \tag{31c}$$

Hence one also obtains

$$\tilde{Q}_t^{t,x,d} = \sigma \nabla_x \tilde{y}_t(x, d), \tag{32a}$$

$$\tilde{V}_t^{t,x,d} = d \,\partial_d \tilde{y}_t(x, d) \, h(x), \tag{32b}$$

and we can define the deterministic control function

$$\tilde{u}_t(x, d) := -d \,\partial_d \tilde{y}_t(x, d) \, h(x) \tag{33}$$

and minimum variance control $\tilde{U}_t = \tilde{u}_t(X_t, \tilde{D}_t)$.

To simplify further, we make the separation *ansatz* $\tilde{y}_t(x, d) = d\, y_t(x)$ and find that this assumption implies

$$\tilde{\mathcal{L}} \tilde{y}_t(x, d) = d \,\mathcal{L}^x \, y_t(x), \tag{34}$$

where $\mathcal{L}^x$ denotes the generator of (9a). Hence $y_t(x)$ has to satisfy the backward Kolmogorov equation

$$- \partial_t y_t(x) = \mathcal{L}^x y_t(x), \qquad y_T(x) = f(x). \tag{35}$$

Furthermore,

$$u_t(x, d) = -d \, y_t(x) \, h(x) \tag{36}$$

and (19a) becomes

$$S_T^\sigma[f] = y_0(X_0) + \int_0^T \tilde{D}_t \, y_t(X_t) \, h(X_t)^\mathsf{T} dZ_t. \tag{37}$$

A closely related estimator can be found in [15] (Section 2.3.3), where it has been used to derive Zakai's equation. The difference is that [15] considers directly the averaged estimator (43) not relying on an FBSDE formulation.

Furthermore, one can derive a BSDE representation for $y_t(x)$ directly. Upon introducing Y_t such that $\tilde{Y}_t = \tilde{D}_t Y_t$ and utilizing (15b), (7b), and (21), one obtains the BSDE

$$dY_t = Q_t^\mathsf{T} dB_t + V_t^\mathsf{T} dW_t, \qquad Y_T = f(X_T), \tag{38}$$

along the FSDEs (7). Please be aware that (Q_t, V_t) are not the same as in the BSDE (21). In fact, since

$$d\tilde{Y}_t = \tilde{D}_t dY_t + Y_t \tilde{D}_t h(X_t)^\mathsf{T} dZ_t + \tilde{D}_t h(X_t)^\mathsf{T} V_t dt \tag{39a}$$

$$= \tilde{D}_t Q_t^\mathsf{T} dB_t + \tilde{D}_t (V_t + Y_t h(X_t))^\mathsf{T} dZ_t, \tag{39b}$$

one finds that $\tilde{Q}_t = \tilde{D}_t Q_t$ and $\tilde{V}_t = \tilde{D}_t V_t + \tilde{D}_t Y_t h(X_t)$. Furthermore, since Y_t does not depend on $Z_{t:T}$, $V_t \equiv 0$ and we obtain

$$dY_t = Q_t^\mathsf{T} dB_t \tag{40}$$

in line with the backward Kolmogorov equation (35). Let us summarise our findings in the following lemma.

Lemma 1 *The estimator (19a) becomes unbiased for $\tilde{Y}_0 = Y_0$ with Y_t, $t \in [0, T]$, defined by the FBSDEs*

$$dX_t = b(X_t)dt + \sigma dB_t, \qquad X_0 \sim \mu, \tag{41a}$$

$$dY_t = Q_t^\mathsf{T} dB_t, \qquad\qquad Y_T = f(X_T). \tag{41b}$$

The control $\tilde{U}_t$, $t \in [0, T]$, is provided by

$$\tilde{U}_t = -\tilde{D}_t Y_t h(X_t). \tag{42}$$

We end this section with a brief discussion related to the problem of filter stability. Let us introduce the abbreviation

$$\mathcal{S}_T^\sigma[f] := \tilde{\mathbb{E}}_Z^*[\mathcal{S}_T^\sigma[f]] = \mu[Y_0] - \int_0^T \tilde{\mathbb{E}}_Z^*[\tilde{U}_t]dZ_t \tag{43}$$

The optimal choice $\tilde{U}_t = -\tilde{V}_t$ leads to

$$\tilde{Y}_T - \mathcal{S}_T^\sigma[f] = \tilde{Y}_0 - \mu[\tilde{Y}_0] + \int_0^T \tilde{Q}_t^\mathsf{T} dB_t + \int_0^T (\tilde{V}_t - \tilde{\mathbb{E}}_Z^*[\tilde{V}_t])dZ_t \tag{44}$$

from which we can recover

$$\mathcal{S}_T^\sigma[f] = \tilde{\mathbb{E}}_Z^*[\tilde{Y}_T] = \sigma_T[f] \tag{45}$$

(unbiasedness) as well as

$$\tilde{\mathbb{E}}^* \left[\left(\tilde{Y}_T - \sigma_T[f] \right)^2 \right] = \mu \left[\left(\tilde{Y}_0 - \mu[\tilde{Y}_0] \right)^2 \right] \tag{46a}$$

$$+ \tilde{\mathbb{E}}^* \left[\int_0^T \left(\|\tilde{Q}_t\|^2 + \|\tilde{V}_t - \tilde{\mathbb{E}}_Z^*[\tilde{V}_t]\|^2 \right) dt \right] \tag{46b}$$

(variance propagation). The last identity gives rise to the differential equation

$$\frac{d}{dt} \tilde{\mathbb{E}}^* \left[\left(\tilde{Y}_t - \sigma_t[f] \right)^2 \right] = \tilde{\mathbb{E}}^* \left[\|\tilde{Q}_t\|^2 + \|\tilde{V}_t - \tilde{\mathbb{E}}_Z^*[\tilde{V}_t]\|^2 \right] \tag{47}$$

and the variance of $\tilde{Y}_t$ is non-increasing as time goes backward from $t = T$ to $t = 0$. The expression on the right hand side of (47) can be viewed as a (non-stationary) Dirichlet form and *carré du champ* operator [2], respectively, associated with the FSDE (15). Furthermore, under an assumed Poincaré-type inequality

$$c \tilde{\mathbb{E}}^* \left[\left(\tilde{Y}_t - \sigma_t[f] \right)^2 \right] \le \tilde{\mathbb{E}}^* \left[\|\tilde{Q}_t\|^2 + \|\tilde{V}_t - \tilde{\mathbb{E}}_Z^*[\tilde{V}_t]\|^2 \right], \tag{48}$$

$c > 0$, the decay becomes exponential. We note that

$$\|\tilde{Q}_t\|^2 + \|\tilde{V}_t - \tilde{\mathbb{E}}_Z^*[\tilde{V}_t]\|^2 = \sigma^2 \|\tilde{D}_t \nabla_x y_t(X_t)\|^2 + \|\tilde{D}_t y_t(X_t) h(X_t) - \sigma_t[y_t h]\|^2, \tag{49}$$

which provides an explicit expression for the decay of variance via (47). It would be of interest to relate these considerations to the problem of filter stability. See the closely related work [18]. An obvious observation is that an exponential decay in the variance of Y_t implies an exponential decay of variance in $\tilde{Y}_t$.

3.2   Innovation-Based Estimator

We now turn to the estimator (19b). The required BSDE for $\check{Y}_t$ is given by

$$d\check{Y}_t = \check{Q}_t^T dB_t + \check{V}_t^T dI_t, \qquad \check{Y}_T = D_T f(X_T), \tag{50}$$

along solutions of the FSDE (9). We note again that

$$D_T f(X_T) - S_T^\pi[f] = D_T f(X_T) - \check{Y}_0 + \int_0^T \check{U}_t^T dI_t \tag{51a}$$

$$= \int_0^T \check{Q}_t^T dB_t + \int_0^T (\check{U}_t + \check{V}_t)^T dI_t. \tag{51b}$$

Similar to the previous case, we take conditional expectation $\mathbb{E}_I^*[\cdot]$ on both sides to conclude

$$\pi_T[f] - \mathbb{E}_I^*[S_T^\pi[f]] = \mathbb{E}_I^*\left[\int_0^T (\breve{U}_t + \breve{V}_t)^\mathrm{T}\mathrm{d}I_t\right], \tag{52}$$

which again yields the optimal choice $\breve{U}_t = -\breve{V}_t$ and the associated estimator (19b) is of minimum variance (see Remark 1).

The time-dependent generator $\mathcal{L}_t$ associated with the FSDE (9) is provided by

$$\mathcal{L}_t g = b^\mathrm{T}\nabla_x g + \frac{\sigma^2}{2}\nabla_x^2 g + \frac{d^2}{2}\|h - \pi_t[h]\|^2 \partial_d^2 g. \tag{53}$$

Let the function $\breve{y}_t(x, d)$ satisfy the associated BSPDE

$$-\mathrm{d}_t\breve{y}_t = \mathcal{L}_t\breve{y}_t\,\mathrm{d}t - \breve{V}_t^\mathrm{T}\mathrm{d}I_t, \qquad \breve{y}_T(x, d) = d\,f(x), \tag{54}$$

where the $\breve{V}_t$ term is needed to make $\breve{y}_t(x, d)$ adapted to the forward process. However, as shown below, $\partial_d^2\breve{y}_t = 0$ and we may conclude that $\breve{V}_t \equiv 0$. Then Itô's formula implies

$$\mathrm{d}\breve{y}_t(X_t, D_t) = (\mathrm{d}_t\breve{y}_t(X_t, D_t) + \mathcal{L}_t\breve{y}_t(X_t, D_t)\mathrm{d}t) + \sigma\nabla_x\breve{y}_t(X_t, D_t)\mathrm{d}B_t \tag{55a}$$

$$+ D_t\partial_d\breve{y}_t(X_t, D_t)(h(X_t) - \pi_t[h])^\mathrm{T}\mathrm{d}I_t \tag{55b}$$

$$= \breve{Q}_t^\mathrm{T}\mathrm{d}B_t + \breve{V}_t^\mathrm{T}\mathrm{d}I_t = \mathrm{d}\breve{Y}_t. \tag{55c}$$

Hence, we find that $\breve{q}_t(x, d) = \sigma\nabla_x\breve{y}_t(x, d)$ as well as

$$\breve{v}_t(x, d) = d\,\partial_d\breve{y}_t(x, d)\,(h(x) - \pi_t[h]). \tag{56}$$

Again, we make the *ansatz* $\breve{y}_t(x, d) = d\,y_t(x)$ and find that $y_t(x)$ has to satisfy the previously stated backward Kolmogorov equation (35). Therefore, we also find that $\breve{Q}_t = \breve{q}_t(X_t, D_t)$ with $\breve{q}_t(x, d) := d\,\sigma\,\nabla_x y_t(x)$ as well as $\breve{V}_t = \breve{v}_t(X_t, D_t)$ with $\breve{v}_t(x, d) = d\,v_t(x)$ where

$$v_t(x) = y_t(x)\,(h(x) - \pi_t[h]). \tag{57}$$

The optimal control is given by

$$\breve{U}_t = -D_t\,y_t(X_t)\,(h(X_t) - \mathbb{E}_I^*[D_t h(X_t)]) \tag{58}$$

and (19b) becomes

$$S_T^\pi[f] = y_0(X_0) + \int_0^T D_t\,y_t(X_t)\,(h(X_t) - \mathbb{E}_I^*[D_t h(X_t)])^\mathrm{T}\mathrm{d}I_t. \tag{59}$$

An alternative derivation is to use the *ansatz* $\check{Y}_t = D_t Y_t$ as in the previous section, where $(Y_t, (Q_t, V_t))$ is the solution of the BSDE (38) along with the FSDE (7). The two BSDEs (50) and (38) are now related by $\check{Q}_t = D_t Q_t$ as well as

$$\check{V}_t = D_t V_t + D_t Y_t \left(h(X_t) - \mathbb{E}_t^*[D_t h(X_t)] \right) \tag{60}$$

and one finds again that $V_t \equiv 0$. Hence the BSDE (38) reduces to (40), which corresponds to the backward Kolmogorov equation (35). Let us summarise our findings in the following lemma.

Lemma 2 *The estimator (19b) becomes unbiased for $\check{Y}_0 = Y_0$ with Y_t, $t \in [0, T]$, defined by the FBSDEs (41) and the control $\check{U}_t$, $t \in [0, T]$, is provided by*

$$\check{U}_t = -D_t Y_t (h(X_t) - \mathbb{E}_t^*[D_t h(X_t)]). \tag{61}$$

The considerations on filter stability, as put forward in Section 3.1, extend naturally to the innovation-based estimator (19b). In particular, (47) becomes

$$\frac{\mathrm{d}}{\mathrm{d}t} \mathbb{E}^* \left[\left(\check{Y}_t - \pi_t[f] \right)^2 \right] \tag{62a}$$

$$= \mathbb{E}^* \left[\sigma^2 \| D_t \nabla_x y_t(X_t) \|^2 + \| D_t y_t(X_t)(h(X_t) - \pi_t[h]) - \pi_t[y_t(h - \pi_t[h])] \|^2 \right] \tag{62b}$$

with $\check{Y}_t = D_t y_t(X_t)$ and an appropriate Poincaré inequality would again establish an exponential decay of variance.

Example 1 The linear-Gaussian case is a special case of the model (7) where $b(x) = A^\mathsf{T} x$, $h(x) = H^\mathsf{T} x$ for some $A \in \Re^{d_x \times d_x}$ and $H \in \Re^{d_x \times d_z}$, and μ is Gaussian with mean m_0 and covariance matrix Σ_0. Fix $\bar{f} \in \Re^{d_x}$ and consider a linear function $f(x) = \bar{f}^\mathsf{T} x$. In this case, the solution to the backward Kolmogorov equation is also a linear function $y_t(x) = \bar{y}_t^\mathsf{T} x$ with

$$-\frac{\mathrm{d}\bar{y}_t}{\mathrm{d}t} = A\bar{y}_t, \quad \bar{y}_T = \bar{f}. \tag{63}$$

Upon introducing the filtered state covariance matrix

$$\Sigma_t := \mathbb{E}_t^*[D_t X_t X_t^\mathsf{T}] - \mathbb{E}_t^*[D_t X_t] \mathbb{E}_t^*[D_t X_t]^\mathsf{T}, \tag{64}$$

the optimal control (58) takes the form

$$\mathbb{E}_t^*[U_t] = -H^\mathsf{T} \Sigma_t \bar{y}_t \tag{65}$$

and the estimator (59) becomes

$$\mathcal{S}_T^\pi[f] := \mathbb{E}_I^*[\mathcal{S}_T^\pi[f]] \tag{66a}$$

$$= \bar{y}_0^\mathsf{T} m_0 + \int_0^T \bar{y}_t^\mathsf{T} \Sigma_t H \mathrm{d}I_t \tag{66b}$$

$$= \bar{f}^\mathsf{T} e^{TA^\mathsf{T}} m_0 + \int_0^T \bar{f}^\mathsf{T} e^{(T-t)A^\mathsf{T}} \Sigma_t H \mathrm{d}I_t. \tag{66c}$$

Define the filtered mean

$$m_t := e^{tA^\mathsf{T}} m_0 + \int_0^t e^{(t-s)A^\mathsf{T}} \Sigma_s H \mathrm{d}I_s, \tag{67}$$

and observe that $\mathcal{S}_T^\pi[f] = \bar{f}^\mathsf{T} m_T$. The Kalman–Bucy filter is obtained by differentiate (67) with respect to t:

$$\mathrm{d}m_t = A^\mathsf{T} m_t \mathrm{d}t + \Sigma_t H \mathrm{d}I_t. \tag{68}$$

Please note that, contrary to (63), the BSPDE formulations (2) leads to the controlled backward Kolmogorov equation

$$-\frac{\mathrm{d}\bar{y}_t}{\mathrm{d}t} = A\bar{y}_t + H\bar{u}_t, \quad \bar{y}_T = \bar{f} \tag{69}$$

instead [17]. We explore this connection further in the following subsection. □

3.3 Observation-Based Estimator II

We now state a BSDE required to define the estimator (19c). The key observation is again that the backward process $\hat{Y}_t$ is chosen such that

$$D_T f(X_T) - \hat{\mathcal{S}}_T^\pi[f] = \hat{Y}_T - \hat{Y}_0 + \int_0^T \hat{U}_t^\mathsf{T} \mathrm{d}Z_t \tag{70}$$

becomes independent of the innovation process $I_{0:T}$ under an optimal choice of the control $\hat{U}_{0:T}$. These considerations lead immediately to the BSDE

$$\mathrm{d}\hat{Y}_t = \hat{Q}_t^\mathsf{T} \mathrm{d}B_t + \hat{V}_t^\mathsf{T} \mathrm{d}I_t - \hat{U}_t^\mathsf{T} \mathbb{E}_I^*[D_t h(X_t)]\mathrm{d}t, \quad \hat{Y}_T = D_T f(X_T), \tag{71}$$

along the FSDE (9) since then

$$D_T f(X_T) - \hat{\mathcal{S}}_T^\pi[f] = \int_0^T \hat{Q}_t^\mathsf{T} \mathrm{d}B_t + \int_0^T (\hat{U}_t + \hat{V}_t)^\mathsf{T} \mathrm{d}I_t \tag{72}$$

and the desired solution is provided by $\hat{U}_t = -\hat{V}_t$. In the context of minimum variance estimation, the conclusion can be drawn from the associated cost function

$$\hat{\mathcal{J}}_T(\hat{U}_{0:T}) := \mathbb{E}^* \left[\int_0^T \left(\|\hat{Q}_t\|^2 + \|\hat{U}_t + \hat{V}_t\|^2 \right) dt \right]. \tag{73}$$

However, contrary to the previous two estimators, the BSDE (71) does no longer lead to a standard backward Kolmogorov-type PDE. This is due to the appearance of the

$$\hat{U}_t^\mathsf{T} \mathbb{E}_I^*[D_t h(X_t)] = -\hat{V}_t^\mathsf{T} \mathbb{E}_I^*[D_t h(X_t)] \tag{74}$$

drift term in (71). Let us summarise our findings in the following lemma.

Lemma 3 *The estimator (19c) becomes unbiased for $\hat{Y}_0$ defined by the FBSDEs*

$$dX_t = b(X_t)dt + \sigma dB_t, \qquad\qquad X_0 \sim \mu, \tag{75a}$$

$$dD_t = D_t \left(h(X_t) - \mathbb{E}_I^*[D_t h(X_t)] \right)^\mathsf{T} dI_t, \qquad D_0 = 1, \tag{75b}$$

$$d\hat{Y}_t = \hat{Q}_t^\mathsf{T} dB_t + \hat{V}_t^\mathsf{T}(dI_t + \mathbb{E}_I^*[D_t h(X_t)]dt), \qquad \hat{Y}_T = D_T f(X_T). \tag{75c}$$

The control $\hat{U}_t$, $t \in [0, T]$, is provided by $\hat{U}_t = -\hat{V}_t$.

Let us again discuss some implications of the proposed estimator and introduce the abbreviation

$$\hat{\mathcal{S}}_T^\pi[f] := \mathbb{E}_I^*[\hat{S}_T^\pi[f]] = \mu[\hat{Y}_0] - \int_0^T \mathbb{E}_I^*[\hat{U}_t]^\mathsf{T} dZ_t, \tag{76}$$

which implies

$$\hat{Y}_T - \hat{\mathcal{S}}_T^\pi[f] = \hat{Y}_0 - \mu[\hat{Y}_0] + \int_0^T \hat{Q}_t^\mathsf{T} dB_t \tag{77a}$$

$$+ \int_0^T (\hat{V}_t + \mathbb{E}_I^*[\hat{U}_t])^\mathsf{T} dI_t + \int_0^T (\hat{U}_t - \mathbb{E}_I^*[\hat{U}_t])^\mathsf{T} \pi_t[h]dt. \tag{77b}$$

The choice $\hat{U}_t = -\hat{V}_t$ leads to

$$\hat{\mathcal{S}}_T^\pi[f] = \mathbb{E}_I^*[\hat{Y}_T] = \pi_T[f] \tag{78}$$

as well as

$$\mathbb{E}^* \left[\left(\hat{Y}_T - \hat{\mathcal{S}}_T^\pi[f] \right)^2 \right] = \mu \left[\left(\hat{Y}_0 - \mu[\hat{Y}_0] \right)^2 \right] \tag{79a}$$

$$+ \mathbb{E}^* \left[\int_0^T \left(\|\hat{Q}_t\|^2 + \|\hat{V}_t - \mathbb{E}_I^*[\hat{V}_t]\|^2 \right) dt \right] \tag{79b}$$

$$+ \mathbb{E}^* \left[\left(\int_0^T (\hat{V}_t - \mathbb{E}_I^*[\hat{V}_t])^T \pi_t[h] dt \right)^2 \right]. \tag{79c}$$

Furthermore, if one applies the averaged control

$$\hat{U}_t = -\mathbb{E}_I^*[\hat{V}_t] \tag{80}$$

in the BSDE (71) directly, then (77) turns into

$$\hat{Y}_T - \hat{\mathcal{S}}_T^\pi[f] = \hat{Y}_0 - \mu[\hat{Y}_0] + \int_0^T \hat{Q}_t^T dB_t + \int_0^T (\hat{V}_t - \mathbb{E}_I^*[\hat{V}_t])^T dI_t. \tag{81}$$

Hence the integral term (79c) vanishes and the resulting (79) reveals the more common minimum variance/optimal control aspect of the averaged control (80). We also note that the resulting estimator (19c) remains unbiased under the averaged control (80).

Let us finally connect the discussion so far to the work of [15, 17], which is based on the estimator (1) and the BSPDE (2). We note that Z_t is not Brownian motion with respect to the path measure $\mathbb{P}$ generated by the FSDEs (7a)–(7b). Hence we first reformulate (2) into the BSDE

$$dY_t = Q_t^T dB_t + V_t^T dW_t - U_t^T h(X_t) dt, \qquad Y_T = f(X_T). \tag{82}$$

along the FSDE (7). In a next step, we obtain

$$Y_T - \mathcal{S}_T = Y_0 - \mu[Y_0] + \int_0^T Q_t^T dB_t + \int_0^T (V_t + \mathcal{U}_t)^T dW_t \tag{83}$$

and

$$\mathbb{E} \left[(Y_T - \mathcal{S}_T)^2 \right] = \mathbb{E} \left[(Y_0 - \mu[Y_0])^2 \right] + \mathbb{E} \left[\int_0^T \left(\|Q_t\|^2 + \|V_t + \mathcal{U}_t\|^2 \right) dt \right]. \tag{84}$$

Please note that the optimal control is *not* provided by $\mathcal{U}_t = -\mathbb{E}[V_t|Z_{0:t}]$. In fact, we introduce

$$\check{Y}_t := D_t Y_t \tag{85}$$

and, using (9) and (82), obtain the BSDE

$$d\check{Y}_t = \check{Q}_t^{\mathrm{T}} dB_t + \check{V}_t^{\mathrm{T}} dI_t - D_t \mathcal{U}_t^{\mathrm{T}} h(X_t) dt, \qquad \check{Y}_T = D_T f(X_T), \tag{86}$$

with $\check{V}_t$ and $\check{Q}_t$ satisfying

$$\check{V}_t = D_t \{V_t + Y_t(h(X_t) - \pi_t[h])\} \tag{87}$$

and $\check{Q}_t = D_t Q_t$, respectively. Note that the BSDE (86) implies

$$\pi_T[f] = \mu[\check{Y}_0] + \int_0^T \mathbb{E}_t^*[\check{V}_t]^{\mathrm{T}} dI_t - \int_0^T \mathcal{U}_t^{\mathrm{T}} \pi_t[h] dt \tag{88}$$

and we can characterise the dependence of the estimation error on $\mathcal{U}_t$:

$$\mathcal{S}_T - \pi_T[f] = -\int_0^T \left(\mathcal{U}_t + \mathbb{E}_t^*\left[\check{V}_t\right]\right)^{\mathrm{T}} dI_t. \tag{89}$$

The error becomes zero (unbiased estimator) for

$$\mathcal{U}_t = -\mathbb{E}_t^*[\check{V}_t] = -\mathbb{E}_t^*[D_t V_t + D_t Y_t(h(X_t) - \pi_t[h])], \tag{90}$$

which also provides the minimiser of (84) (minimum variance estimator) in agreement with the results from [17].

Please also compare the BSDE formulations (71) to the BSDE formulation (86), which are both along the FSDE (9). In particular, $\hat{Q}_t$ corresponds formally to $\check{Q}_t$ and $\hat{V}_t$ to $\check{V}_t$, respectively. There remains a difference in the use of either $\hat{U}_t^{\mathrm{T}} \mathbb{E}_t^*[D_t h(X_t)]$ or $\mathcal{U}_t^{\mathrm{T}}(D_t h(X_t))$, respectively, as the additional drift term. However, since the representation of an unbiased estimator is unique (Proposition 2.31 in [1]), the resulting estimators are equivalent, that is, $\hat{\mathcal{S}}_T^\pi[f] = \mathcal{S}_T[f]$.

Example 2 In the linear Gaussian case, $V_t \equiv 0$ in (82) and the BSDE gives rise to the controlled backward Kolmogorov equation (69). Furthermore, the control (90) reduces to

$$\bar{u}_t = -H^{\mathrm{T}} \Sigma_t \bar{y}_t, \tag{91}$$

where $\bar{y}_t$ now satisfies the closed loop backward equation

$$-\frac{d\bar{y}_t}{dt} = A\bar{y}_t - HH^{\mathrm{T}} \Sigma_t \bar{y}_t, \qquad \bar{y}_T = \bar{f} \tag{92}$$

in line with the dual optimal control perspective of [13–15].   □

3.4 Observation Error Based Estimator

We finally discuss the rather unconventional estimator (19d) with the observation error defined by (5). We again employ the FSDE (15) and introduce the associated BSDE

$$d\bar{Y}_t = \bar{Q}_t^\mathsf{T} dB_t + \bar{V}_t^\mathsf{T} dZ_t + \bar{U}_t^\mathsf{T} h(X_t) dt, \qquad \bar{Y}_T = \tilde{D}_T f(X_T). \tag{93}$$

Hence

$$\bar{Y}_T - \bar{S}_T^\sigma[f] = \int_0^T \bar{Q}_t^\mathsf{T} dB + \int_0^T (\bar{V}_t + \bar{U}_t)^\mathsf{T} dZ_t \tag{94}$$

from which we conclude that

$$\tilde{\mathbb{E}}_Z^*[\bar{S}_T^\sigma[f]] = \tilde{\mathbb{E}}_Z^*[\bar{Y}_T] = \sigma_T[f] \tag{95}$$

provided that $\bar{U}_t = -\bar{V}_t$.

Let us introduce a second BSDE for Y_t defined by $\bar{Y}_t = \tilde{D}_t Y_t$. The BSDE is given by

$$dY_t = Q_t^\mathsf{T} dB_t + V_t^\mathsf{T} dW_t + U_t^\mathsf{T} h(X_t) dt, \qquad Y_T = f(X_T), \tag{96}$$

and the following identities hold:

$$\bar{Q}_t = \tilde{D}_t Q_t, \quad \bar{V}_t = \tilde{D}_t (V_t + Y_t h(X_t)), \quad \bar{U}_t = \tilde{D}_t U_t. \tag{97}$$

Using the optimal control $\bar{U}_t = -\bar{V}_t$, we obtain the closed FBSDE system consisting of the FSDE (7) and the BSDE

$$dY_t = Q_t^\mathsf{T} dB_t + V_t^\mathsf{T} dW_t - (V_t + Y_t h(X_t))^\mathsf{T} h(X_t) dt, \qquad Y_T = f(X_T). \tag{98}$$

It holds that $V_t \equiv 0$ and one obtains the Feynman-Kac type BSDE [7, 19]

$$dY_t = Q_t^\mathsf{T} dB_t - Y_t h(X_t)^\mathsf{T} h(X_t) dt, \qquad Y_T = f(X_T) \tag{99}$$

along the FSDE (7a) with associated backward PDE

$$-\partial_t y_t = \mathcal{L}^x y_t - \|h\|^2 y_t, \qquad y_T = f. \tag{100}$$

Let us summarise our findings in the following lemma.

Lemma 4 *The estimator (19d) becomes unbiased for $\bar{Y}_0$ defined by $\bar{Y}_0 = y_0(X_0)$, where y_t satisfies the backward PDE (100) and the control $\bar{U}_t$, $t \in [0, T]$, is provided by*

$$\bar{U}_t = -\tilde{D}_t y_t(X_t) h(X_t). \tag{101}$$

4 FBSDE Estimators for Optimal Control Application

In this section, we discuss an extension of the estimator (19b) to partially observed
stochastic optimal control problems [3, 5]. More specifically, consider the controlled
diffusion process

$$dX_t = b(X_t)dt + G\alpha_t dt + \sigma dB_t. \tag{102}$$

The observation model is kept identical to (7b). The objective is to minimise the
cost function

$$V_T(\alpha_{0:T}) = \mathbb{E}\left[\int_0^T \left(c_t(X_t) + \frac{1}{2}\|\alpha_t\|^2\right) dt + f(X_T)\right] \tag{103}$$

for given $c_t(x)$ and $f(x)$ over all admissible controls $\alpha_{0:T}$.

In literature, the optimal control problem is reshaped into a fully observed
problem of the unnormalised filter, and then the stochastic maximum principle
applies. See Chapter 8 of [3] for a detailed discussion. We present the FBSDE
approach to obtain the adjoint equation using the nonlinear filter in this section.

Since we wish to condition the control on the available data, we introduce the
cost function for a given observation trajectory

$$V_T(\alpha_{0:T}|I_{0:T}) = \mathbb{E}_I^*\left[\int_0^T D_t \left(c_t(X_t) + \frac{1}{2}\|\alpha_t\|^2\right) dt + D_T f(X_T)\right]. \tag{104}$$

Note that (104) corresponds to the motivating example (6) for a fixed control input
$\alpha_{0:T}$.

Upon extending the techniques developed in Sect. 3.2, we construct an estimator
of the form

$$V_T(\alpha_{0:T}|I_{0:T}) = \mu[\check{Y}_0] + \int_0^T \mathbb{E}_I^*[\check{V}_t]^\mathsf{T} dI_t, \tag{105}$$

where $(\check{Y}_t, \check{V}_t)$ satisfy an appropriate generalization of the BSDE (50); namely

$$d\check{Y}_t = \check{Q}_t^\mathsf{T} dB_t + \check{V}_t^\mathsf{T} dI_t - D_t \left(c_t(X_t) + \frac{1}{2}\|\alpha_t\|^2\right) dt, \qquad \check{Y}_T = D_T f(X_T), \tag{106}$$

along the FSDE (9) with the drift term in (9a) now being given by $b(X_t, \alpha_t)$. Note
that $\mu[\check{Y}_0] = V_T(\alpha_{0:T})$.

Following the weak formulation of stochastic optimal control [4, 7], we introduce

$$\bar{B}_t = B_t + \sigma^{-1} \int_0^t G\alpha_s \mathrm{d}s \tag{107}$$

and recall that $\bar{B}_t$ is Brownian motion under a modified probability measure $\mathbb{P}^*$ according to Girsanov's theorem [19]. Hence the FSDEs (9a)–(9b) become

$$\mathrm{d}\bar{X}_t = b(\bar{X}_t)\mathrm{d}t + \sigma\mathrm{d}\bar{B}_t, \qquad\qquad \bar{X}_0 \sim \mu, \tag{108a}$$

$$\mathrm{d}\bar{D}_t = \bar{D}_t \left(h(\bar{X}_t) - \pi_t[h]\right)^{\mathrm{T}}\mathrm{d}I_t + \sigma^{-1}\bar{D}_t(G\alpha_t)^{\mathrm{T}}\mathrm{d}\bar{B}_t, \qquad \bar{D}_0 = 1. \tag{108b}$$

The associated BSDE is provide by

$$\mathrm{d}\bar{Y}_t = \bar{Q}_t^{\mathrm{T}}\mathrm{d}\bar{B}_t + \bar{V}_t^{\mathrm{T}}\mathrm{d}I_t - \bar{D}_t \left(c_t(\bar{X}_t) + \frac{1}{2}\|\alpha_t\|^2\right)\mathrm{d}t, \quad \bar{Y}_T = \bar{D}_T f(\bar{X}_T). \tag{109}$$

In order to find the desired optimal control, we make the *ansatz* $\bar{Y}_t = \bar{D}_t Y_t$, which turns the BSDE (109) into the BSDE

$$\mathrm{d}Y_t = Q_t^{\mathrm{T}}\mathrm{d}\bar{B}_t + V_t^{\mathrm{T}}\mathrm{d}I_t \tag{110a}$$

$$- \left\{c_t(\bar{X}_t) + \frac{1}{2}\|\alpha_t\|^2 + V_t^{\mathrm{T}}(h(\bar{X}_t) - \pi_t[h]) + \sigma^{-1}Q_t^{\mathrm{T}}G\alpha_t\right\}\mathrm{d}t, \tag{110b}$$

with final condition $Y_T = f(\bar{X}_T)$. Compare the previously derived BSDE (38) and the discussion from Sect. 3.2. Recall that $\mu[\check{Y}_0] = \mu[\bar{Y}_0] = \mu[Y_0]$, $\bar{Q}_t = \bar{D}_t(Q_t + \sigma^{-1}Y_t G\alpha_t)$, and $\bar{V}_t = \bar{D}_t(V_t + Y_t(h(\bar{X}_t) - \pi_t[h]))$.

Assuming sufficient spatial regularity, we now consider the BSPDE reformulation of (110). However, contrary to the discussion in Sect. 3.2, the auxiliary term V_t no longer vanishes since α_t is a random process and we are led to consider the BSPDE

$$-\mathrm{d}Y_t = \mathcal{L}^{x,\alpha_t}Y_t\mathrm{d}t + \left(c_t + \frac{1}{2}\|\alpha_t\|^2 + V_t^{\mathrm{T}}(h - \pi_t[h])\right)\mathrm{d}t - V_t^{\mathrm{T}}\mathrm{d}I_t, \, Y_T = f, \tag{111}$$

instead. This SPDE formulation leads to a dual optimal control problem [21] with the forward SPDE provided by the KS equation

$$\mathrm{d}\pi_t = (\mathcal{L}^{x,\alpha_t})^\dagger\pi_t\mathrm{d}t + \pi_t(h - \pi_t[h])^{\mathrm{T}}\mathrm{d}I_t \tag{112}$$

[1, 11]. Here $(\mathcal{L}^{x,\alpha_t})^\dagger$ denotes the adjoint of $\mathcal{L}^{x,\alpha_t}$. Upon calculating $\mathrm{d}\pi_t[Y_t]$, it is easy to verify that

$$\mathcal{V}_T(\alpha_{0:T}|I_{0:T}) = \mu[\mathbf{Y}_0] + \int_0^T \pi_t \left[V_t + \mathbf{Y}_t(h - \pi_t[h])\right]^\mathsf{T} dI_t, \tag{113}$$

which corresponds to (105).

Lemma 5 *The control in the KS equation (112) and the BSPDE (111) has to satisfy*

$$\alpha_t = \arg\min_\alpha \pi_t \left[\mathcal{L}^{x,\alpha}\mathbf{Y}_t + \frac{1}{2}\|\alpha\|^2\right] \tag{114}$$

in order to minimise $\mathcal{V}_T(\alpha_{0:T}) = \mu[\mathbf{Y}_0]$.

Proof See Theorem 8.2.1. in [3] for the related optimality criterion in terms of Zakai's equation and its adjoint BSPDE formulation. A formal argument can be stated as follows. Let $\alpha_{0:T}^{(0)}$ denote some control and $\pi_{0:T}^{(0)}$, $\mathbf{Y}_{0:T}^{(0)}$ the associated solutions of the KS equation (112) and the BSPDE (111), respectively. Assume that the control does not satisfy (114). Then determine a new $\mathbf{Y}_{0:T}^{(1)}$ by solving (111) subject to a new control law $\alpha_{0:T}^{(1)}$ determined by (114) with $\pi_{0:T} = \pi_{0:T}^{(0)}$. It holds by the comparison theorem that $\mu[\mathbf{Y}_0^{(1)}] < \mu[\mathbf{Y}_0^{(0)}]$. Next determine $\pi_{0:T}^{(1)}$ by solving (112) with the new control law $\alpha_{0:T}^{(1)}$. This procedure is to be repeated until a control $\alpha_{0:T}^*$ has been found which cannot be improved upon further.

The SPDE pair (112) and (111) can be replaced by the FSDEs (108) together with the BSDE (110). The optimality condition (114) turns then into

$$\alpha_t = -\sigma^{-1}G^\mathsf{T}\bar{\mathbb{E}}_I^*\left[\bar{D}_t Q_t\right]. \tag{115}$$

Note that one can follow the procedure in this section under the modified measure $\tilde{\mathbb{P}}$ and consider the estimator of the form (19a). See also Eq. 8.2.40 in [3] for a related BSPDE formulation associated with Zakai's evolution equation for the unnormalised filtering distribution σ_t.

Remark 2 We close this section by highlighting a link to the observation-based estimator (1) and its derivation in Sect. 3.3. Let us formally introduce the following partially observed stochastic control problem. We set $G = 0$ in (102) and introduce the running cost

$$c_t(x, \alpha) = h(x)^\mathsf{T}\alpha. \tag{116}$$

Hence the cost functional (104) becomes

$$\mathcal{V}_T(\alpha_{0:T}|I_{0:T}) = \mathbb{E}_I^*\left[\int_0^T D_t h(X_t)\alpha_t\, dt + D_T f(X_T)\right] \tag{117}$$

and its estimator is given by (105) and (113), respectively. Then (113) and (117) together imply

$$\pi_T[f] = \mu[Y_0] - \int_0^T \pi_t[h]^T \alpha_t \mathrm{d}t + \int_0^T \pi_t [V_t + Y_t(h - \pi_t[h])]^T \mathrm{d}I_t. \quad (118)$$

The definition (4) of the innovation process dictates that the estimator (1) corresponds to the choice

$$\mathcal{U}_t = \alpha_t = -\pi_t [V_t + Y_t(h - \pi_t[h])]. \quad (119)$$

However, please be aware that $\alpha_{0:T} = \mathcal{U}_{0:T}$ is not a minimiser of the cost functional (117) [16].

5 Conclusions

Building on the previous work [15–17] on optimal estimation for nonlinear filtering problems, in this chapter we approach the problem from an FBSDE approach, as also widely used in the context of optimal control problems [7]. Our approach sheds new light on the underlying structure of conditional estimation by carefully selecting the set of FBSDEs. In particular, we have strictly followed the classical FBSDE framework by allowing only Brownian noise in the FBSDE formulations. We have also demonstrated that the two estimation formulations for $\sigma_T[f]$ actually lead to a deterministic control law. This fact had been previously been used in [15] in the context of scenario (ii) for deriving Zakai's equation; but its full implications only emerge in the context conditional estimation. A further application could include the study of filter stability using (100).

It will also be of interest to further explore numerical approximations of the proposed estimators and control laws based, for example, on [8–10] and their relation to mean-field filtering equations [6, 20, 22].

Acknowledgments This work has been funded by Deutsche Forschungsgemeinschaft (DFG) - Project-ID 318763901 - SFB1294. The authors thank Prashant Mehta for sharing his insight into the subject of this chapter. The authors would also like to thank the Isaac Newton Institute for Mathematical Sciences, Cambridge, for support and hospitality during the programme *The Mathematical and Statistical Foundation of Future Data-Driven Engineering* where work on this chapter was undertaken. The programme was supported by EPSRC grant no EP/R014604/1.

References

1. A. Bain and D. Crisan. *Fundamentals of Stochastic Filtering*, volume 60. Springer Science & Business Media, 2008.
2. D. Bakry, I. Gentil, and M. Ledoux. *Analysis and Geometry of Markov Diffusion Operators*. Springer Verlag, Switzerland, 2014.

3. A. Bensoussan. *Stochastic Control of Partially Observable Systems*. Cambridge University Press, Cambridge, 1992.
4. A. Bensoussan. *Estimation and Control of Dynamical Systems*. Springer, Cham, 2018.
5. V. S. Borkar. *Optimal control of diffusion processes*. Longman Scientific & Technical, Harlow, 1989.
6. E. Calvello, S. Reich, and A. M. Stuart. Ensemble Kalman methods: A mean field perspective. *arXiv preprint arXiv:2209.11371*, 2022.
7. R. Carmona. *Lectures on BSDEs, Stochastic Control, and Stochastic Differential Games with Financial Applications*. SIAM, Philadelphia, 2016.
8. J. Chessari, R. Kawai, Y. Shinozaki, and T. Yamada. Numerical methods for backward stochastic differential equations: A survey. *arXiv preprint arXiv:2101.08936*, 2021.
9. W. E, J. Han, and A. Jentzen. Deep learning-based numerical methods for high-dimensional parabolic partial differential equations and backward stochastic differential equations. *Communications in Mathematics and Statistics*, 5:349–380, 2017.
10. W. E, J. Han, and A. Jentzen. Algorithms for solving high-dimensional PDEs: From nonlinear Monte Carlo to machine learning. *Nonlinearity*, 35:278, 2021.
11. A. H. Jazwinski. *Stochastic Processes and Filtering Theory*. Courier Corporation, 2007.
12. R. E. Kalman. On the general theory of control systems. In *Proceedings First International Conference on Automatic Control, Moscow, USSR*, pages 481–492, 1960.
13. R. E. Kalman. A new approach to linear filtering and prediction problems. *Journal of Basic Engineering*, 82(1):35–45, 03 1960. ISSN 0021-9223.
14. R. E. Kalman and R. S. Bucy. New results in linear filtering and prediction theory. *Journal of Basic Engineering*, 83(1):95–108, 03 1961. ISSN 0021-9223.
15. J. W. Kim. *Duality for Nonlinear Filtering*. PhD thesis, University of Illinois Urbana-Champaign, 2022.
16. J. W. Kim and P. G. Mehta. Duality for nonlinear filtering I: Observability. *IEEE Transactions on Automatic Control*, 69(2):699–711, 2024.
17. J. W. Kim and P. G. Mehta. Duality for nonlinear filtering II: Optimal control. *IEEE Transactions on Automatic Control*, 69(2):712–725, 2024.
18. J. W. Kim and P. G. Mehta. Variance decay property for filter stability. *arXiv preprint arXiv:2305.12850*, 2024.
19. G. A. Pavliotis. *Stochastic Processes and Applications*. Springer Verlag, New York, 2016.
20. A. Taghvaei and P. G. Mehta. A survey of feedback particle filter and related controlled interacting particle systems (CIPS). *Annual Reviews in Control*, 55:356–378, 2023.
21. B. Wittenmark. Stochastic adaptive control methods: A survey. *International Journal of Control*, 21:705–730, 1975.
22. T. Yang, P. G. Mehta, and S. Meyn. Feedback particle filter. *IEEE Trans. Automat. Control*, 58(10):2465–2480, 2013. ISSN 0018-9286.

Data Assimilation for the Stochastic Camassa-Holm Equation Using Particle Filtering: A Numerical Investigation

Colin J. Cotter, Dan Crisan, and Maneesh Kumar Singh

1 Introduction

Data assimilation (DA) is a set of methodologies that integrate past knowledge, represented as numerical models of a system, with newly acquired observational data from the same system [24, 26]. This tool is used in many domains, including meteorology, oceanography, and environmental research, to merge observational data with numerical models to improve forecast and simulation accuracy [26]. A concise overview of DA is presented in [2], including the references contained within. For stochastic systems, data assimilation in the context of the filtering problem can be rigorously formulated as stochastic filtering. In this work, we emphasise a stochastic filtering problem where a hidden stochastic process (signal) is observed at discrete times with noise. The nonlinear filtering problem consists of computing the law of the signal, given observations that are collected sequentially. More details on stochastic filtering can be found in [3] and references therein.

In this study, we investigate the data assimilation method for a nonlinear stochastic partial differential equation that corresponds to a viscous shallow water equation. In particular, we examine the particle filter methodology for a stochastic Camassa–Holm (CH) model with transport noise (referred to as Stochastic Advection by Lie Transport, or SALT). A detailed outlook on SALT-type noise can be found in [10, 18]. The deterministic CH equation [6] admits solutions with singularities, which possess a sharp peak at the apex of their velocity profile. By following the variation principle approach in stochastic fluid dynamics [18], the stochastic Camassa–Holm (SCH) equation in the SALT framework is derived in [13]. The interaction of peakons (peaked soliton solutions) with the stochastic transport in

C. J. Cotter · D. Crisan · M. K. Singh (✉)
Department of Mathematics, Imperial College London, London, UK
e-mail: maneesh-kumar.singh@imperial.ac.uk

© The Author(s) 2025
B. Chapron et al. (eds.), *Stochastic Transport in Upper Ocean Dynamics III*,
Mathematics of Planet Earth 13, https://doi.org/10.1007/978-3-031-70660-8_7

the SCH model is investigated in [4]. This model is useful in providing a 1+1 (one space and one time dimension) SPDE example where the behaviour of SALT in data assimilation can be easily investigated. However, there is a disadvantage: the solutions become rougher as time progresses. Numerical solutions behave like a space-time random field at longer times, leading to an atypical data assimilation problem, in that it is easy to solve numerically since the long-time solution of the filtering problem is a steady state distribution describing this random field. In this chapter, we incorporate a viscous dissipation to the equations to control the regularity of the solution at longer times.

Particle filtering in high dimensions typically requires adaptation to deal with the curse of dimensionality, which otherwise leads to a loss of particle diversity as explained in Sect. 3. In this chapter, we continue along the recent line of work to combine tempering, jittering and nudging techniques to avoid this diversity loss. In [23], adaptation to particle filtering is discussed for the stochastic Navier–Stokes equation with linear additive type noise. Data assimilation using an adapted particle filter for the two-dimensional Euler equation and quasi-geostrophic model is investigated in [8, 9]. A tempering-based adaptive particle filter to infer from a partially observed stochastic rotating shallow water (SRSW) model is studied in [21]. Recently, a lagged particle filter has been introduced for stable filtering of high-dimensional state-space models in [25].

This article aims to investigate the potential for jittering, tempering and nudging for stochastic PDEs by focusing on identical twin experiments using an SPDE with one space dimension (SCH), where the effect on the whole solution can be easily visualised. Our investigations also demonstrate the emerging capability of our parallel library for particle filtering with SPDEs, built around the Firedrake automated code generation system. In this chapter, we focus on numerical experiments testing the stability of the particle filters available in our library. A full evaluation of accuracy will follow in future work.

The organization of this chapter is as follows. In Sect. 2, we introduce the stochastic Camassa–Holm equation and discuss the numerical approximation of the model problem. In Sect. 3, we briefly review the bootstrap particle filter and its consequent augmentations with jittering, tempering and nudging procedures. In Sect. 4, a numerical study is conducted, illustrating the behaviour of data assimilation methods. Finally, we summarize with some concluding remarks in Sect. 5.

2   The SCH Model and Its Discretisation

The deterministic viscous Camassa–Holm is expressed as the following evolution equation,

$$
\begin{aligned}
m &= u - \alpha^2 \partial_{xx} u, \\
\mathrm{d}m &- \mu \partial_{xx} m \, \mathrm{d}t + (\partial_x m + m \partial_x) u \mathrm{d}t,
\end{aligned}
\tag{1}
$$

for the evolution in time $t \in [0, T)$ of the fluid momentum density $m(x, t)$ and the velocity $u(x, t)$ solved on the spatial domain $[0, L]$ with periodic boundary conditions and initial conditions $u(x, 0) = u^0(x)$, with $\alpha > 0$ some chosen constant parameter.

In the SALT framework, this deterministic PDE is transformed into a stochastic PDE by replacing $du \mapsto udt + dU$, where U is some stochastic process. Instead of the noise expansions of the form $dU = \sum_{k=1}^{K} \xi^k dW^k$ used in some previous works on SALT, in this chapter we use Gaussian random space-time fields obtained from the Matérn formula,

$$(I - \kappa^{-2}\nabla^2)^k dU(x, t) = \eta dW(x, t), \tag{2}$$

where $dW(x, t)$ is a space-time white noise that is cylindrical in space and Stratonovich (in keeping with the SALT approach) in time. The coefficients η and κ determine the expected smoothness of the process U. For this work, we consider $\eta = 1$, and $k = 3$.

Henceforth, we consider the SCH equation,

$$m = u - \alpha^2 \partial_{xx}u, \, dm - \mu\partial_{xx}m \, dt + (\partial_x m + m\partial_x)(udt + \circ dU), \tag{3}$$

with the same setup as for (1). Here $\circ$ is used to indicate that the integral is Stratonovich in time rather than Ito.

For spatial discretisation, we will be interested in approximating solutions of the SCH that are periodic functions in the spatial variable. We will consider a finite element discretisation on a uniform mesh of the interval $I = [0, L]$ with N cells of width $h = L/N$.

The space-time white noise $dW(x, t)$ is approximated by a process $dW_h(x, t)$ depending on a finite number of Brownian motions according to

$$W_h(x, t) = \sum_{i=1}^{N} \frac{1}{A_i^{1/2}} \phi_i(x) W_i(t), \tag{4}$$

where A_i is the width of cell i (which is equal to h for a uniform grid), ϕ_i is the indicator function of cell i, and $\{W_i(t)\}_{i=1}^{N}$ are N iid standard Brownian motions. In other words, $\int_{t_A}^{t_B} dW_h \in Q_h$ for any times $t_A < t_B$, where Q_h is the piecewise constant finite element space (DG0). This is a low-order spatial approximation of the space-time white noise; higher-order approximations can be obtained. However, the required square root factorisation of the finite element mass matrix is not diagonal and we did not implement this. Croci et al. [14] provided an efficient formulation for continuous finite element fields using a generalised form of the square root that exploits the local assembly procedure.

We then approximate U with $U_h := U_{h,k}$, for $n = 0, 1, 2, \ldots$, where $\{U_{h,j}\}_{j=1}^{k}$ are space-time Gaussian random fields in the continuous linear Lagrange (P1) finite element space on the mesh, denoted here as V_h. Then we approximate the solution of (2) by solving k approximated second-order elliptic problems, according to

$$(\Delta U_{h,j}^n, v) + \kappa^{-2}(\nabla \Delta U_{h,j}^n, \nabla v)$$

$$= \begin{cases} \eta(\Delta^n W_h, v), \ j = 1 \\ (\Delta U_{h,j-1}^n, v), \ j > 1 \end{cases}, \quad \forall v \in V_h, \quad j = 1, 2, \ldots, k, \tag{5}$$

where $(\cdot, \cdot)$ represents the usual L^2 inner product for functions on I, $\Delta U_{h,j}^n = \int_{t_n}^{t_{n+1}} dU_{h,j}(t)$, $\Delta W_h^n = \int_{t^n}^{t^{n+1}} dW_h(t)$ for two consecutive time levels $t^n < t^{n+1}$. In this chapter, we use $k = 3$.

Then, the semidiscrete numerical scheme seeks $m(t) \in V_h$ and $u(t) \in V_h$ such that

$$\begin{aligned}(u, \psi) + \alpha^2(\partial_x u, \partial_x \psi) - (m, \psi) = 0, \quad \forall \psi \in V_h, \\ (m_t, \phi) + \mu(\partial_x m, \partial_x \phi) + (m \partial_x v, \phi) - (mv, \partial_x \phi) = 0, \quad \forall \phi \in V_h.\end{aligned} \tag{6}$$

This is equivalent to a finite-dimensional stochastic differential equation on $\mathbb{R}^N$. More details on the selection of the initial conditions are given in a later section.

For the time discretization, we select a uniform time step $\Delta t = T/M$ and $t_n = n\Delta t$, $n = 1, 2, \ldots, N_T$, and solve for $m^n \approx m(x, t_n)$ and $u^n \approx u(x, t_n)$.

Then we use ΔU_h^n in an implicit midpoint rule discretisation (leading to a Stratonovich method since ΔU_h^n multiplies $m^{n+1/2}$, not m^n.), finding $m^{n+1}, u^n \in V_h$

$$\begin{aligned} (u^{n+1}, \psi) + \alpha^2(\partial_x u^{n+1}, \partial_x \psi) - (m^{n+1}, \psi) = 0, \quad \forall \psi \in V_h, \\ (m^{n+1} - m^n, \phi) + \mu \Delta t (\partial_x m^{n+1/2}, \partial_x \phi) \\ + (m^{n+1/2} \partial_x (\Delta t u^{n+1/2} + \Delta U_h^n), \phi) \\ - (m^{n+1/2}(\Delta t u^{n+1/2} + \Delta U_h^n), \partial_x \phi) = 0, \quad \forall \phi \in V_h, \end{aligned} \tag{7}$$

and $m^{n+1/2} = (m^{n+1} + m^n)/2$, etc.

3   Data Assimilation Methods

In Sect. 2, we defined the SPDE providing the unknown signal, i.e., the system we are interested in performing Bayesian inference upon. In this work, we will use the language of stochastic filtering to provide the background framework for a Bayesian inference case study for the SCH model.

Let X and Y be two processes defined on the probability space $(\Omega, \mathcal{F}, \mathbb{P})$. The process X is usually called the signal process or the "truth", with range in a specified function space V in the SPDE case (approximated by a finite element space in this work) and Y is the observation process, with range $\mathbb{R}^M$. The pair of processes (X, Y) forms the basis of the nonlinear filtering problem: find the best approximation of the posterior distribution of the signal X_t, denoted by π_t given the observations $Y_1, Y_2, \ldots, Y_t$. In our context, the observations consist of noisy measurements of the true state recorded at discrete times and they are taken at locations on a data grid $\mathcal{G}_d$, defined later. The data assimilation is performed at these times, which we call the assimilation times.

In this work, we discuss the approximation of the posterior distribution of the signal by *particle filters*. These sequential Monte Carlo methods generate approximations of the posterior distribution using sets of *particles*, which represent samples from the conditional distribution of X. Particle filters are employed to make inferences about the signal process. This involves utilizing Bayes' theorem, considering the time-evolution induced by the signal X_t, and taking into account the observation process Y_t. The observation data Y_t is, in our case, an M-dimensional process that consists of noisy measurements of the (one dimensional) velocity field u taken at a point belonging to the data grid $\mathcal{G}_d$,

$$Y_t := \mathcal{P}_d^s(X_t) + V_t, \tag{8}$$

where the observation operator $\mathcal{P}_d^s : V \to \mathbb{R}^M$ is a projection operator corresponding to interpolation of X to the points on the data grid $\mathcal{G}_d$, and $V_t \sim \mathcal{N}(\mathbf{0}, I_\sigma)$, where $I_\sigma = \mathrm{diag}(\sigma_1^2, \sigma_2^2, \ldots, \sigma_M^2)$. While we assumed standard normal distributions for V_t, the methodology presented is valid for any observation likelihood with a computable pdf. The ensemble of particles evolved between assimilation times according to the law of the signal.

Next, we explain briefly the various types of particle filters used in this article. Before going into the details of particle filters, we introduce some technical terms used to explain filters. The likelihood weight function is defined as

$$\mathcal{W}(X_{t_n}, Y_n) = \exp\left(-\frac{1}{2}\sum_{i=1}^{M}\left\|\frac{\mathcal{P}_d^s(X_{t_n})_i - Y_{t_n,i}}{\sigma_i}\right\|_2^2\right), \tag{9}$$

where $Y_{t_1}, \ldots Y_{t_M}$ are the components of Y_t (one for each of the M observation points). One can then write the observation likelihood at instance t_n as

$$p(Y(t_n)|X(t_n)) \propto \mathcal{W}(X_{t_n}, Y_{t_n}).$$

We enumerate the particles $X_t^1, X_t^2, \ldots, X_t^{N_p}$. To measure the variability of the weights (9) of particles at time t, we use the *effective sample size* (ESS),

$$\text{ESS}(\overline{\mathbf{w}}) \stackrel{\Delta}{=} \|\mathbf{w}\|_{\ell^2}^2 = \left(\sum_{n=1}^{N_p} (\overline{w}_n)^2\right)^{-1},$$

$$\overline{w}_n := \frac{w_n}{\sum_{n=1}^{N_p} w_n}, \quad \text{where } w_n = \mathcal{W}(X_t^{(n)}, Y_t, \tag{10}$$

where N_p is the number of particles. ESS quantifies the distribution of weights. The ESS value approaches N_p if the particle weights are almost uniform, and it is close to one when fewer particles have large weights and the remaining particles have small weights. We resample whenever $N_p^* = \text{ESS}/N_p$ drops below a specified threshold. For all the numerical experiments, we used the threshold 0.8.

3.1 Particle Filters: Basic Terminology

In this subsection, we briefly discuss the basics of particle filters, so that we can present our results in context. We mostly describe the methodology; more details on why it works can be found in the references [8, 9, 23].

3.1.1 Bootstrap Particle Filter

The bootstrap particle filter is the basic particle filter, also called a Sampling Importance Resampling filter. In the bootstrap filter, given an initial distribution of particles (obtained as samples from a prior distribution for the initial state), each particle is propagated forward according to the signal equation (the spatially discretised SCH equation in our case), with independent realisations of the Brownian motions. Here, and in the more sophisticated particle filter formulations later, we consider intermittent data assimilation in intervals of length $\Delta\tau$, subdivided into model timesteps Δt with $\Delta\tau/\Delta t = N_s$ some positive integer. When the discrete time is discussed, we will use the suffix u^n, m^n to indicate the solution n timesteps after the last assimilation time, i.e. the index n resets to 0 after the most recent data has been assimilated, for the purposes of presentation here.

The empirical distribution,

$$d\mu_F = \sum_{i=1}^{N_p} \frac{1}{N_p} \delta(X - X_i), \tag{11}$$

where δ is the Dirac measure, is an approximation of the prior (forecast) distribution for the signal, before receiving the observations.

Subsequently, utilizing partial observations, weights for new particles are calculated. This is done by computing the likelihood weight function (9) for each particle, and then renormalising so that the complete set of weights sums to 1. The weighted empirical distribution,

$$d\mu_A = \sum_{i=1}^{N_p} \overline{w}_i \delta(X - X_i), \tag{12}$$

where $\{\overline{w}_i\}_{i=1}^{N_p}$ are the normalised weights, is an approximation of the posterior (analysis) distribution for the signal, conditional on the received observations.

Next, a selection process is used on the weighted particles. This is a statistical procedure that aims to find a new equally weighted set of particles that approximate the same distribution as the old nonequally weighted set of particles. On average, the particles with larger weights will be duplicated, while the particles which have smaller weights will be eliminated. In the simplest case, this is done by sampling with replacement from the ensemble of particles using the multinomial distribution described by the weights. In this work, we use the systematic resampling algorithm [17], to reduce the sampling error.

The ESS is a crude diagnostic that measures how far the weights are from being uniform. The value of ESS typically drops fast for higher dimensional problems because of the curse of dimensionality: the particles are relatively sparse in observation space (i.e. under the mapping of $\mathcal{P}_d^s$) and a small number of particles will have much higher weights than the others with high probability. This causes a loss of diversity amongst the particle population after resampling. As a result, the particles fail to give a better approximation of the posterior distribution. To overcome this situation, one would require a huge number of particles.

To resolve the filter degeneracy, we replace the direct resampling from the weighted predictive approximations by a *tempering* procedure combined with *jittering* and *nudging* procedures, described below.

3.1.2 Tempering and Jittering

As discussed above, due to the sample degeneracy, the ESS value will rapidly fall below the tolerance N_p^*. The purpose of *tempering* is to take incremental steps between the approximations of the prior and posterior distributions, resampling on each step, to maintain a high ESS value. In each tempering step $k = 1, \ldots, N_\theta$, the particle likelihood weights are evaluated and scaled by $0 < \Delta\theta_k < 1$, with $\sum_{k=1}^{N_k} \Delta\theta_k = 1$, where $\Delta\theta_k$ is chosen so that ESS $> N_p^*$ for that step. In our work we use an adaptive tempering procedure as discussed in [5, 19, 20]. This procedure repeatedly reduces $\Delta\theta_k$ until the condition holds, keeping the number of tempering steps to a minimum.

After each tempering step, the particles are resampled according to the scaled (and normalised) weights. This alone is insufficient to prevent the accumulation of a large number of duplicates in the particle ensemble. To remove the duplication in the resulting ensembles, we employ *jittering*. This can be understood by using the fact that distributions on the state at assimilation time τ_n are equivalent to distributions on the joint distribution of the state at assimilation time τ_{n-1} together with the Brownian increments from τ_{n-1} to τ_n. The equivalence comes because given a sample from the latter, we can solve the signal equation forwards with the initial condition given by the state value $X_{\tau_{n-1}}$ at time τ_{n-1}, using the realisation of the Brownian increments ΔW,[1] to obtain a state value X_{τ_n} at time τ_n. Each particle can thus be represented by $(X_{\tau_{n-1}}, \Delta W)$, which may be duplicated after resampling. Using the disintegration formula $\pi(X_{\tau_{n-1}}, \Delta W | \hat{Y}_n) = \pi(X_{\tau_{n-1}} | \hat{Y}_n) \pi(\Delta W | X_{\tau_{n-1}}, \hat{Y}_n)$, where $\hat{Y}_n$ is a shorthand for all the observations $Y_1, Y_2, \ldots, Y_n$, we see that keeping $X_{\tau_{n-1}}$ the same, but obtaining a new ΔW sample from $\pi(\Delta W | X_{\tau_{n-1}}, \hat{Y}_n)$, produces another consistent sample from $\pi(X_{\tau_{n-1}}, \Delta W | \hat{Y}_n)$. This is the jittering technique that allows duplicates to be replaced by different samples from the same distribution. This is achieved by performing tempering using the $(X_{\tau_{n-1}}, \Delta W)$ representation, and only updating to $X_{\tau_{n-1}}$ once the tempering step is complete. At each tempering step, after resampling, the noise realisation ΔW for each particle is moved using a Monte Carlo Markov Chain (MCMC) method with the target being the tempered posterior distribution with the likelihood function scaled by $\theta_k = \sum_k \Delta \theta_k$ at step k. Since the samples are from the conditional distribution $\pi(\Delta W | X_{\tau_{n-1}}, \hat{Y}_n)$, this can be done independently for each particle.

The MCMC method describes a sequence of samples ΔW from $\pi(\Delta W | X_{\tau_{n-1}})$, namely $\Delta W^0, \Delta W^1, \Delta W^2, \ldots$, where ΔW^0 is the sample of ΔW given after resampling. After space and time discretisation, this is a finite array of numbers (of dimension $N_s \times N$ for our discretisation choices) whose prior distribution is iid $\mathcal{N}(0, \Delta t)$. We use the Preconditioned Crank Nicholson (PCN) algorithm [12] to move the particles. Given a previous old sample ΔW_0, PCN proposes a new sample $\Delta W_1 = (2 - \delta)/(2 + \delta)\Delta W_0 + (8\delta)^{1/2}/(2 + \delta)\Delta \hat{W}$, where $\Delta \hat{W}$ is a new sample from the prior distribution, and $\delta > 0$ is a time-stepping parameter. The proposal is accepted with probability

$$a = \max\left(1, \frac{\mathcal{W}(X_{\tau_n}(\Delta W_0), Y_n)^{\theta_k}}{\mathcal{W}(X_{\tau_n}(\Delta W_1), Y_n)^{\theta_k}}\right), \tag{13}$$

where $X_{\tau_n}(\Delta W_0)$ and $X_{\tau_n}(\Delta W_0)$ denote the solutions generated from ΔW_0 and ΔW_1, respectively. Otherwise, the old value of ΔW is repeated.[2] Larger δ means

---

[1] Note that here ΔW represents *all* of the Brownian increments from τ_n to τ_{n+1}.

[2] Note that a numerically stable implementation should deal with log weights.

that the proposal is moved further away and smaller δ means that the proposal is more likely to be accepted. The number of MCMC iterations (we call them "jittering steps") per tempering steps is fixed. The MCMC algorithm doesn't need to converge in statistics, just that the duplicated particles are sufficiently spread. The accept-reject criteria ensure statistical consistency as proved in [5].

In the completed algorithm, one selects several jittering steps N_j per tempering step. Thus the full assimilation step consists of $N_j \times N_t$ tempering steps, where N_t is the number of tempering steps which are selected adaptively. Each jittering step requires one forward model run from τ_{n-1} to τ_n, and each tempering step requires one resampling step (which requires parallel communication). Thus the total cost is $N_t \times N_j \times C_f + N_t \times C_r$ where C_f is the cost of one forward model run and C_r is the cost of one resampling step.

3.1.3 Nudging

In the nudging particle filter framework, we introduce a time-dependent control variable that we may choose, so that particles are "nudged" towards regions where the observations suggest the signal is likely to be. This *nudging* is done in a way that preserves the consistency of the particle filter, i.e. the particle ensemble remains a consistent set of samples from the prior distribution, after appropriate modification of the weights. A similar technique is discussed in [23] where the likelihood-informed proposals are used. Implicit particle filters [7] are related and rely on the principle of pushing particles to high-probability regions to prevent the collapse of the filter in high-dimensional state spaces. We also mention the equal/equivalent weights particle filter which aims to keep the particle weights uniform in high dimensional problems [1, 28]. Presently, we justify this at the level of the spatial semidiscretisation, which is interpretable as an SDE, of the form

$$dx = f(x)dt + G(x)dW, \ x(0) = x_0, \tag{14}$$

where $x \in \mathbb{R}^N$, $f : \mathbb{R}^N \to \mathbb{R}^N$, $G : \mathbb{R}^N \to \mathbb{R}^{N \times Q}$ (i.e., $G(x)$ is an $N \times Q$ matrix for each $x \in \mathbb{R}^N$), and $W(t)$ is a Q-dimensional Brownian motion. If one instead solves the modified SDE,

$$d\hat{x} = f(\hat{x})dt + G(\hat{x})(\lambda(t)dt + dW), \ \hat{x}(0) = x_0, \tag{15}$$

where $\lambda(t) \in \mathbb{R}^Q$, then the joint probability measure for $\hat{x}(t)$ for $t \in (0, T)$ is absolutely continuous with respect to the joint probability measure for $x(t)$ on the same range, and the Radon-Nikodym derivative from one measure to the other is

$$G = \exp\left(-\int_0^T \frac{1}{2}|\lambda(t)|^2 dt + \int_0^T \lambda(t) \cdot dW\right), \tag{16}$$

subject to appropriate regularity conditions on $f(x)$ and $G(x)$ which we assume here. This means that we can choose λ to reduce the likelihood weight, but we must pay the price of multiplying this weight by G.

In particular, in this work, we use the Girsanov formula to correct the solution of the spatially discrete SCH Eq. (6) to keep the particles closer to the true state. To implement this, we update the SCH model (3) with a 'nudging term', replacing the $j = 1$ case of (5) with

$$(\Delta U_j, v) + \kappa^{-2}(\nabla \Delta U_j, \nabla v) = \eta(\Delta W_h + \Delta \Lambda, v), \quad \forall v \in V_h, \tag{17}$$

where $\Delta \Lambda = \Lambda(t)(t_B - t_A)$, and $\Lambda(t) \in Q_h$. The Girsanov formula (16) can be rewritten as

$$G = \exp\left(-\int_0^T \frac{1}{2}(\Lambda^2(t), 1/A)\mathrm{d}t + \int_0^T (\Lambda(t), \mathrm{d}W_h/A)\right), \tag{18}$$

where $A \in Q_h$, such that $\int_{e_i} A \mathrm{d}x = 1$, for each cell e_i. In our discrete time approximation of this, we use $\Delta U_h = \Delta U_k$ from (17) with $t_A = t_n$ and $t_B = t_{n+1}$ in (7), and we use the approximated time integral,

$$G \approx G_{\Delta}t = \exp\left(-\sum_{n=0}^{N_s}\left(\frac{1}{2}((\Lambda^n)^2, 1/A)\Delta t + (\Lambda^n, \Delta W_h^n/A)\right)\right). \tag{19}$$

After choosing $\Lambda(t)$, the particles will have a new weights according to Girsanov's theorem, given by

$$\tilde{W}(u, Y, \Lambda) = W(u, Y) + G(\Lambda).$$

Hence, it makes sense to maximise $\tilde{W}$ over Λ. However, a critical aspect is that Girsanov's theorem only holds if $\lambda(t)$ only depends on $W(s)$ for $s < t$. Our strategy is to incrementally optimise $\lambda(t)$ as we reveal $W(t)$; we have to adapt $\lambda(t)$ to past noise. After time discretisation, this means that we first initialise $\Lambda^n = 0$ and $\Delta W^n = 0$ for $n = 0, 1, \ldots, N_s$. Then we optimise $\tilde{W}$ over Λ^0, keeping the other values of Λ fixed. Next, we randomly sample ΔW_h^0 from the Q-dimensional distribution $\mathcal{N}(0, \Delta t I)$. Then we optimise $\tilde{W}$ over Λ^1, keeping the other values fixed, and randomly sample ΔW_h^1 from $\mathcal{N}(0, \Delta t I)$, and so on until we reach $n = N_s$. This approach was investigated in [8] in application to data assimilation for a quasigeostrophic ocean model, but Λ was only nonzero in the final stage of the splitting method in the last timestep $n = N_s$, which allows the optimisation problem to be solved using linear least squares. This simplification was due to the lack of an available nonlinear optimisation algorithm within that code framework. This is something that we have addressed in the current work.

In general, these optimisation problems are nonlinear since the observations at τ_n depend on the entire history of dW from $t = \tau_{n-1}$ to τ_n through the nonlinear SDE. We solve these problems numerically, using a gradient descent algorithm (BFGS), with the gradient of the functional computed using the adjoint technique. In our implementation, this is automated using Firedrake [15], which is built according to the methodology of [16].[3] Briefly, when the forward model is run for the first time, Firedrake records the sequence of solves and arithmetic operations in a graph structure. Since all of these operations are expressed symbolically in Firedrake, from which code is automatically generated, the symbols can be automatically differentiated, leading to a symbolic representation of the adjoint model from which more code can be automatically generated and run.

After applying the nudging step, it may still be necessary to use tempering and jittering as above. In that case, the θ_k adjusted weight formulas described above need to be modified by replacing W with $\tilde{W}$, as described in [8].

3.1.4 Ensemble Parallelism

Our present implementation using Firedrake allows us to combine spatial domain decomposition for each particle with ensemble parallelism across particles. The algorithms discussed above involve independent calculations for each particle, with the exception of the resampling step, when particle states (and noise increments) need to be replaced with copies from others. The weight normalisation step also requires inter-particle communication, but only for one floating point number per particle, which is insignificant for large models. Ensemble parallelism means dividing the ensemble of particles into batches and executing the independent calculations for each batch, before updating the particles from copies, which may come from other batches. In our implementation, we use distributed memory parallelism using the Message Passing Interface (MPI) protocol, see Fig. 1.

Our present algorithm for resampling is quite naive: we just compute which particles need to be replaced by copies of which other ones, and send and receive from batches as necessary. Since communication of entire model states between ranks is costly (and likely to dominate the algorithm cost for large ensemble sizes), a more sophisticated approach should optimise the order of the particles after resampling to minimise communication. Alternatively, algorithms such as the Islands particle filter should be considered [27]. We also mention the parallel resampling of [22], although this is more appropriate for shared memory parallelism than the distributed memory parallelism that we require for SPDEs. In the present work, we do not investigate

---

[3] Since we need to solve repeated optimisation problems with different data, this required minor extensions to Firedrake, namely the `family derivative_components` argument to `family adjoint.ReducedFunctional`, which allows to ignore derivatives concerning the observed data in the minimisation calculation; these changes are now in the main branch of Firedrake.

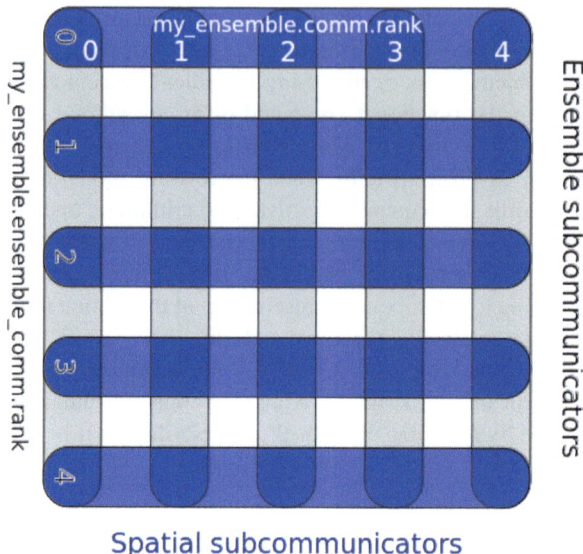

Fig. 1 Spatial and ensemble parallelism for an ensemble with 5 batches of particles, each executed in parallel over 5 processors, using 25 ranks in total. Spatial subcommunicators are used for the domain decomposition algorithm for the iterative solvers involved in solving the forward equations for each particle, and ensemble subcommunicators are used to transfer particle states (and noises) during resampling

parallel performance. We just note that this combined parallelism is possible in our code framework, and will present a thorough investigation in future work on more challenging problems in 2D and 3D.

4 Numerical Investigations

In our numerical experiments of particle filtering applied to the SCH model, we use an ensemble of 150 particles. This is motivated by (the upper end of) typical ensemble sizes of ensemble uncertainty quantification and data assimilation systems for operational weather forecasting, where ensemble sizes are limited by the large computational cost of the forward models. In the standard context, the velocity of the SCH model (7) is observed every 5 time steps. For all numerical experiments, we choose the length L of the spatial domain to be 40, and we take $\alpha = 1$. There are $N = 100$ equispaced cells in the decomposition of the interval $[0, L]$, and the model time step is $\Delta t = 0.025$. We explore multiple scenarios to evaluate stochastic filtering. In the initial scenario, observation data is gathered from the entire spatial domain, while in the second scenario, only half of the spatial domain is observed.

All the numerical simulations are conducted using our general-purpose particle filter library [11] which is built upon Firedrake [15].

4.1 Experiment 1: Full Domain Observed

In this experiment, we took measurements at $M = 80$ equispaced grid points in the interval $[0, L]$. These grid points constitute $\mathcal{G}_d$. The observations were perturbed with iid $\mathcal{N}(0, 0.5)$ measurement errors.

4.1.1 Initialization of Particles and Truth

In this experiment the initialization of particles and the true solution of the model problem (7) is constructed in the following way.

For particle $n = 1, 2, \ldots, N_p$, we solve the (finite element discretisation of the) following elliptic problem on the periodic domain $[0, L]$ with zero boundary,

$$\begin{cases} (I - \nabla^2)U_n^{0,1} = |W_n|, & n = 1, 2, \ldots, N_p, \\ (I - \nabla^2)U_n^{0,j+1} = U_n^{0,j}, & n = 1, 2, \ldots, N_p, \quad j = 1, 2, \end{cases} \tag{20}$$

where W_n is the DG0 function where each basis coefficient (i.e. constant cell value) is sampled from $\mathcal{N}(0, h)$. We then calculate initial conditions for the particles as $u_n^{p,0} = ((\alpha_n)^2 U_n^{0,3} + (\beta_n)^2)$ for $n = 1, 2, \ldots, N_p$, where random parameters $\alpha_n, \beta_n \sim \mathcal{N}(0, 1)$ for $n = 1, 2, \ldots, N_p$. Then, the initial condition for the truth u_0 is sampled from the same distributions as the particles.

For the validation of various filtering procedures, we use the ensemble mean l_2-norm relative error (EMRE), the relative bias (RB) and the relative forecast ensemble spread (RES). These are defined as follows,

$$EMRE(u^a, u^p) := \frac{1}{N_p} \sum_{n=1}^{N_p} \frac{\|u^a - u_n^p\|_2}{\|u^a\|_2},$$

$$RB(u^a, u^p) := \frac{\|u^a - \overline{u}^p\|_2}{\|u^a\|_2}, \quad \text{ensemble mean } \overline{u}^p := \frac{1}{N_p} \sum_{n=1}^{N_p} u_n^p,$$

$$RES(u^p, \overline{u}^p) := \frac{1}{N_p - 1} \sum_{n=1}^{N_p} \frac{\|u_n^p - \overline{u}^p\|_2}{\|u^a\|_2}.$$

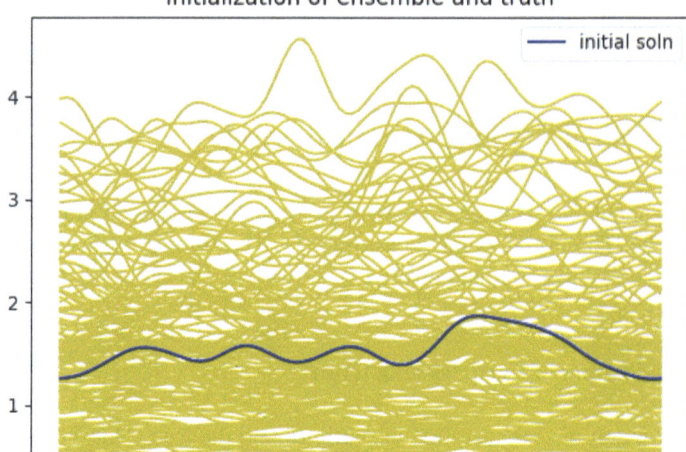

Fig. 2 Initialization of all 150 particles and true state (the latter shown in blue). In this experiment there is a broad spread of the initial ensemble, expressing a wide distribution of possible states that envelopes the true value

These quantities are computed for each ensemble step. The purpose of these statistics is to demonstrate that the particle filters are stable (or not).

In Fig. 2, the initialization of the ensemble of particles and the true state is displayed. We have coloured the ensemble of 150 particles yellow in all the figures depicting the trajectories of the ensemble and true values. We observe that the cloud of particles is diverse. The initial ensemble gives a good description of the initial uncertainty with $EMRE(u^a, u^{p,0}) = 0.48$ and $RB(u^a, u^{p,0}) = 0.12$.

4.1.2 Bootstrap Filter

Firstly, we discuss how the bootstrap filter performs for the SCH model. With the above initialization, trajectories of the truth and particles are displayed for the different observation (data assimilation) steps in Fig. 3. In this and all evaluations in this chapter, we focus only on stability, not accuracy. As time evolves, the ensemble spread reduces gradually but the cloud of particles does not track the truth. This phenomenon can be confirmed by Fig. 3. The statistics of the difference between the truth and ensemble are discussed with the lenses of EMRE, RB and RES. These terms are calculated and displayed in Fig. 4 against assimilation steps. One can see the RES gradually decreases but the mean square error EMRE and relative bias RB diverge and saturate, indicating filter divergence.

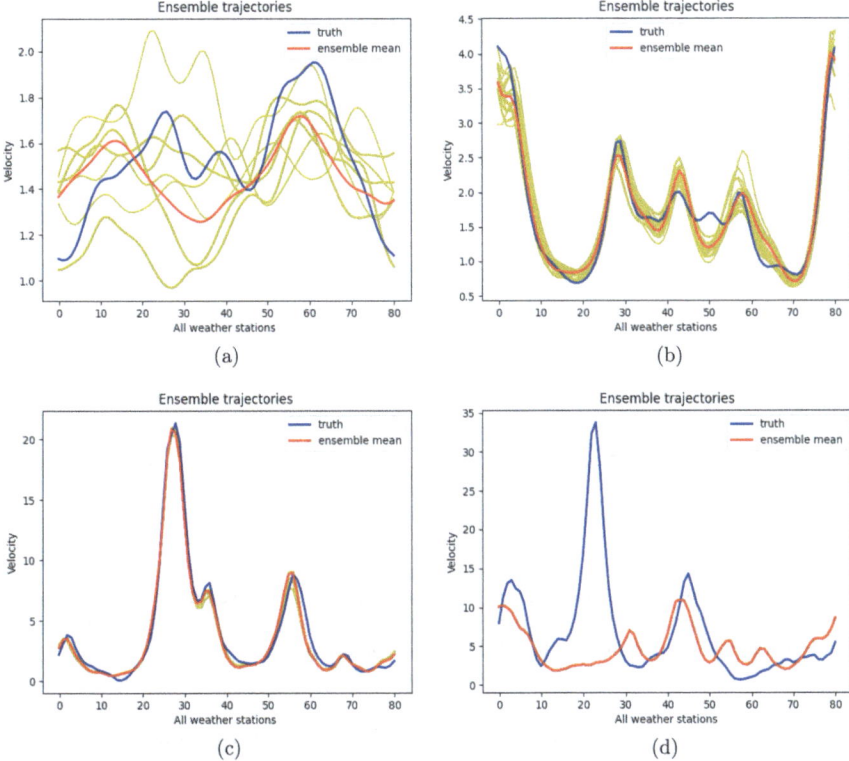

Fig. 3 Comparison of the evolution of the true state vs posterior ensemble and ensemble mean at data grids (weather stations). In order to assimilate data we use the bootstrap particle filter and the outcome is displayed for the specified assimilation steps. We observe that the filter is diverging by the 1000th assimilation step. (**a**) DA step 1. (**b**) DA step 100. (**c**) DA step 500. (**d**) DA step 1000

4.1.3 Tempering and Jittering

Now, we discuss the data assimilation algorithm that uses tempering and jittering. The jittering parameter δ is equal to 0.15, and there are five jittering steps per tempering step. Given that our framework allows for adapted tempering, the number of tempering steps needed for our numerical experiment falls within the range of 7 to 10. In Fig. 5, we exhibit a few instances to emphasize the reduction of uncertainty resulting from applying the particle filter with tempering and jittering, as described in the previous section. Additionally, it is possible to compare the ensemble after a single DA step using the tempering-jittering filter with the outcome from the bootstrap filter. In particular, for the bootstrap filter, almost all particles are replaced by duplicates of a small number of particles, and the ensemble is not very diverse.

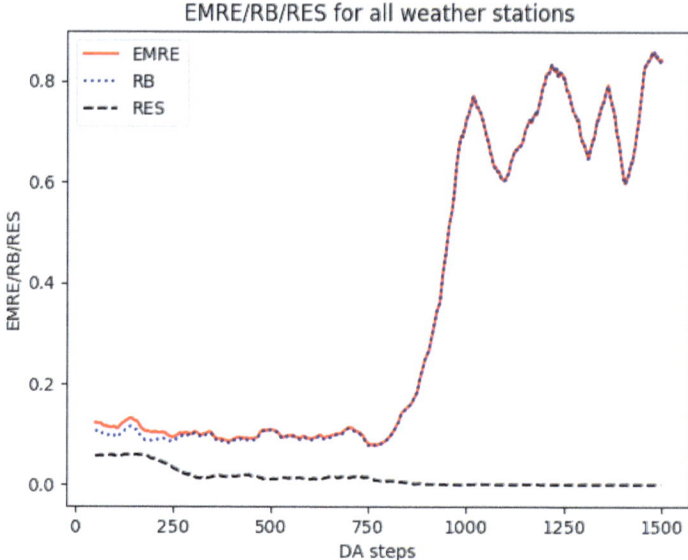

Fig. 4 Evolution of the relative ensemble mean error (EMRE), relative bias (RB) and ensemble spread (RES) associated with bootstrap particle filter. We observe that the filter appears to be diverging (and then saturating) by the 1000th assimilation step

However, the tempering allows the ensemble to be rediversified from those particles using the noise.

We plot the EMRE, RB and RES in Fig. 6. One can see that the EMRE and the RES are comparable and stable as time evolves. In contrast to the previous case, the particle filter follows the truth much better.

4.1.4 Nudging

We will now look at the performance of the data assimilation methodology which includes nudging before using tempering and jittering. In Fig. 7, we exhibit a few instances to emphasize the reduction of uncertainty resulting from applying the particle filter with nudging, tempering and jittering. We plot the EMRE, RB and RES in Fig. 8. In these preliminary results, we observe that this particle filter is stable but does not yet provide a dramatic improvement over the filter without nudging.

To conclude the results regarding Experiment 1, We have summarized the comparison of particle filters by displaying time-averaged EMRE, RB and RES values in Table 1.

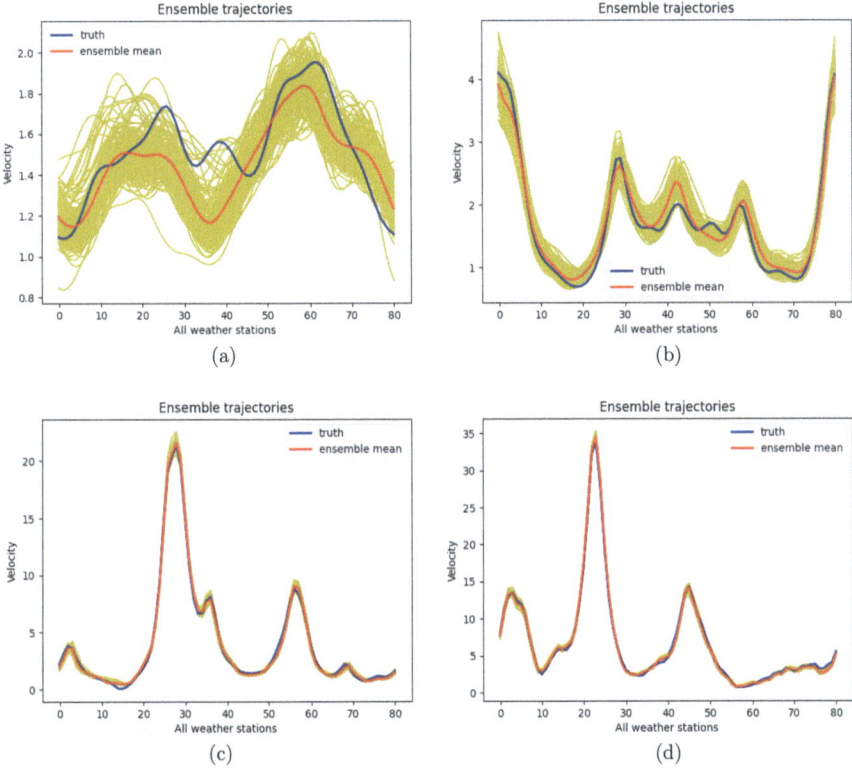

Fig. 5 Comparison of the evolution of the true state vs posterior ensemble and ensemble mean at data grids (weather stations). To assimilate data we use *tempering* and *jittering* and the outcome is displayed for the specified assimilation steps. We observe that the true solution is contained within the spread of the ensemble despite the low spread, indicating that the particle filter is dealing well with this filtering problem. (**a**) DA step 1. (**b**) DA step 100. (**c**) DA step 500. (**d**) DA step 1000

4.2   Experiment 2: Half Domain Observed

In this experiment, the observation data is taken from half of the spatial domain, *i.e* $[0, L/2]$, at 40 equispaced points. With this modification, we now examine the bootstrap filter for the SCH model. From Fig. 9, we see that truth is within the spread of the particles after initial data assimilation steps but particles lost track of truth even in the observed domain $[0, L/2]$. Also, we have plotted the values of EMRE, RB and RES in Fig. 11, where one can observe that the error and bias (EMRE and RB) significantly increase as time evolves.

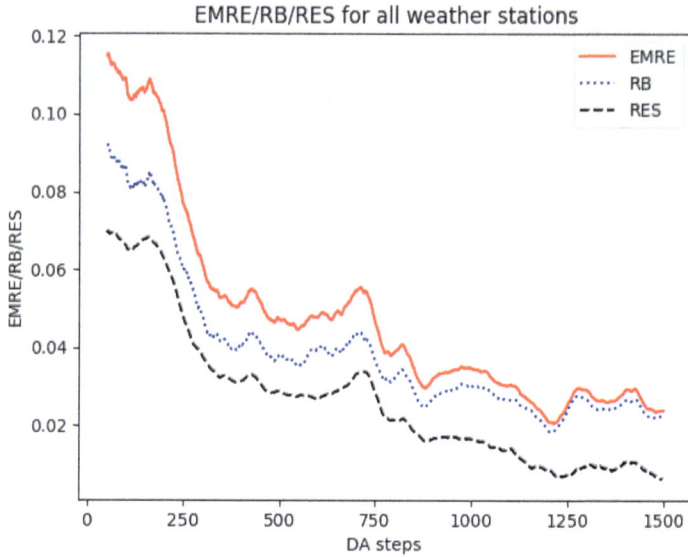

Fig. 6 Evolution of the relative ensemble mean error (EMRE), relative bias (RB) and ensemble spread (RES) associated with the filter using *tempering* and *jittering*. We observe that the filter is stable

Next, we discuss the particle filter using tempering and jittering. We have used the same tempering-jittering procedure as discussed in the previous example. From Fig. 10, we see that the ensemble surrounds the true value after some data assimilation steps in $[0, L/2]$. This does not occur initially; this is because it is not possible to reach all possible states of the posterior through modifications of the SALT noise. The ensemble does end up surrounding the true solution in the displayed results from the 100, 500, 1000 and 2000th data assimilation steps in this figure. This shows that the particle filter is stable, and this is confirmed by the EMRE, RB and RES measures in Fig. 11. We conclude our comparison of particle filters by presenting time-averaged EMRE, RB and RES in Table 2. We leave the incorporation of nudging for this example to further work.

5 Concluding Remarks

In this work, we investigated adaptive tempering, jittering and nudging techniques applied to the stochastic Camassa-Holm equation with SALT noise and viscosity, demonstrating our new capability that can be applied to arbitrary stochastic models

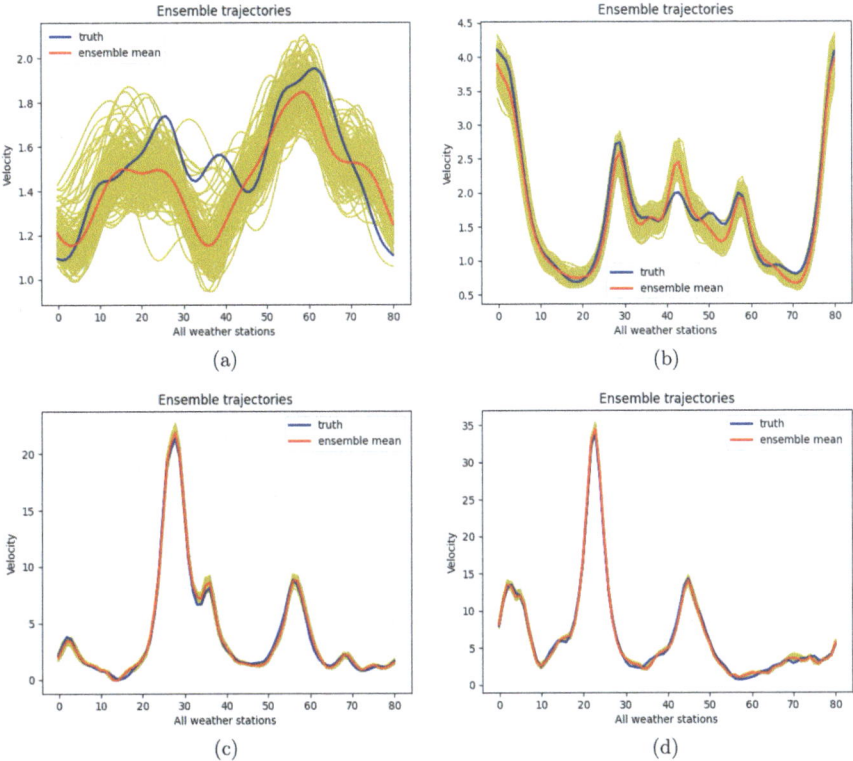

Fig. 7 Comparison of the evolution of the true state vs posterior ensemble and ensemble mean at data grids (weather stations). To assimilate data we use *nudging*, *tempering* and *jittering* and outcome is displayed for the specified assimilation steps. We observe that the true solution is contained within the spread of the ensemble despite the low spread, indicating that the particle filter is dealing well with this filtering problem. (**a**) DA step 1. (**b**) DA step 100. (**c**) DA step 500. (**d**) DA step 1000

written in Firedrake, made possible using MPI parallelism across particles. The nudging filter involves solving nonlinear optimisation problems that are enabled with Firedrake's automated adjoint system. We demonstrated that these approaches lead to particle filters that are stable with relatively few (150) particles in cases where the classical bootstrap filter fails.

In forthcoming work, we will undertake a detailed investigation of the accuracy of these filters when applied to 2D and 3D problems, making use of high-

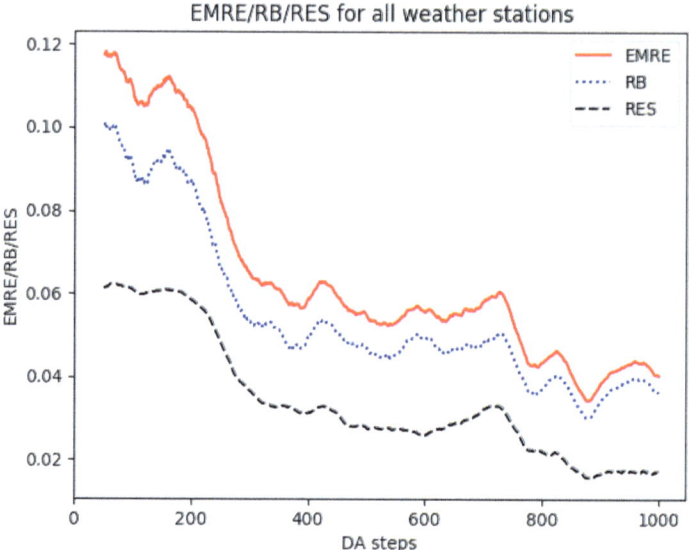

Fig. 8 Evolution of the relative ensemble mean error (EMRE), relative bias (RB) and ensemble spread (RES) associated with the filter using *nudging*, *tempering* and *jittering*. We observe that the particle filter is dealing well with this filtering problem.

Table 1 Time-averaged EMRE, RB and RES associated with Experiment 1

Time-averaged	BS particle filter	tempering and jittering	tempering and jittering with nudging
EMRE	0.3513	0.0485	0.0507
RB	0.3488	0.0398	0.0401
RES	0.0141	0.0270	0.0271

performance computing. It is necessary to go beyond metrics such as ESS to properly determine the optimal value for δ and the optimal number of jittering steps; the gold standard is to compare against MCMC estimates of statistics. We will also investigate whether the required accuracy can be more efficiently reached using Metropolis Adjusted Langevin (MALA) or Hybrid Monte Carlo (HMC) samplers in the jittering steps.

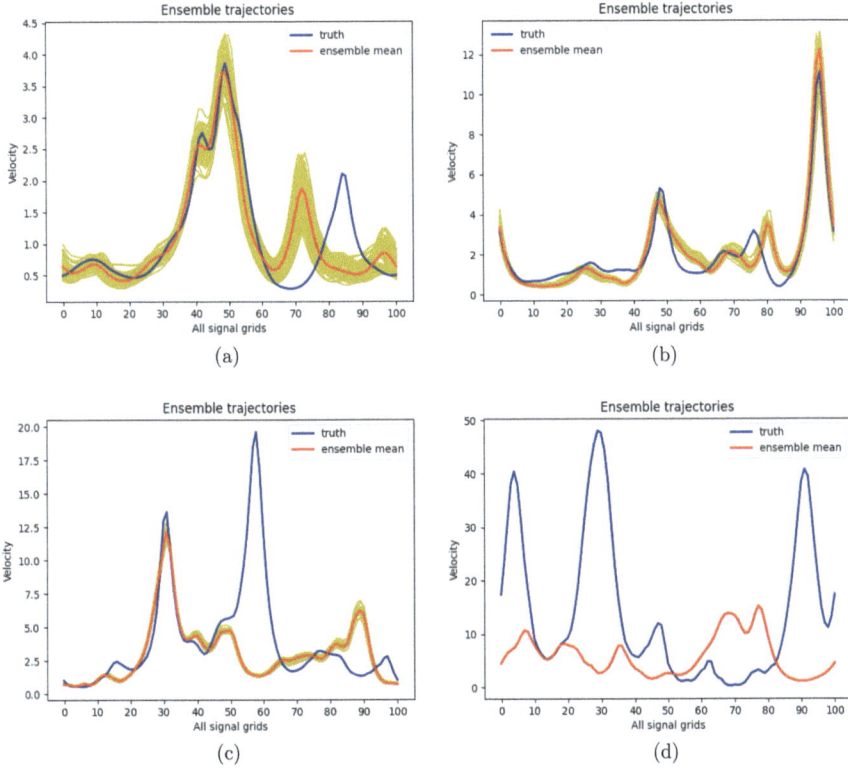

Fig. 9 Comparison of the evolution of the true state vs posterior ensemble and ensemble mean at signal grids. To assimilate data we use a bootstrap particle filter and the outcome is displayed for the specified assimilation steps. We observe that the particle filter has diverged, with a narrow ensemble spread that does not envelope the true solution. (**a**) DA step 100. (**b**) DA step 500. (**c**) DA step 1000. (**d**) DA step 2000

We will investigate the parallel performance of our implementation, and if necessary will develop more sophisticated ordering or subgrouping algorithms for resampling to achieve better parallel scalability.

Once we have established this capability, it is our goal to use data assimilation to calibrate the SALT parametrisation, modifying the Gaussian Matérn field to model the differences between fine grid and coarse grid simulations, to facilitate faster data assimilation approaches.

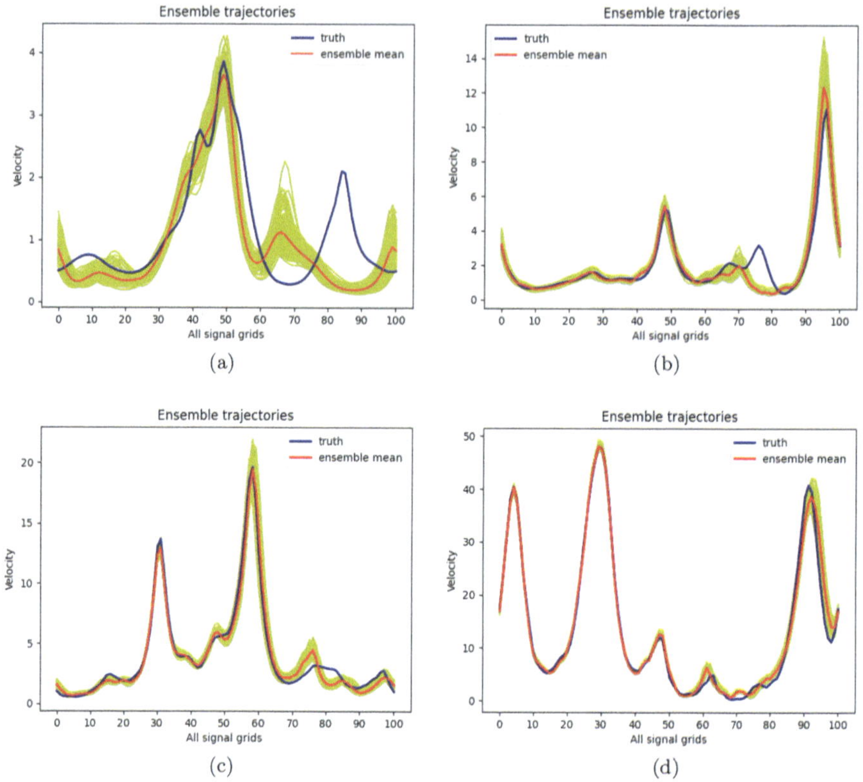

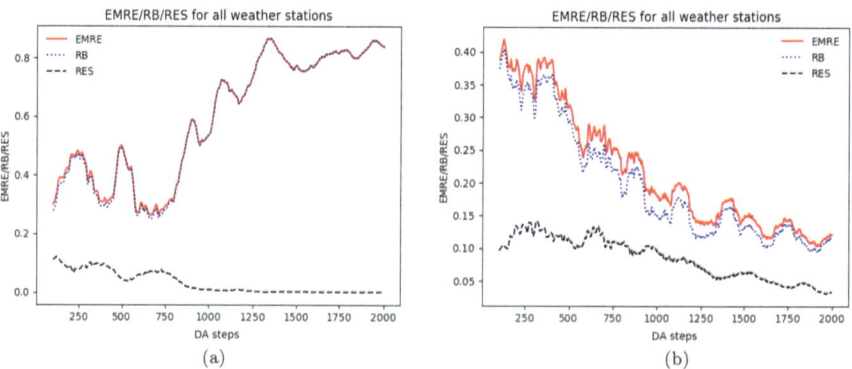

Fig. 10 Comparison of the evolution of the true state vs posterior ensemble and ensemble mean at signal grids. To assimilate data we use *tempering* and *jittering* and the outcome is displayed for the specified assimilation steps. We observe that the spread of the ensemble does not include the true solution in the early stages. However, the filter is stable and later the spread of the ensemble does include the true solution. (**a**) DA step 100. (**b**) DA step 500. (**c**) DA step 1000. (**d**) DA step 2000

Fig. 11 Evolution of the relative ensemble mean error (EMRE), relative bias (RB) and ensemble spread (RES) for all signal grid points for the partially observed case. We observe that the bootstrap filter diverges, while the tempering and jittering filter is stable. (**a**) Bootstrap filter. (**b**) *tempering* and *jittering*

Table 2 Time-averaged
EMRE, RB and RES
associated with Experiment 2

Time-averaged	*BS particle filter*	*tempering* and *jittering*
EMRE	0.5923	0.2198
RB	0.5873	0.1997
RES	0.0327	0.0840

References

1. M. Ades and P. J. Van Leeuwen. The equivalent-weights particle filter in a high-dimensional system. *Quarterly Journal of the Royal Meteorological Society*, 141(687):484–503, 2015.
2. M. Asch, M. Bocquet, and M. Nodet. *Data assimilation: methods, algorithms, and applications*. SIAM, 2016.
3. A. Bain and D. Crisan. *Fundamentals of stochastic filtering*, volume 3. Springer, 2009.
4. T. M. Bendall, C. J. Cotter, and D. D. Holm. Perspectives on the formation of peakons in the stochastic Camassa–Holm equation. *Proceedings of the Royal Society A*, 477(2250):20210224, 2021.
5. A. Beskos, D. Crisan, and A. Jasra. On the stability of sequential Monte Carlo methods in high dimensions. *The Annals of Applied Probability*, 24(4):1396–1445, 2014.
6. R. Camassa and D. D. Holm. An integrable shallow water equation with peaked solitons. *Physical review letters*, 71(11):1661, 1993.
7. A. Chorin, M. Morzfeld, and X. Tu. Implicit particle filters for data assimilation. *Communications in Applied Mathematics and Computational Science*, 5(2):221–240, 2010.
8. C. J. Cotter, D. Crisan, D. D. Holm, W. Pan, and I. Shevchenko. Data assimilation for a quasi-geostrophic model with circulation-preserving stochastic transport noise. *Journal of Statistical Physics*, 179(5–6):1186–1221, 2020.
9. C. J. Cotter, D. Crisan, D. D. Holm, W. Pan, and I. Shevchenko. A particle filter for stochastic advection by Lie transport: A case study for the damped and forced incompressible two-dimensional Euler equation. *SIAM/ASA Journal on Uncertainty Quantification*, 8(4):1446–1492, 2020.
10. C. J. Cotter, G. A. Gottwald, and D. D. Holm. Stochastic partial differential fluid equations as a diffusive limit of deterministic lagrangian multi-time dynamics. *Proceedings of the Royal Society A: Mathematical, Physical and Engineering Sciences*, 473(2205):20170388, 2017.
11. C. J. Cotter and M.K. Singh. The "nudging" particle filter library, 2024. https://github.com/colinjcotter/nudging.
12. S. L. Cotter, G. O. Roberts, A. M. Stuart, and D. White. MCMC Methods for Functions: Modifying Old Algorithms to Make Them Faster. *Statistical Science*, 28(3):424–446, 2013.
13. D. Crisan and D. D. Holm. Wave breaking for the stochastic Camassa–Holm equation. *Physica D: Nonlinear Phenomena*, 376:138–143, 2018.
14. M. Croci, M. B. Giles, M. E. Rognes, and P. E. Farrell. Efficient white noise sampling and coupling for multilevel Monte Carlo with nonnested meshes. *SIAM/ASA Journal on Uncertainty Quantification*, 6(4):1630–1655, 2018.
15. Ham D. A. et al. *Firedrake User Manual*. Imperial College London and University of Oxford and Baylor University and University of Washington, first edition, 5 2023.
16. P. E. Farrell, D. A. Ham, S. W. Funke, and M. E. Rognes. Automated derivation of the adjoint of high-level transient finite element programs. *SIAM Journal on Scientific Computing*, 35(4):C369–C393, 2013.
17. N. J. Gordon, D. J. Salmond, and A. F. M. Smith. Novel approach to nonlinear/non-Gaussian Bayesian state estimation. In *IEE proceedings F (radar and signal processing)*, volume 140(2), pages 107–113. IET, 1993.
18. D. D. Holm. Variational principles for stochastic fluid dynamics. *Proceedings of the Royal Society A: Mathematical, Physical and Engineering Sciences*, 471(2176):20140963, 2015.

19. A. Jasra, D. A. Stephens, A. Doucet, and T. Tsagaris. Inference for Lévy-driven stochastic volatility models via adaptive sequential monte carlo. *Scandinavian Journal of Statistics*, 38(1):1–22, 2011.
20. N. Kantas, A. Beskos, and A. Jasra. Sequential monte carlo methods for high-dimensional inverse problems: A case study for the navier–stokes equations. *SIAM/ASA Journal on Uncertainty Quantification*, 2(1):464–489, 2014.
21. O. Lang, P.-J. Van Leeuwen, D. Crisan, and R. Potthast. Bayesian inference for fluid dynamics: a case study for the stochastic rotating shallow water model. *Frontiers in Applied Mathematics and Statistics*, 8:949354, 2022.
22. L. M. Murray, A. Lee, and P. E. Jacob. Parallel resampling in the particle filter. *Journal of Computational and Graphical Statistics*, 25(3):789–805, 2016.
23. F. Pons Llopis, N. Kantas, A. Beskos, and A. Jasra. Particle filtering for stochastic Navier–Stokes signal observed with linear additive noise. *SIAM Journal on Scientific Computing*, 40(3):A1544–A1565, 2018.
24. S. Reich and C. Cotter. *Probabilistic forecasting and Bayesian data assimilation*. Cambridge University Press, 2015.
25. H. Ruzayqat, A. Er-Raiy, A. Beskos, D. Crisan, A. Jasra, and N. Kantas. A lagged particle filter for stable filtering of certain high-dimensional state-space models. *SIAM/ASA Journal on Uncertainty Quantification*, 10(3):1130–1161, 2022.
26. P. J. Van Leeuwen, H. R. Künsch, L. Nerger, R. Potthast, and S. Reich. Particle filters for high-dimensional geoscience applications: A review. *Quarterly Journal of the Royal Meteorological Society*, 145(723):2335–2365, 2019.
27. C. Vergé, C. Dubarry, P. Del Moral, and E. Moulines. On parallel implementation of sequential Monte Carlo methods: the island particle model. *Statistics and Computing*, 25(2):243–260, 2015.
28. M. Zhu, P. J. Van Leeuwen, and J. Amezcua. Implicit equal-weights particle filter. *Quarterly Journal of the Royal Meteorological Society*, 142(698):1904–1919, 2016.

Some Properties of a Non-hydrostatic Stochastic Oceanic Primitive Equations Model

Arnaud Debussche, Etienne Mémin, and Antoine Moneyron

In this chapter, we study how relaxing the classical hydrostatic balance hypothesis affects theoretical aspects of the LU primitive equations well-posedness. We focus on models that sit between incompressible 3D LU Navier-Stokes equations and standard LU primitive equations, aiming for numerical manageability while capturing non-hydrostatic phenomena. Our main result concerns the well-posedness of a specific stochastic interpretation of the LU primitive equations. This holds with rigid-lid type boundary conditions, and when the horizontal component of noise is independent of z, see [1, 2]. In fact these conditions can be related to the dynamical regime in which the primitive equations remain valid. Moreover, under these conditions, we show that the LU primitive equations solution tends toward the one of the deterministic primitive equations for a vanishing noise, thus providing a physical coherence to the LU stochastic model.

1 Introduction

Stochastic modelling for large-scale fluid dynamics is essential but challenging due to the intrinsic complexity of geophysical flows, as they are chaotic systems with fully developed turbulence. These features induce computational limitations, which classically require to use physical approximations. However, in the last years, stochastic modelling has emerged as a powerful setting for deriving suitable representations [4, 18, 19, 27], allowing greater variability than deterministic large-

A. Debussche
Univ Rennes, CNRS, IRMAR UMR 6625, Rennes, France

E. Mémin · A. Moneyron (✉)
Univ Rennes, INRIA, IRMAR UMR 6625, Rennes, France
e-mail: antoine.moneyron@inria.fr

© The Author(s) 2025   161
B. Chapron et al. (eds.), *Stochastic Transport in Upper Ocean Dynamics III*,
Mathematics of Planet Earth 13, https://doi.org/10.1007/978-3-031-70660-8_8

scale representation. These models aim for plausible forecasts together with efficient uncertainty quantification.

Specifically, the location uncertainty approach (LU) has been developed in the past decade for deriving reliable stochastic models, using stochastic principles for mass, momentum, and energy conservation [30, 38]. It is applied successfully in geophysical and reduced-order models, and LU versions of fluid flow dynamics models have shown promising properties—on classical geophysical models [3, 32–34], stochastic reduced order models [35–37] and large eddy simulation models [7, 8, 24]. The LU formalism is based on transport noises, which are intensively investigated by the mathematical community [1, 2, 5, 10, 13, 14, 16, 17, 22, 26, 31].

Furthermore, the deterministic primitive equations—which assume hydrostatic equilibrium—are standard yet limited [39], especially in representing important phenomena for climate such as deep convection. Remarkably, this model is known to be well-posed with rigid-lid boundary conditions [6]. Moreover, a recent work showed that a stochastic representation of primitive equations with transport noise is well-posed under a "deterministic-like" hydrostatic hypothesis, using water world type boundary conditions [1]. However, the authors assumed that the horizontal noise is independent of the vertical axis, which makes the barotropic and baroclinic noises tractable. In this chapter, we explore weaker hydrostatic assumptions in the LU representation of the primitive equations to model non-hydrostatic phenomena more accurately.

Our chapter investigates the well-posedness of these LU primitive equations under modified hydrostatic balance, offering potential models bridging stochastic primitive equations and 3D Navier-Stokes equations. Our main result concerns the global solutions and continuity under specific noise assumptions, aiming to better represent non-hydrostatic phenomena.

The chapter is organised as follows: first we detail the assumptions made in the LU framework, then we define the function spaces to derive the abstract mathematical problem. Finally, we state our results on existence and uniqueness of solutions for the model we proposed. In particular, this model admits global pathwise solutions under regular enough conditions, and with an additional noise structure assumption. We only give a sketch of proof for our main result; a more complete one will be submitted subsequently.

2 Oceans Dynamics Models in the LU Framework

The LU formulation separates the flow displacement into large-scale dynamics and highly oscillating unresolved motion, as follows,

$$dX_t = u(X_t, t)dt + \sigma(X_t, t)dW_t. \tag{2.1}$$

Here, X denotes the (3D) *Lagrangian* displacement, defined in a bounded cylindrical domain $\mathscr{S} = \mathscr{S}_H \times [0, -h] \subset \mathbb{R}^3$, where $\mathscr{S}_H$ is a subset of $\mathbb{R}^2$ with smooth boundary. In this formalism, $u : \mathscr{S} \times [0, T] \rightarrow \mathbb{R}^3$ is the *Eulerian* velocity of the fluid flow. The large-scale velocity $u(X_t, t)$ correlates in both space and time, while the unresolved small-scale velocity $\sigma(X_t, t)dW$ is uncorrelated-in-time yet correlated-in-space. We further refer to this second component as a noise term, which must be interpreted in the Itô sense.

Let us define this noise term more precisely: consider a cylindrical Wiener process W on the space of square integrable functions $\mathscr{W} := L^2(\mathscr{S}, \mathbb{R}^3)$. Thus, there exists a Hilbert orthonormal basis $(e_i)_{i \in \mathbb{N}}$ of $\mathscr{W}$ and a sequence of independent standard Brownian motions $(\hat{\beta}^i)_{i \in \mathbb{N}}$ on a filtered probability space $(\Omega, \mathscr{F}, (\mathscr{F}_t)_t, \mathbb{P})$ such that,

$$W = \sum_{i \in \mathbb{N}} \hat{\beta}^i e_i.$$

Note that the sum $\sum_{i \in \mathbb{N}} \hat{\beta}^i e_i$ does *not* converge *a priori* in $\mathscr{W}$. Hence we interpret the previous identity in a space $\mathscr{U}$ including $\mathscr{W}$, such that the embedding $\mathscr{W} \hookrightarrow \mathscr{U}$ is Hilbert-Schmidt. Typically, $\mathscr{U}$ can be the dual space of any reproducing kernel Hilbert subspace of $\mathscr{W}$ for the inner product $(\cdot, \cdot)_{\mathscr{W}}$, that is for instance $H^{-s}(\mathscr{S}, \mathbb{R}^3)$ with $s > \frac{3}{2}$.

Then, we define the noise through a deterministic time-dependent correlation operator σ_t: let $\hat{\sigma} : [0, T] \rightarrow L^2(\mathscr{S}^2, \mathbb{R}^3)$ a bounded symmetric kernel, and define

$$(\sigma_t f)(x) = \int_{\mathscr{S}} \hat{\sigma}(x, y, t) f(y) dy, \quad \forall f \in \mathscr{W}.$$

With this definition, σ_t is a Hilbert-Schmidt operator mapping $\mathscr{W}$ into itself, so that the noise can be defined as

$$\sigma_t W_t = \sum_{i \in \mathbb{N}} \hat{\beta}_t^i \sigma_t e_i,$$

where the previous series converges in the sense of $L^2(\Omega, \mathscr{W})$. Here we interpret $\mathscr{W}$ as the space carrying the process $\sigma_t W_t$, while the notation $L^2(\mathscr{S}, \mathbb{R}^3)$ is kept for denoting the space of tridimensional velocities. Importantly, $\sigma_t W_t$ is an abuse of notation as σ_t is an operator is defined on $\mathscr{W}$ while W_t does not converge on $\mathscr{W}$—that is to say "$\sigma_t(W_t)$" *is not* properly defined.

Moreover, we consider a family of eigenfunctions $(\phi_k)_k$ of the operator σ_t, which we scale by their corresponding eigenvalues, and such that $(\phi_k)_k$ is a Hilbert basis of $\mathscr{W}$. By writing the previous series in terms of $(\phi_k)_k$, it can be shown that there exists a sequence of independent standard Brownian motions $(\beta_t^k)_k$, defined on the previously introduced filtered space $(\Omega, \mathscr{F}, (\mathscr{F}_t)_t, \mathbb{P})$, so that

$$\sigma_t W_t = \sum_{k \in \mathbb{N}} \beta_t^k \phi_k(t).$$

As such, $(\Omega, \mathcal{F}, (\mathcal{F}_t)_t, \mathbb{P}, W)$ is a stochastic basis. In addition, the previous series converges in $\mathcal{W}$ almost surely, and in the sense of $L^p(\Omega, \mathcal{W})$ for all $p \in \mathbb{N}$ [11, 20].

Furthermore, we may define the variance tensor as follows, which is the diagonal part of the covariance tensor,

$$a(x, t) = \int_{\mathcal{S}} \hat{\sigma}(x, y, t)\hat{\sigma}(y, x, t)dy = \sum_{k=0}^{\infty} \phi_k(x, t)\phi_k(x, t)^T. \qquad (2.2)$$

Let us note that, in full generality—that is when $\hat{\sigma}_t$ is itself a random function—the operator-valued process σ_t is subject to an integrability condition

$$\mathbb{P}\left(\int_0^T \|a(\cdot, t)\|_{L^2(\mathcal{S}, \mathbb{R}^{3\times3})}dt < \infty\right) = 1,$$

where $\|\cdot\|_{L^2(\mathcal{S}, \mathbb{R}^{3\times3})}$ is the Hilbert norm associated to $L^2(\mathcal{S}, \mathbb{R}^{3\times3})$, the matrix space $\mathbb{R}^{3\times3}$ being equipped with the Frobenius norm. As such, the integral $\int_0^t \sigma_s dW_s$ is a $\mathcal{W}$-valued Gaussian process with expectation zero and bounded variance: $\mathbb{E}[\|\int_0^t \sigma_s dW_s\|_{L^2}^2] < \infty$. The quadratic variation of $\int_0^t (\sigma_t dW_s)(x)$ is given by the finite variation process $\int_0^t a(x, s)ds$.

In a similar way as deriving the classical Navier-Stokes equations, the LU Navier-Stokes equations emerge through a stochastic version of the Reynolds transport theorem (SRTT) [30]. Let q be a random scalar, within a volume $\mathcal{V}(t)$ transported by the flow. Then, for incompressible unresolved flows—that is $\nabla \cdot \sigma_t = 0$—the SRTT reads

$$d\left(\int_{\mathcal{V}(t)} q(x, t)dx\right) = \int_{\mathcal{V}(t)} \left(\mathbb{D}_t q + q\nabla \cdot (u - u_s)dt\right)dx, \qquad (2.3)$$

$$\mathbb{D}_t q = d_t q + (u - u_s)\cdot \nabla q \, dt + \sigma dW_t \cdot \nabla q - \frac{1}{2}\nabla \cdot (a\nabla q)dt, \qquad (2.4)$$

involving an additional drift $u_s = \frac{1}{2}\nabla \cdot a$, coined as the Itō-Stokes drift in [3]. Consequently, u_s is a vector field from $\mathcal{S} \times [0, T]$ to $\mathbb{R}^3$, alike any trajectory of u. In addition, $d_t q(x, t) = q(x, t + dt) - q(x, t)$ is the forward time increment at a fixed spatial point x, and $\mathbb{D}_t q$ is a stochastic transport operator introduced in [30, 32], playing the role of the material derivative. The Itō-Stokes drift is directly related to the divergence of the variance tensor a, representing the effects of noise inhomogeneity on large-scale dynamics. Such advection terms are often added in large-scale ocean dynamics models —under the name of bollus velocity— to take account for surface waves and Langmuir turbulence [9, 23, 29]. As shown in [3], the LU framework holds similar features which accounts for the effects of the small-scale inhomogeneity on the large-scale flow. Additionally, the stochastic transport operator represents physically interpretable terms for large-scale flows representation. Namely, Eq. (2.4) gathers the following four terms

- an evolution term $d_t q$,
- a transport term $(u - u_S) \cdot \nabla q \, dt$, that is the advection of the large-scale quantity q by the large-scale Itō-Stokes drift corrected velocity $u - u_S$,
- a transport term $\sigma d W_t \cdot \nabla q$, that is the advection of q by the unresolved velocity,
- an inhomogeneous diffusion term $-\frac{1}{2} \nabla \cdot (a \nabla q) dt$ representing small-scale mixing effects on q.

Remarkably, the energy associated to the backscattering term $\sigma d W_t \cdot \nabla q$ is exactly compensated by the stochastic diffusion term $-\frac{1}{2} \nabla \cdot (a \nabla q) dt$ [32]. This equilibrium can be interpreted as an instance of a fluctuation-dissipation theorem.

2.1 The LU Primitive Equations

Let us derive the LU primitive equations. Assuming that the flow is isochoric with constant material density, and that the noise is divergence-free alongside with a divergence-free corresponding Itō-Stokes drift, i.e.

$$\nabla \cdot u = \nabla \cdot u_S = 0, \qquad \nabla \cdot \sigma_t d W_t = 0, \tag{2.5}$$

we deduce that $\mathbb{D}_t q = 0$, for any conservative scalar quantity q. In addition, we model density through a linear law of state involving salinity and temperature,

$$\rho = \rho_0 \left(1 + \beta_T (T - T_r) + \beta_S (S - S_r) \right), \tag{2.6}$$

with ρ_0 the reference density of the ocean at a typical temperature T_r and salinity S_r, assuming that the thermodynamic parameters $\beta_T := \frac{1}{\rho_0} \frac{\partial \rho}{\partial T}$ and $\beta_S := \frac{1}{\rho_0} \frac{\partial \rho}{\partial T}$ are constant. Plus, we define three anisotropic diffusion operators that is used further: for $i \in \{v, T, S\}$, provided that the viscosities μ_i, ν_i are given *a priori*, denote

$$A^i = -\mu_i (\partial_{xx} + \partial_{yy}) - \nu_i \partial_{zz}. \tag{2.7}$$

Applying the SRTT to the conservation of momentum principle in rotating frame—see [13] in non rotating frame—we derive the following stochastic equations of motion,

$$\mathbb{D}_t u + f k \times (u \, dt + \sigma d W_t) = -\frac{1}{\rho_0} \nabla (p \, dt + dp_t^\sigma) - A^v (u \, dt + \sigma d W_t). \tag{2.8}$$

Importantly, we have introduced a martingale noise pressure term dp_t^σ in addition to the classical pressure term $p \, dt$, due to stochastic modelling. Applying the SRTT similarly to the conservation of energy and saline mass, we find

$$\mathbb{D}_t T = -A^T T, \quad \mathbb{D}_t S = -A^S S. \tag{2.9}$$

Now, define $u^* = u - u_s$, and note that by assumption u^*, u and u_s are divergence free. Therefore, if we denote by v^*, v and v_s their respective horizontal components, we can express their vertical components via the integro-differential operator

$$w(\cdot) = \int_z^0 \nabla_H \cdot (\cdot), \tag{2.10}$$

under the hypothesis that $w = 0$ when $z = 0$ or $z = -h$. The horizontal gradient operator is denoted by $\nabla_H = (\partial_x \ \partial_y)^T$, and we use further $\Delta_H(\cdot) = \nabla_H \cdot (\nabla_H(\cdot))$ to denote the horizontal Laplace operator. Also, write $\sigma^H dW_t$ and $\sigma^w dW_t$ the horizontal and vertical components of σdW_t, respectively. Thus, the horizontal and vertical momentum equations read

$$\mathbb{D}_t v + \Gamma(v\, dt + \sigma^H dW_t) = -A^v(v\, dt + \sigma^H dW_t) - \frac{1}{\rho_0}\nabla_H(p\, dt + dp_t^\sigma), \tag{2.11}$$

$$\mathbb{D}_t w = -A^v(w\, dt + \sigma^w dW_t) - \frac{1}{\rho_0}\partial_z(p\, dt + dp_t^\sigma) - \frac{\rho}{\rho_0}g\, dt, \tag{2.12}$$

where $\Gamma((a \ b)^T) = f(-b \ a)^T$ is the horizontal projection of the Coriolis term.

A recent study—see article [1]—derived stochastic primitive equations using a "deterministic-like" hydrostatic hypothesis: assuming that the vertical acceleration is negligible compared to the gravity term, the vertical momentum equation boils down to

$$\partial_z p + \rho g = 0, \quad \text{and} \quad \partial_z dp_t^\sigma = 0. \tag{2.13}$$

We further call this assumption the *strong hydrostatic hypothesis*. The validity of this hydrostatic balance corresponds to a regime with small ratio between the squared aspect ratio $\alpha^2 = h^2/L^2$—with h and L denoting vertical and horizontal scales, respectively—and the Richardson number $Ri = \frac{N^2}{(\partial_z v)^2} \sim N^2 h^2/U^2$, defined from the ratio between the stratification given by the Brunt-Väsäilä frequency $N^2 = -\frac{g}{\rho_0}\partial_z \rho$ and the squared vertical shear of the horizontal velocity [28]. In the stochastic setting, the strong hydrostatic balance holds if the noise is small enough not to unbalance this regime.

Gathering all the equations described previously and assuming that the strong hydrostatic hypothesis holds, we obtain the following problem

$$\mathbb{D}_t v + \Gamma(v\, dt + \sigma^H dW_t) = -A^v(v\, dt + \sigma^H dW_t) - \frac{1}{\rho_0}\nabla_H(p\, dt + dp_t^\sigma),$$

$$\mathbb{D}_t T = -A^T T dt, \quad \mathbb{D}_t S = -A^S S dt,$$

$$\nabla_H \cdot v + \partial_z w = 0,$$

$$\partial_z p + \rho g = 0, \quad \partial_z dp_t^\sigma = 0,$$

$$\rho = \rho_0 (1 + \beta_T (T - T_r) + \beta_S (S - S_r)).$$

This system is closely aligned with the deterministic primitive system, since the stochastic transport operator is interpreted as a material derivative. The aforementioned model studied in [1] corresponds to the model above, with the stochastic diffusion term $\frac{1}{2}\nabla \cdot (a\nabla(\cdot))$ being replaced by $\nu_\sigma \Delta(\cdot)$, where $\nu_\sigma > 0$ is a constant. Well-posedness results were established assuming that the initial condition is smooth enough, with periodic horizontal boundary conditions and rigid-lid type vertical boundary conditions—termed *water world* or *aqua planet*. Their key results highlighted local-in-time well-posedness, and global-in-time well-posedness when the horizontal component of the noise is barotropic, i.e. independent of the z-coordinate. We are aiming to extend these well-posedness proofs to more general models, introducing different assumptions about the vertical momentum equation, which we term *weak hydrostatic hypotheses*. These assumptions accommodate relaxed hydrostatic equilibria, touching regimes at the edge of the deterministic hydrostatic assumption validity. This is expected to capture better non-hydrostatic phenomena (like wind or buoyancy-driven turbulence), and deep oceanic convection, influenced by strong noise disrupting the strong hydrostatic regime. We further present one possible interpretation of the *weak hydrostatic hypothesis*.

To derive our new model, we neglect only the large-scale contribution of the vertical acceleration in the vertical momentum Eq. (2.12),

$$-\frac{1}{\rho_0} \partial_z (p\, dt + dp_t^\sigma) - \frac{\rho}{\rho_0} g\, dt = \mathbb{D}_t w + A^v (w\, dt + \sigma^w dW_t)$$

$$= \underbrace{d_t w + u \cdot \nabla w\, dt + A^v (w\, dt)}_{\approx 0} - u_s \cdot \nabla w dt + \sigma dW_t \cdot \nabla w$$

$$- \frac{1}{2} \nabla \cdot (a\nabla w) dt + A^v (\sigma^w dW_t)$$

$$\approx -u_s \cdot \nabla w dt + \sigma dW_t \cdot \nabla w - \frac{1}{2} \nabla \cdot (a\nabla w) dt + A^v (\sigma dW_t).$$

that is,

$$\frac{1}{\rho_0} \partial_z p = -\frac{\rho}{\rho_0} g + \frac{1}{2} \nabla \cdot (a\nabla w) + u_s \cdot \nabla w dt, \text{ and}$$

$$\frac{1}{\rho_0} \partial_z dp_t^\sigma = -\sigma dW_t \cdot \nabla w - A^v (\sigma^w dW_t). \tag{2.14}$$

Notice that the stochastic advection and diffusion terms have been kept in this interpretation of the hydrostatic hypothesis. In its deterministic version, this hypothesis neglects the whole vertical acceleration term and molecular diffusion term, arguing that they are negligible compared to the gravity term $\frac{\rho}{\rho_0}g$. Transitioning to the stochastic case, we claim the same for the large-scale vertical terms $d_t w + u \cdot \nabla w dt$ and $A^v w dt$. However, due to stochastic modelling, three terms remain: the back scattering noise advection term $\sigma dW_t \cdot \nabla w$, the Itō-Stokes drift advection term $u_s \cdot \nabla w dt$, and the diffusion term $\frac{1}{2}\nabla \cdot (a\nabla w)dt$. Differently from $u \cdot \nabla w dt$, the term $\sigma dW_t \cdot \nabla w$ cannot be neglected compared to $\frac{\rho}{\rho_0}g dt$ since the latter is a pure bounded variation term. In addition, the terms $u_s \cdot \nabla w dt$ and $\frac{1}{2}\nabla \cdot (a\nabla w)dt$ are kept since they depend directly on the noise—if $\Upsilon^{1/2}$ is the noise scaling, then they scale like Υ. This formulation leads to the following problem

$$\mathbb{D}_t v + \Gamma(v\,dt + \sigma^H dW_t) = -A^v(v\,dt + \sigma^H dW_t) - \frac{1}{\rho_0}\nabla_H(p\,dt + dp_t^\sigma),$$

$$\mathbb{D}_t T = -A^T T dt, \quad \mathbb{D}_t S = -A^S S dt,$$

$$\nabla_H \cdot v + \partial_z w = 0,$$

$$\frac{1}{\rho_0}\partial_z p + \frac{\rho}{\rho_0}g = \frac{1}{2}\nabla \cdot (a\nabla w) + u_s \cdot \nabla w, \quad \frac{1}{\rho_0}\partial_z dp_t^\sigma = -\sigma dW_t \cdot \nabla w - A^v(\sigma^w dW_t),$$

$$\rho = \rho_0(1 + \beta_T(T - T_r) + \beta_S(S - S_r)).$$

Considering such a problem, a major difficulty arises: the presence of the transport noise $\sigma dW_t \cdot \nabla w$ in the stochastic pressure term induces the following terms in horizontal velocity dynamics,

$$\nabla_H \int_z^0 \sigma dW_t \cdot \nabla w\,dz', \quad \nabla_H \int_z^0 u_s \cdot \nabla w\,dz'\,dt.$$

We don't know how to make a suitable energy estimate for global pathwise existence with such terms, notably because three derivatives on the horizontal velocity v are involved—since $w(v) = \int_z^0 \nabla_H \cdot v$. To tackle this issue, we propose one way to regularise the previous system.

Weak Low-Pass Filtered Hydrostatic Hypothesis

To enforce greater regularity for the vertical transport noise term, we define a regularising convolution kernel K, and replace Eq. (2.14) by

$$\frac{1}{\rho_0}\partial_z p = -\frac{\rho}{\rho_0}g + \frac{1}{2}\nabla \cdot (a^K \nabla w) + K * [u_s \cdot \nabla w],$$

$$\frac{1}{\rho_0}\partial_z dp_t^\sigma = -K * [\sigma dW_t \cdot \nabla w] - A^v(\sigma^w dW_t). \qquad (2.15)$$

In the previous formulae, we denote by a^K the operator $f \longmapsto \sum_{k=0}^{\infty} \phi_k \mathscr{C}_K \mathscr{C}_K^*$ $(\phi_k^T f)$, and by $\mathscr{C}_K$ the operator $f \longmapsto K * f$. The regularising kernel only affects the vertical transport noise, and not potential vertical additive noises. Also, the stochastic diffusion operator $\frac{1}{2} \nabla \cdot (a^K \nabla(\cdot))$ is chosen to be the covariation correction term associated to $K * [\sigma dW_t \cdot \nabla w]$. We refer to this assumption as the *weak low-pass filtered hydrostatic hypothesis*.

This approach consists in filtering the vertical transport noise and disregarding the vertical acceleration of the resolved component of velocity. The noise terms, alongside the stochastic diffusion term, account for the deviation from a strong hydrostatic equilibrium. By convolving the vertical transport noise with K, we effectively truncate its highest frequencies. This new hypothesis relaxes the strong hydrostatic balance, allowing for more general stochastic pressures and extending the validity of the system dynamical regime beyond the strong hydrostatic case. Comparing this methodology to the one presented in [2], authors introduced a temperature noise impacting the pressure equation, that is a perturbation of *thermodynamic* origin. Unlike theirs, our model involves transport noise of the vertical velocity component, that is a perturbation of *mechanical* origin. This approach retains more terms linked to the vertical velocity w, accounting for the influence of unresolved small-scale velocity—like turbulence or submesoscale components— on the large-scale vertical velocity. Hence, the structure of this problem resembles a more genuine tridimensional problem, justifying the additional regularization through a filtering kernel. Using a filtering kernel is a common practice in defining numerical models for primitive equations. It is also prevalent in establishing well-posedness for specific subgrid models like the Gent-McWilliam model, particularly in the field of mesoscale dynamics, as highlighted in [25]. Thus, assuming that this weak hydrostatic hypothesis holds rather than the strong one, one derives the following problem

$$\mathbb{D}_t v + \Gamma(v \, dt + \sigma^H dW_t) = -A^v(v \, dt + \sigma^H dW_t) - \frac{1}{\rho_0} \nabla_H(p \, dt + dp_t^\sigma),$$

(2.16)

$$\mathbb{D}_t T = -A^T T dt, \tag{2.17}$$

$$\mathbb{D}_t S = -A^S S dt, \tag{2.18}$$

$$\nabla_H \cdot v + \partial_z w = 0, \tag{2.19}$$

$$\frac{1}{\rho_0} \partial_z p + \frac{\rho}{\rho_0} g = \frac{1}{2} \nabla \cdot (a^K \nabla w) + K * [u_s \cdot \nabla w], \tag{2.20}$$

$$\frac{1}{\rho_0} \partial_z dp_t^\sigma + K * [\sigma dW_t \cdot \nabla w] + A^v(\sigma^w dW_t) = 0, \tag{2.21}$$

$$\rho = \rho_0 \Big(1 + \beta_T(T - T_r) + \beta_S(S - S_r) \Big). \tag{2.22}$$

In addition, decompose the boundary as $\partial S = \Gamma_u \cup \Gamma_b \cup \Gamma_l$—respectively the upper, bottom and lateral boundaries—and equip this problem with the following rigid-lid type boundary conditions [5, 12]

$$\partial_z v = 0, \quad w = 0, \quad v_T \partial_z T + \alpha_T T = 0, \quad \partial_z S = 0 \quad \text{on } \Gamma_u, \quad (2.23)$$

$$v = 0, \quad w = 0, \quad \partial_z T = 0, \quad \partial_z S = 0 \quad \text{on } \Gamma_b,$$

$$v = 0, \quad \partial_{n_H} T = 0, \quad \partial_{n_H} S = 0 \quad \text{on } \Gamma_l,$$

and initial conditions

$$v(t = 0) = v_0 \in H^1(\mathscr{S}, \mathbb{R}^2), \, T(t = 0) = T_0 \in H^1(\mathscr{S}, \mathbb{R}),$$

$$S(t = 0) = S_0 \in H^1(\mathscr{S}, \mathbb{R}), \quad (2.24)$$

fulfilling the previous boundary conditions. We further assume that the noise, its variance tensor and their gradients cancel on the horizontal boundary Γ_l. Also, we assume that the vertical noise $\sigma_z dW_t$ cancels on the vertical upper and bottom boundaries $\Gamma_u \cup \Gamma_b$. Furthermore, the noise is assumed to be regular enough in the following sense,

$$\sup_{t \in [0,T]} \sum_{k=0}^{\infty} \|\phi_k\|^2_{H^4(\mathscr{S}, \mathbb{R}^3)} < \infty, \quad u_S \in L^\infty\Big([0, T], H^4(\mathscr{S}, \mathbb{R}^3)\Big), \quad (2.25)$$

$$d_t u_S \in L^\infty\Big([0, T], H^3(\mathscr{S}, \mathbb{R}^3)\Big), \quad a \nabla u_S \in L^\infty\Big([0, T], H^2(\mathscr{S}, \mathbb{R}^3)\Big).$$

and

$$a \in H^1\Big([0, T], H^3(\mathscr{S}, \mathbb{R}^{3 \times 3})\Big). \quad (2.26)$$

These regularity assumptions are not limiting in practice because most ocean models consider spatially smooth noises [38], since these are the physically observed ones. Also, remark that the weak low-pass filtered hydrostatic hypothesis adds a stochastic contribution to the pressure terms compared to the strong hypothesis— see Eqs. (2.20) and (2.21)—

$$\frac{1}{\rho_0} \partial_z p + \frac{\rho}{\rho_0} g = \frac{1}{2} \nabla \cdot (a^K \nabla w) + K * [u_S \cdot \nabla w],$$

$$\text{and} \quad \frac{1}{\rho_0} \partial_z dp_t^\sigma + K * [\sigma dW_t \cdot \nabla w] + A^v(\sigma^w dW_t) = 0.$$

Therefore, this impacts the horizontal momentum equation via the horizontal pressure gradients, so one has

$$\frac{1}{\rho_0}\nabla_H p = -\nabla_H \int_z^0 \frac{\rho}{\rho_0} g \, dz' + \nabla_H \int_z^0 \frac{1}{2}\nabla \cdot (a^K \nabla w)$$

$$+ K * [u_s \cdot \nabla w] \, dz' + \frac{1}{\rho_0}\nabla_H p^s$$

$$=: -\nabla_H \int_z^0 \frac{\rho}{\rho_0} g \, dz' + \frac{1}{\rho_0}\nabla_H p^{weak} + \frac{1}{\rho_0}\nabla_H p^s,$$

$$\frac{1}{\rho_0}\nabla_H dp_t^\sigma = -\nabla_H \int_z^0 K * [\sigma dW_t \cdot \nabla w] - A^v(\sigma^w dW_t) \, dz' + \frac{1}{\rho_0}\nabla_H dp_t^{\sigma,s}$$

$$=: \frac{1}{\rho_0}\nabla_H dp_t^{\sigma,weak} + \frac{1}{\rho_0}\nabla_H dp_t^{\sigma,s},$$

where p^{weak} and $p^{\sigma,weak}$ denote respectively the bounded variation and martingale pressures associated to the weakening of the hydrostatic hypothesis. Introducing new additive noise terms does not pose a challenge for a sufficiently regular σdW_t. However, the key difficulty arises from the presence of the term $\nabla_H \int_z^0 \sigma dW_t \cdot \nabla w \, dz'$ in the expression of $\frac{1}{\rho_0}\nabla_H dp_t^\sigma$. This term represents the horizontal impact of the vertical transport noise $\sigma dW_t \cdot \nabla w$. Regularising this term using a smoothing filter inherently enlarges its spatial scale, causing the spatial scale of the vertical transport noise to remain above the resolution cutoff scale without aliasing artifacts. Likewise, $\nabla_H \int_z^0 \nabla \cdot (a \nabla w) \, dz'$ represents the horizontal influence of the covariation correction originating from the LU Navier-Stokes equations. Establishing the well-posedness of such a model, subject to appropriate regularisation and structural conditions, stands as our primary outcome.

2.2 Definition of the Spaces

In this subsection, we define the function spaces that are used further. Remind that the spatial domain is denoted by $\mathscr{S} = \mathscr{S}_H \times [-h, 0] \subset \mathbb{R}^3$. First, we introduce the following inner products

$$(v, v^\sharp)_{H_1} = (v, v^\sharp)_{L^2(\mathscr{S},\mathbb{R}^2)}, \quad (v, v^\sharp)_{V_1} = (\nabla v, \nabla v^\sharp)_{L^2(\mathscr{S},\mathbb{R}^2)}$$

$$(T, T^\sharp)_{H_2} = (T, T^\sharp)_{L^2(\mathscr{S},\mathbb{R})}, \quad (T, T^\sharp)_{V_2} = (\nabla T, \nabla T^\sharp)_{L^2} + \frac{\alpha_T}{v_T}(T, T^\sharp)_{L^2(\Gamma_u,\mathbb{R})},$$

$$(S, S^\sharp)_{H_3} = (S, S^\sharp)_{L^2(\mathscr{S},\mathbb{R})}, \quad (S, S^\sharp)_{V_3} = (\nabla S, \nabla S^\sharp)_{L^2(\mathscr{S},\mathbb{R})},$$

and let

$$(U, U^\sharp)_H = (v, v^\sharp)_{H_1} + (T, T^\sharp)_{H_2} + (S, S^\sharp)_{H_3}, \quad (U, U^\sharp)_V = (v, v^\sharp)_{V_1}$$

$$+ (T, T^\sharp)_{V_2} + (S, S^\sharp)_{V_3}, \tag{2.27}$$

for all $U, U^\sharp \in L^2(\mathscr{S}, \mathbb{R}^4)$. Also we denote by $\|\cdot\|_H, \|\cdot\|_{H_i}$ and $\|\cdot\|_V, \|\cdot\|_{V_i}$ the associated norms. With a slight abuse of notation, we may write $\|\cdot\|_H, \|\cdot\|_V$ in place of $\|\cdot\|_{H_i}, \|\cdot\|_{V_i}$ respectively, and similarly use $(\cdot, \cdot)_H, (\cdot, \cdot)_V$ rather than $(\cdot, \cdot)_{H_i}, (\cdot, \cdot)_{V_i}$. Moreover, notice that we distinguished the inner products $(\cdot, \cdot)_{H_2}$ and $(\cdot, \cdot)_{H_3}$, even if they denote the same operation. This is for consistency with the following definitions of the spaces H_2 and H_3.

Then, denote by $\mathscr{V}_1$ the space of functions $C^\infty(\mathscr{S}, \mathbb{R}^2)$ with a compact support strictly included in $\mathscr{S}$, such that for all $v \in \mathscr{V}_1$, $\nabla_H \cdot \int_{-h}^0 v = 0$. Plus, define $\mathscr{V}_2 = \mathscr{V}_3$ the space of functions $C^\infty(\mathscr{S}, \mathbb{R})$. Denote by H_i the closure of $\mathscr{V}_i$ for the norm $\|.\|_{H_i}$, and V_i its closure by $\|.\|_{V_i}$. Eventually, define $H = H_1 \times H_2 \times H_3$ and $V = V_1 \times V_2 \times V_3$, which are also the closures of $\mathscr{V}_1 \times \mathscr{V}_2 \times \mathscr{V}_3$ by $\|.\|_H$ and $\|.\|_V$, respectively. Often, by abuse of notation, we write $(\cdot, \cdot)_H$ instead of $(\cdot, \cdot)_{V' \times V}$. More generally, if K is a subspace of H and K' its dual space, we write $(\cdot, \cdot)_H$ instead of $(\cdot, \cdot)_{K' \times K}$. Moreover, we define $\mathscr{D}(A) = V \cap H^2(\mathscr{S}, \mathbb{R}^4)$, where $A = (A^v A^T A^S)^T$ gathers the three diffusion operators. As such, $A : \mathscr{D}(A) \to H$ is an unbounded operator. In addition, for any Hilbert spaces $\mathscr{H}_1$ and $\mathscr{H}_2$, we define $\mathscr{L}_2(\mathscr{H}_1, \mathscr{H}_2)$ the space of Hilbert-Schmidt operators from $\mathscr{H}_1$ to $\mathscr{H}_2$, and $\|\cdot\|_{\mathscr{L}_2(\mathscr{H}_1, \mathscr{H}_2)}$ its associated norm.

Again, we distinguished the spaces H_2 and H_3, even though they formally denote the same space. However, V_2 and V_3 *are* different spaces since they are not equipped with the same inner products due to different boundary conditions on the temperature and salinity —Robin and Neumann respectively. This distinction allows to interpret H_2 and V_2 as temperature spaces, and H_3 and V_3 as salinity spaces. In addition, H_1 and V_1 are interpreted as horizontal velocity spaces —since their elements are $\mathbb{R}^2$-valued processes. Using this formalism, the vertical velocity w is written as a functional of the horizontal velocity $v \in H_1$ through the continuity equation, namely $w(v) = \int_z^0 \nabla_H \cdot v \, dz'$.

Eventually, we define the barotropic and baroclinic projectors $\mathscr{A}_2 : \mathbb{R}^3 \to \mathbb{R}^2$, $\mathscr{A} : \mathbb{R}^3 \to \mathbb{R}^3$ and $\mathscr{R} : \mathbb{R}^3 \to \mathbb{R}^3$ of the velocity component as follows. For $v \in \mathscr{V}_1$, h being the depth of the ocean, let

$$\mathscr{A}_2[v](x, y) = \frac{1}{h} \int_{-h}^0 v(x, y, z') dz', \quad \mathscr{A}[v](x, y, z) = \mathscr{A}_2[v](x, y),$$

$$\mathscr{R}[v] = v - \mathscr{A}[v]. \tag{2.28}$$

Remark that $\mathscr{A}$ and $\mathscr{R}$ are orthogonal projectors with respect to the inner product $(\cdot, \cdot)_H$. To simplify notations, we may use $\bar{v}$ in place of $\mathscr{A}_2[v]$ or $\mathscr{A}[v]$, and $\tilde{v}$ in place of $\mathscr{R}[v]$.

2.3 Abstract Formulation of the Problems

In this section, we aim to express, in abstract form, the previous problem under the weak low-pass filtered hydrostatic hypothesis (2.14). First, define the 4D vector U, representing the state of the system, and the correction U^* of U by the Itō-Stokes drift, as

$$U = (v, T, S)^T, \quad U^* = (v^*, T, S)^T.$$

Then, denote the advection operator by

$$B(U^*, \cdot) = B(v^*, \cdot) := (v^* \cdot \nabla_H)(\cdot) + w(v^*)\partial_z(\cdot).$$

Moreover, define two Leray type projectors $\mathbf{P}^v$ and $\mathbf{P}$ as follows:

$$\mathbf{P}^v(v) = \mathbf{P}_{2D}\mathscr{A}(v) + \mathscr{R}(v), \quad \mathbf{P}(U) = (\mathbf{P}^v(v), T, S)^T, \tag{2.29}$$

where $\mathbf{P}_{2D}$ is the standard 2D Leray projector, which is associated to the barotropic component $\mathscr{A}(v)$. Notice that the baroclinic component $\mathscr{R}(v)$ is left unchanged by the projector $\mathbf{P}^v$, that is $\mathbf{P}$ only affects the barotropic component of velocity. For notational convenience we keep the same notations for the composition of the following operators with the Leray projector:

$$AU = \mathbf{P}(A^v v, A^T T, A^S S)^T, \quad CU$$
$$= \mathbf{P}(Cv, 0, 0)^T,$$
$$B(U^*, U) = \mathbf{P}(B(v^*, v), B(v^*, T), B(v^*, S))^T. \tag{2.30}$$

Thus, if $U \in \mathscr{D}(A)$, then $AU \in H$, and we obtain the relation

$$d_t U + \left[AU + B(U^*, U) + CU + \frac{1}{\rho_0}\mathbf{P}\begin{pmatrix} \nabla_H p \\ 0 \\ 0 \end{pmatrix} \right.$$

$$\left. - \frac{1}{2}\mathbf{P}\nabla\cdot(a\nabla U) \right] dt = -\mathbf{P}(\sigma dW_t \cdot \nabla U)$$

$$- (A + \Gamma)(\sigma^H dW_t) - \frac{1}{\rho_0}\mathbf{P}\begin{pmatrix} \nabla_H dp_t^\sigma \\ 0 \\ 0 \end{pmatrix}. \tag{2.31}$$

Now write the problem in terms of U^* with a change of variable, to get

$$d_t U^* + [AU^* + B(U^*) + CU^* + \frac{1}{\rho_0} \mathbf{P} \begin{pmatrix} \nabla_H p \\ 0 \\ 0 \end{pmatrix}$$

$$+ F_\sigma(U^*)]dt = G_\sigma(U^*)dW_t - \frac{1}{\rho_0} \mathbf{P} \begin{pmatrix} \nabla_H dp_t^\sigma \\ 0 \\ 0 \end{pmatrix}, \qquad (2.32)$$

where the operators F_σ and G_σ are defined as

$$F_\sigma(U^*)dt = \mathbf{P}\Big[d_t U_S + [B(U^*, U_S) - \frac{1}{2}\nabla\cdot(a\nabla U_S)]dt + A U_S$$

$$+ \Gamma U_S - \frac{1}{2}\nabla\cdot(a\nabla U^*)dt\Big],$$

$$G_\sigma(U^*)dW_t = \mathbf{P}\Big[-(\sigma dW_t \cdot \nabla)U^* - (\sigma dW_t \cdot \nabla)U_S$$

$$- A(\sigma^H dW_t) - \Gamma(\sigma^H dW_t)\Big],$$

with $(v_s, w_s)^T = u_S = \frac{1}{2}(\nabla\cdot a)$, $U_S = (v_s, 0, 0)^T$. Remind that u_S is divergence free, so $w_s = w(v_s)$ from the definition of operator $w(v)$ —that is Eq. (2.10). In addition, we can derive the following relations for the pressure terms, using Eqs. (2.20) and (2.21),

$$\frac{1}{\rho_0}\nabla_H p = -g\,\nabla_H \int_z^0 (\beta_T T + \beta_S S)dz'$$

$$+ \nabla_H \int_z^0 \Big[K * [u_S \cdot \nabla(w(v^*) + w_s)]$$

$$- \frac{1}{2}\nabla\cdot(a^K \nabla(w(v^*) + w_s))\Big]dz'$$

$$-\frac{1}{\rho_0}\nabla_H(dp_t^\sigma) = \nabla_H \int_z^0 \Big[K * [\sigma dW_t \cdot \nabla(w(v^*) + w_s)]$$

$$+ A(\sigma^w dW_t)\Big]dz' + \frac{1}{\rho_0}\nabla_H dp_t^{\sigma,s}.$$

The quantities $p^s\,dt$ and $dp_t^{\sigma,s}$ are respectively the bounded variation and the martingale contributions to the surface pressure. As they are independent on the z-axis, we have for all $v^\sharp \in V_1$,

$$\left(\frac{1}{\rho_0}\mathbf{P}^v\nabla_H p^s, v^\sharp\right)_H = -\left(\frac{1}{\rho_0}p^s, \nabla_H\cdot\mathscr{A}v^\sharp\right)_H = 0,$$

using the boundary conditions on $\mathscr{A}v^\sharp$, which are $\mathscr{A}v^\sharp = 0$ on $\mathscr{S}_H$ and $\mathscr{A}v^\sharp\cdot\mathbf{n} = \frac{\partial}{\partial\mathbf{n}}\mathscr{A}v^\sharp\times\mathbf{n} = 0$ on $\partial\mathscr{S}_H$. This shows that $\mathbf{P}^v\nabla_H p^s = 0$. Similarly, we get $\mathbf{P}^v\nabla_H dp_t^{\sigma,s} = 0$. Therefore, we get the following relation,

$$\frac{1}{\rho_0}\mathbf{P}^v[\nabla_H p] = \mathbf{P}^v\left[-g\nabla_H\int_z^0(\beta_T T + \beta_S S)dz'\right]$$

$$+ \mathbf{P}^v\left[\nabla_H\int_z^0\left[K*[u_s\cdot\nabla(w(v^*)+w_s)]\right.\right.$$

$$\left.\left.-\frac{1}{2}\nabla\cdot(a^K\nabla(w(v^*)+w_s))\right]dz'\right],$$

$$-\frac{1}{\rho_0}\mathbf{P}^v[\nabla_H(dp_t^\sigma)] = \mathbf{P}^v\left[\nabla_H\int_z^0\left[K*[\sigma dW_t\cdot\nabla(w(v^*)+w_s)]\right.\right.$$

$$\left.\left.+ A(\sigma^w dW_t)\right]dz'\right].$$

On the one hand, the bounded variation surface pressure p^s corresponds to a Lagrange multiplier associated with the constraint $\nabla_H\cdot\mathscr{A}v = 0$. This reminds of Cao and Titi's proof in the deterministic setting [6], where they showed that, up to some coupling terms, the barotropic mode follows a 2D Navier-Stokes dynamics while the baroclinic mode follows a 3D Burgers dynamics. Crucially, the bounded variation surface pressure does not influence the baroclinic dynamics, given the divergence-free nature of the barotropic mode under vertical boundary conditions. On the other hand, the martingale surface pressure $p^{s,\sigma}$ emerges from our proposed stochastic modelling, acting as a perturbation to the pressure p^s. This additional term $p^{s,\sigma}$ may impact both the barotropic and baroclinic dynamics. However, under the strong hydrostatic hypothesis, the martingale pressure equation simplifies to $\partial_z dp_t^\sigma = 0$ from equation (2.13), meaning $dp_t^\sigma = dp_t^{s,\sigma}$. Hence, the martingale pressure term solely affects the barotropic dynamics in this scenario, akin to the deterministic case, enabling the use of analogous methodologies. Our derivations lead to the following formulation of the filtered problem.

Low-Pass Filtered Problem ($\mathscr{P}_K$)

For any $K\in H^3(\mathscr{S},\mathbb{R})$, we define ($\mathscr{P}_K$), the abstract primitive equations problem with weak low-pass filtered hydrostatic hypothesis, as follows,

$$\begin{cases} d_t U^* + [AU^* + B(U^*) + CU^* + \frac{1}{\rho_0}P\nabla_H p + F_\sigma(U^*)]dt = G_\sigma(U^*)dW_t \\ \quad - \frac{1}{\rho_0}P\nabla_H dp_t^\sigma, \\ - \frac{1}{\rho_0}P\nabla_H(dp_t^\sigma) = \mathbf{P}\left[\nabla_H \int_z^0 \left[K * [\sigma dW_t \cdot \nabla(w(v^*) + w_s)] + A(\sigma^w dW_t)\right]dz'\right], \\ \frac{1}{\rho_0}P\nabla_H p = \mathbf{P}\left[\nabla_H \int_z^0 K * \left[u_s \cdot \nabla(w(v^*) + w_s)\right]dz' \\ \quad - \nabla_H \int_z^0 \frac{1}{2}\nabla \cdot (a^K \nabla(w(v^*) + w_s))dz' \\ \quad - g\,\nabla_H \int_z^0 (\beta_T T + \beta_S S)dz'\right], \end{cases}$$

under the condition $\int_{-h}^0 \nabla_H \cdot v^* = 0$. The problem is supplemented with the boundary conditions (2.23) and the initial conditions (2.24). As mentioned before, we assume that the noise, its variance tensor and their gradients cancel on the horizontal boundary Γ_l, and that the vertical noise $\sigma_z dW_t$ cancels on the vertical upper and bottom boundaries $\Gamma_u \cup \Gamma_b$. Moreover, the noise is assumed to follow the regularity conditions (2.25) and (2.26).

3 Main Results

Our main results concern the well-posedness of the weak low-pass filtered problem $(\mathscr{P}_K)$,

Theorem 3.1 *Suppose $K \in H^3(\mathscr{S}, \mathbb{R})$. Then, the following propositions hold,*

1. *The problem $(\mathscr{P}_K)$ admits at least one global-in-time martingale solution, for all $T > 0$, in the space*

$$L^2\left(\Omega, L^2([0, T], V)\right) \cap L^2\left(\Omega, L^\infty([0, T], H)\right).$$

2. *There exists a stopping time $\tau > 0$ such that $(\mathscr{P}_K)$ admits a local-in-time pathwise solution U^*, which fulfils, for all $T > 0$ and for all stopping time $0 < \tau' < \tau$,*

$$U_{\tau' \wedge \cdot}^* \in L^2\left(\Omega, L^2([0, T], \mathscr{D}(A))\right) \cap L^2\left(\Omega, C([0, T], V)\right).$$

This solution is unique up to indistinguishability, that is for all solutions U^* and $\hat{U}^*$ of $(\mathscr{P}_K)$ associated to the stopping times $\tau, \tilde{\tau}$ respectively, the following holds,

$$\mathbb{P}\left(\sup_{[0,T]} \|U_{\tau' \wedge \cdot}^* - \hat{U}_{\tau' \wedge \cdot}^*\|_H^2 = 0; \quad \forall T > 0\right) = 1,$$

for all stopping time $0 < \tau' < \tau \wedge \hat{\tau}$.

Barotropic Horizontal Noise Assumption

In addition, we propose the following assumption on the noise structure, which we refer to as the *barotropic horizontal noise assumption (BHN)*.

$\sigma_H dW_t$ is independent of the variable z, i.e. the horizontal noise is constant over the z axis.

Such assumption is used to demonstrate our results of global pathwise well-posedness, and continuity with respect to the initial data and the noise data. Namely,

Theorem 3.2 *Assume that (BHN) holds, and choose $K \in H^{15/4}(\mathscr{S}, \mathbb{R})$. Then, the problem $(\mathscr{P}_K)$ admits a global-in-time pathwise solution, which is unique up to indistinguishability, in the space*

$$L^2_{loc}\Big([0, +\infty), \mathscr{D}(A)\Big) \cap C\Big([0, +\infty), V\Big).$$

This solution is continuous in the following sense: fix $T > 0$, then

- *for a fixed noise data, U^* is Lispchitz with respect to the initial data in the sense of*

$$V \rightarrow L^2\Big(\Omega, L^2([0, T], \mathscr{D}(A))\Big) \cap L^2\Big(\Omega, C([0, T], V)\Big),$$

- *define Σ, a space of noise operators, as follows,*

$$\Sigma = \Big\{\sigma \in \mathscr{L}_2\Big(\mathscr{W}, H^4(\mathscr{S}, \mathbb{R}^3)\Big) \,\Big|\, a \in H^1\Big([0, T], H^3(\mathscr{S}, \mathbb{R}^{3\times3})\Big)\Big\};$$

then U^* is locally Lipschitz with respect to both the initial data and the noise in the sense of

$$V \times \Sigma \rightarrow L^2\Big(\Omega, L^2([0, T], \mathscr{D}(A))\Big) \cap L^2\Big(\Omega, C([0, T], V)\Big).$$

In particular, for a fixed initial data, if we denote by U^σ the solution of $(\mathscr{P}_K)$ associated to the noise data $\sigma \in \mathscr{N}$, then

$$U^{\Upsilon^{1/2}\sigma} \xrightarrow[\Upsilon \to 0]{} U^0.$$

Here, U^0 denotes the solution to the problem $(\mathscr{P}_K)$ with noise zero—i.e. $\sigma = 0$—that is to say the classical deterministic primitive equations.

We only give a sketch of the proof for Theorem 3.2 below. Theorem 3.1 can be obtained by adapting the work proposed in [13] to the primitive equations, which can be considered as a simplification of the 3D Navier-Stokes equations. It relies on considering a Galerkin approximation of the problem, then using energy estimates and tightness arguments to show that these solutions converge toward a solution of the initial problem, see [5, 12, 13, 15].

Fix $T > 0$. Remind that the barotropic and baroclinic modes of velocity are defined in equation (2.28). Assuming that Theorem 1 hold, we show that the following estimates hold for any stopping times $0 < \eta < \zeta < T$, there exist three constants $C_1, C_2, C_3 > 0$ such that:

- Barotropic velocity estimate in "$H^1 - H^2$"

$$
\mathbb{E}\Big[\|\bar{v}\|_V^2(\zeta) \Big] + \mathbb{E}\Big[\int_\eta^\zeta \|A\bar{v}\|_H^2 \, ds \Big] \leq C_1 \mathbb{E}\Big[\|\bar{v}(\eta)\|_V^2 + 1 \Big]
$$
$$
+ C_1 \mathbb{E}\Big[\int_\eta^\zeta \|U\|_H^2 \|U\|_V^2 \|\bar{v}\|_V^2 + \|U\|_V^2 + \int_{\mathscr{S}} |\tilde{v}|^2 |\nabla_3 \tilde{v}|^2 + \|\partial_z v\|_V^2 ds \Big],
$$

$$(3.1)$$

- Vertical gradient of velocity estimate in "$L^2 - H^1$"

$$
\mathbb{E}\Big[\sup_{[\eta,\zeta]} \|\partial_z v\|_H^2 \Big] + \mathbb{E}\Big[\int_\eta^\zeta \|\partial_z v\|_V^2 \, ds \Big] \leq C_2 \mathbb{E}\Big[\|\partial_z v(\eta)\|_H^2 + 1 \Big]
$$
$$
+ C_2 \mathbb{E}\Big[\int_\eta^\zeta (1 + \|U\|_V^2)(1 + \|\partial_z v\|_H^2) + \int_{\mathscr{S}} |\nabla \tilde{v}|^2 |\tilde{v}|^2 ds \Big].
$$

$$(3.2)$$

- Baroclinic velocity estimate in "$L^4 - W^{1,4}$"

$$
\mathbb{E}\Big[\sup_{[\eta,\zeta]} |\tilde{v}|_{L^4}^4 \Big] + \mathbb{E}\Big[\int_\eta^\zeta \int_{\mathscr{S}} (|\nabla \tilde{v}|^2 |\tilde{v}|^2 + |\nabla |\tilde{v}|^2|^2 + |\partial_z \tilde{v}|^2 |\tilde{v}|^2 + |\partial_z |\tilde{v}|^2|^2) d\mathscr{S} ds \Big]
$$
$$
\leq C_3 \mathbb{E}\Big[1 + \|\tilde{v}(\eta)\|_{L^4}^4 + \int_\eta^\zeta (\|U\|_V^2 + 1) \|\tilde{v}\|_{L^4}^4 \Big] + \frac{1}{16(C_2 \vee 1)} \mathbb{E}\Big[\int_\eta^\zeta \|\partial_z v\|_V^2 ds \Big].
$$

$$(3.3)$$

Once gathered, these estimates lead to the existence of a global pathwise solution to the problem $(\mathscr{P}_K)$, using similar arguments as in [1] —in particular the stochastic Grönwall's lemma, see [21]. Moreover, using similar arguments, we may show the uniqueness of the solution, as well as the continuity of the solution with respect to the initial condition and the noise operator σ. These estimates are related to the proof of the well-posedness of the deterministic primitive equations proposed originally in the article [6]. Essentially, the argument is that $\bar{v}$ follows a 2D Navier Stokes equation, while $\tilde{v}$ follows a 3D Burger equation, up to some coupling terms. Therefore, the idea is to estimate $\bar{v}$ in the strong sense ("$H^1 - H^2$") and $\tilde{v}$ in an L^p-space—namely L^4 and $W^{1,4}$. Moreover, a third estimate is needed on the vertical velocity gradient $\partial_z v$ ("$L^2 - H^1$"). In addition, the noise structure constraints (BHN) we applied for ensuring global existence remind of those proposed in [1, 2], particularly $\sigma_H dW_t = (\sigma_H dW_t)(x, y)$ and $\sigma_z dW_t = (\sigma_z dW_t)(x, y, z)$.

However, the scheme of proof in [1] differs from ours: it initially establishes the existence of a *maximal* solution before exploring its globality-in-time. We emphasize that the pathwise solution we derive—using the Galerkin approximation—is *a priori* non-maximal without the assumption (BHN). Notably, our proof accommodates scenarios where the vertical acceleration $\mathbb{D}_t w$ is entirely neglected, which corresponds to choosing $K = 0$ and neglecting additive noise in equation (2.15). Our well-posedness outcome is similar as the one of [1], except that we transitioned from water world to rigid-lid boundary conditions. Nonetheless, assuming that the noise and its gradient cancel on the boundary remains pivotal for the validity of our integration by parts arguments.

A more detailed proof will be submitted subsequently.

Remark

The barotropic horizontal noise assumption (BHN) aligns with the validity domain of the primitive equations in their deterministic form. These equations hold true when the squared aspect ratio $\alpha^2 := (h/L)^2$ is negligible compared to the Richardson number $Ri := \frac{N^2}{(\partial_z v)^2}$ [28]. Here, v denotes horizontal velocity, h the ocean depth, L the horizontal scale (e.g., $\sqrt{|\mathcal{S}_H|}$), and $N^2 = -\frac{g}{\rho_0}\partial_z \rho$ the Brünt-Väisälä frequency. This condition reads

$$\frac{\alpha^2}{Ri} \ll 1 \quad \text{or equivalently} \quad (\partial_z v)^2 \ll \frac{N^2}{\alpha^2}. \tag{3.4}$$

This particularly holds in the limit of small vertical shear of the horizontal component. In such case, the horizontal velocity becomes almost independent of z, which we call "quasi-barotropic". In stochastic flow contexts, the horizontal noise models a small-scale velocity, denoted by $\Upsilon^{1/2}v'$, where $\Upsilon^{1/2}$ is a scaling factor ensuring v and v' share the same order of magnitude. Hence, condition (3.4) can be rewritten as

$$(\partial_z(v + \Upsilon^{1/2}v'))^2 \ll \frac{N^2}{\alpha^2}, \quad \text{or} \quad (\partial_z v)^2, \ \Upsilon(\partial_z v')^2 \ll \frac{N^2}{\alpha^2}. \tag{3.5}$$

Consequently, the LU stochastic primitive equations remain physically valid under condition (3.5), that is when the horizontal noise modeling $\Upsilon^{1/2}v'$ is either sufficiently small ($\Upsilon \to 0$) or quasi-barotropic (($\partial_z v')^2 \to 0$). In this setup, the noise structure hypothesis in Theorem 1.3 is equivalent to $\partial_z v' = 0$. We anticipate that a slight deviation from this assumption—that is to say considering a noise with a non-zero small enough baroclinic mode—would yield similar well-posedness results. This is because the energy stemming from the baroclinic noise—when small enough—is likely to be balanced by the combined molecular and stochastic diffusions. However, considering a large baroclinic noise component seems to be a serious challenge when proving the LU primitive equations well-posedness.

References

1. A. Agresti et al. "The stochastic primitive equations with transport noise and turbulent pressure". In: *Stochastics and Partial Differential Equations: Analysis and Computations* (Oct. 2022). DOI: 10.1007/s40072-022-00277-3. URL: https://doi.org/10.1007/s40072-022-00277-3.
2. A. Agresti et al. *The stochastic primitive equations with non-isothermal turbulent pressure.* 2023. arXiv: 2210.05973 [math.AP].
3. W. Bauer et al. "Deciphering the role of small-scale inhomogeneity on geophysical flow structuration: a stochastic approach". In: *Journal of Physical Oceanography* 50.4 (2020), pp. 983–1003.
4. J. Berner et al. "Stochastic parameterization: Toward a new view of weather and climate models". In: *Bulletin of the American Meteorological Society* 98.3 (2017), pp. 565–588.
5. Z. Brzeźniak and J. Slavík. "Well-posedness of the 3D stochastic primitive equations with multiplicative and transport noise". In: *Journal of Differential Equations* 296 (2021), pp. 617–676. ISSN: 0022-0396. DOI: https://doi.org/10.1016/j.jde.2021.05.049. URL: https://www.sciencedirect.com/science/article/pii/S0022039621003521.
6. C. Cao and E. S. Titi. "Global well-posedness of the three-dimensional viscous primitive equations of large scale ocean and atmosphere dynamics". In: *Annals of Mathematics* (2007), pp. 245–267.
7. P. Chandramouli, E. Mémin, and D. Heitz. "4D large scale variational data assimilation of a turbulent flow with a dynamics error model". In: *Journal of Computational Physics* 412 (2020), p. 109446.
8. P. Chandramouli et al. "Coarse large-eddy simulations in a transitional wake flow with flow models under location uncertainty". In: *Computers & Fluids* 168 (2018), pp. 170–189.
9. A. D. Craik and S. Leibovich. "A rational model for Langmuir circulations". In: *Journal of Fluid Mechanics* 73.3 (1976), pp. 401–426.
10. D. Crisan, F. Flandoli, and D. Holm. "Solution properties of a 3D stochastic Euler fluid equation". In: *Journal of Nonlinear Science* 29.3 (2019), pp. 813–870.
11. G. Da Prato and J. Zabczyk. *Stochastic equations in infinite dimensions.* Cambridge university press, Second Edition, 2014.
12. A. Debussche, N. Glatt-Holtz, and R. Temam. "Local martingale and pathwise solutions for an abstract fluids model". In: *Physica D: Nonlinear Phenomena* 240.14-15 (July 2011), pp. 1123–1144. DOI: 10.1016/j.physd.2011.03.009. URL: https://doi.org/10.1016%2Fj.physd.2011.03.009.
13. A. Debussche, B. Hug, and E. Mémin. "A Consistent Stochastic Large-Scale Representation of the Navier–Stokes Equations". In: *Journal of Mathematical Fluid Mechanics* 25.1 (Jan. 2023), p. 19. DOI: 10.1007/s00021-023-00764-0. URL: https://doi.org/10.1007/s00021-023-00764-0.
14. F. Flandoli, L. Galeati, and D. Luo. "Delayed blow-up by transport noise". In: *Communications in Partial Differential Equations* 46.9 (2021), pp. 1757–1788. DOI: 10.1080/03605302.2021.1893748. eprint: https://doi.org/10.1080/03605302.2021.1893748. URL: https://doi.org/10.1080/03605302.2021.1893748.
15. F. Flandoli and D. Gatarek. "Martingale and stationary solutions for stochastic Navier–Stokes equations". In: *Probability Theory and Related Fields* 102.3 (Sept. 1995), pp. 367–391. ISSN: 1432-2064. DOI: 10.1007/BF01192467. URL: https://doi.org/10.1007/BF01192467.
16. F. Flandoli, M. Gubinelli, and E. Priola. "Well-posedness of the transport equation by stochastic perturbation". In: *Inventiones mathematicae* 180.1 (Apr. 1, 2010), pp. 1–53. DOI: 10.1007/s00222-009-0224-4. URL: https://doi.org/10.1007/s00222-009-0224-4.

17. F. Flandoli and D. Luo. "High mode transport noise improves vorticity blow-up control in 3D Navier–Stokes equations". In: *Probability Theory and Related Fields* 180.1 (June 1, 2021), pp. 309–363. DOI: 10.1007/s00440-021-01037-5. URL: https://doi.org/10.1007/s00440-021-01037-5.

18. C. L. Franzke and T. J. O'Kane. *Nonlinear and stochastic climate dynamics*. Cambridge University Press, 2017.

19. C. L. Franzke et al. "Stochastic climate theory and modeling". In: *Wiley Interdisciplinary Reviews: Climate Change* 6.1 (2015), pp. 63–78.

20. L. Gawarecki and V. Mandrekar. *Stochastic differential equations in infinite dimensions: with applications to stochastic partial differential equations*. Springer Science & Business Media, 2010.

21. N. Glatt-Holtz and M. Ziane. "Strong pathwise solutions of the stochastic Navier–Stokes system". In: *Advances in Differential Equations* 14.5/6 (2009), pp. 567–600. DOI: 10.57262/ade/1355867260. URL: https://doi.org/10.57262/ade/1355867260.

22. D. Goodair, D. Crisan, and O. Lang. "Existence and uniqueness of maximal solutions to SPDEs with applications to viscous fluid equations". In: *Stochastics and Partial Differential Equations: Analysis and Computations* (2023), pp. 1–64.

23. J.-L. Guermond, J. T. Oden, and S. Prudhomme. "Mathematical perspectives on large eddy simulation models for turbulent flows". In: *Journal of Mathematical Fluid Mechanics* 6 (2004), pp. 194–248.

24. S. K. Harouna and E. Mémin. "Stochastic representation of the Reynolds transport theorem: revisiting large-scale modeling". In: *Computers & Fluids* 156 (2017), pp. 456–469.

25. P. Korn and E. S. Titi. "Global Well-Posedness of the Primitive Equations of Large-Scale Ocean Dynamics with the Gent-McWilliams-Redi Eddy Parametrization Model". In: (2023). arXiv: 2304.03242 [math.AP].

26. O. Lang, D. Crisan, and E. Mémin. "Analytical Properties for a Stochastic Rotating Shallow Water Model Under Location Uncertainty". In: *Journal of Mathematical Fluid Mechanics* 25.2 (Feb. 20, 2023), p. 29. DOI: 10.1007/s00021-023-00769-9. URL: https://doi.org/10.1007/s00021-023-00769-9.

27. A. J. Majda, I. Timofeyev, and E. Vanden E.nden. "Models for stochastic climate prediction". In: *Proceedings of the National Academy of Sciences* 96.26 (1999), pp. 14687–14691.

28. J. Marshall et al. "Hydrostatic, quasi-hydrostatic, and nonhydrostatic ocean modeling". In: *Journal of Geophysical Research: Oceans* 102.C3 (1997), pp. 5733–5752. DOI: https://doi.org/10.1029/96JC02776. eprint: https://agupubs.onlinelibrary.wiley.com/doi/pdf/101029/96JC02776. URL: https://agupubs.onlinelibrary.wiley.com/doi/abs/10.1029/96JC02776.

29. J. C. McWilliams, P. P. Sullivan, and C.-H. Moeng. "Langmuir turbulence in the ocean". In: *Journal of Fluid Mechanics* 334 (1997), pp. 1–30.

30. E. Memin. "Fluid flow dynamics under location uncertainty". In: *Geophysical & Astrophysical Fluid Dynamics* 108.2 (2014), pp. 119–146.

31. R. Mikulevicius and B. L. Rozovskii. "Global L_2-solutions of stochastic Navier–Stokes equations". In: *The Annals of Probability* 33.1 (Jan. 1, 2005), pp. 137–176. DOI: 10.1214/009117904000000630. URL: https://doi.org/10.1214/009117904000000630.

32. V. Resseguier, E. Memin, and B. Chapron. "Geophysical flows under location uncertainty, Part I Random transport and general models". In: *Geophysical & Astrophysical Fluid Dynamics* 111.3 (2017), pp. 149–176.

33. V. Resseguier, E. Mémin, and B. Chapron. "Geophysical flows under location uncertainty, part II quasi-geostrophy and efficient ensemble spreading". In: *Geophysical & Astrophysical Fluid Dynamics* 111.3 (2017), pp. 177–208.

34. V. Resseguier, E. Mémin, and B. Chapron. "Geophysical flows under location uncertainty, Part III SQG and frontal dynamics under strong turbulence conditions". In: *Geophysical & Astrophysical Fluid Dynamics* 111.3 (2017), pp. 209–227.

35. V. Resseguier et al. "Stochastic modelling and diffusion modes for proper orthogonal decomposition models and small-scale flow analysis". In: *Journal of Fluid Mechanics* 826 (2017), pp. 888–917.

36. V. Resseguier et al. "Quantifying truncation-related uncertainties in unsteady fluid dynamics reduced order models". In: *SIAM/ASA Journal on Uncertainty Quantification* 9.3 (2021), pp. 1152–1183.
37. G. Tissot, A. V. Cavalieri, and E. Mémin. "Stochastic linear modes in a turbulent channel flow". In: *Journal of Fluid Mechanics* 912 (2021), A51.
38. F. L. Tucciarone, E. Mémin, and L. Li. "Primitive Equations Under Location Uncertainty: Analytical Description and Model Development". In: *Stochastic Transport in Upper Ocean Dynamics*. Ed. by B. Chapron et al. Cham: Springer International Publishing, 2023, pp. 287–300. ISBN: 978-3-031-18988-3.
39. G. K. Vallis. *Atmospheric and oceanic fluid dynamics*. Cambridge University Press, 2017.

Derivation of Stochastic Models for Coastal Waves

Arnaud Debussche, Etienne Mémin, and Antoine Moneyron

In this chapter, we consider a stochastic nonlinear formulation of classical coastal waves models under location uncertainty (LU). In the formal setting investigated here, stochastic versions of the Serre-Green-Naghdi, Boussinesq and classical shallow water wave models are obtained through an asymptotic expansion, which is similar to the one operated in the deterministic setting. However, modified advection terms emerge, together with advection noise terms. These terms are well-known features arising from the LU formalism, based on momentum conservation principle.

1 Introduction

The ocean surface waves constitute an essential component of ocean dynamics, as they are directly related to a strong energy exchange with the underlying current. However, the mean current and the waves follow different dynamics, operating at different times and spatial scales. Moreover, they are based on significantly different physical assumptions: surface waves rely on potential—or irrotational—flow, whereas the turbulent dynamics of the current is expressed through a non-zero vorticity. As a result, the wave-current coupling is very complex to model numerically. For this purpose, designing simplified stochastic models of surface waves would be advantageous to account for both dynamics. This can be achieved by

A. Debussche (✉)
Univ Rennes, CNRS, IRMAR UMR 6625, Rennes, France

E. Mémin · A. Moneyron
Univ Rennes, INRIA, IRMAR UMR 6625, Rennes, France
e-mail: antoine.moneyron@inria.fr

© The Author(s) 2025
B. Chapron et al. (eds.), *Stochastic Transport in Upper Ocean Dynamics III*,
Mathematics of Planet Earth 13, https://doi.org/10.1007/978-3-031-70660-8_9

adding as a specific noise term in the stochastic representation of ocean circulation [21, 35].

To derive such stochastic wave models, ways to adapt the classical deterministic derivation to the stochastic setting have been widely investigated. It was shown for example that deep-ocean stochastic long waves can arise from a linearised stochastic shallow water system [23] or from a stochastic Hamiltonian formulation [7]. Water waves travelling from deep water areas to shallower regions—where the water depth is much less than their wavelength—undergo significant alterations. Swift variations in height, velocity, and direction lead to substantial modifications of the free water surface profiles. Initially resembling almost perfect sine waves, these profiles evolve into an asymmetric shape. Mathematically, coastal waves are described by different dynamical models corresponding to various approximations of the irrotational Euler equations—which are averaged along the water column ultimately.

More specifically, we use the location uncertainty principle developed in [22], which is based on the addition of a noise term on the Lagrangian formulation of the displacement. In fact, this approach is a critical aspect in ocean stochastic modelling and has been widely discussed: an SDE based stochastic generalisation of the deterministic Lagrangian expression of the flow was proposed in [30, 31] for instance, from which the authors derive an Eulerian expression. Such idea also exists in turbulence modelling: for example McWilliam and Berloff proposed stochastic parametrization built upon Langevin models of turbulence in [3], devised in the wake of Kraichnan's seminal work [17].

In this chapter, we aim to derive a location uncertainty interpretation of non-linear coastal waves models, based on the modelling principle introduced in [22]. These models are naturally accompanied by some nonlinear variabilities associated with numerical or modelling uncertainties. More fundamentally, this stochastic formalism provides a way to incorporate the effect of the non-resolved vertical turbulent component. In this work, we focus on the family of models associated with the Serre-Green-Naghdi equations [12, 33], which allow capturing the non-hydrostatic phenomena related to vertical acceleration. In particular, we aim to assess the behavior of the associated numerical simulations. Our main findings are that the LU interpretations of the Saint-Venant, Boussinesq and Serre-Green-Naghdi wave models allow to break the symmetry of the wave and introduce variability compared to their deterministic counterparts. Nevertheless, they do enjoy the same conservation principles with slight conditions on the noise structure. Hence, the LU setting would allow to explore the influence of a "conservative randomness" on the classical wave model dynamics. Plus, we provide some graphical observations of this influence using numerical simulations.

The article is organised as follows: after a brief summary on the location uncertainty principle (LU), we derive a stochastic representation of the Serre-Green-Nagdhi shallow water waves model following the same physical approximations used in the deterministic setting [1]. Then, we show how simpler models can be obtained from further approximations, and discuss the conservation of usual quantities such as mass, momentum and energy. The last section describes and assesses some numerical results associated with different models.

2 Stochastic Flow and Transport

Let $\mathscr{D} \subset \mathbb{R}^d$ (with $d = 2$ or 3) denote a bounded spatial domain. The LU principle relies on a stochastic formulation of the Lagrangian trajectory (X_t) of the form

$$dX_t(x) = v(X_t(x), t)dt + (\sigma_t dB_t)(x), \quad X_0(x) = x \in \mathscr{D}. \tag{2.1}$$

This stochastic differential equation (SDE) decomposes the flow in terms of a smooth- in-time velocity component v that is both spatially and temporally correlated, and a fast unresolved flow component σdB_t (called noise term in the following). This latter component must be understood in the Itō sense, and is uncorrelated in time yet correlated in space. Importantly, since the Itō integral is of null expectation (as a martingale), the relation (2.1) decomposes the flow unequivocally as a large-scale displacement component and a null mean fluctuation, the mean displacement corresponding to the large-scale component expectation.

Let us now provide a precise definition of this random fluctuation component. The noise term takes values in the Hilbert space $H := (L^2(\mathscr{D}))^d$ and is defined from the Wiener process (also called cylindrical Brownian motion) [26] on a stochastic basis $(\Omega, \mathscr{F}_t, (\mathscr{F}_t)_{t \in [0,T]}, \mathbb{P})$. By definition, this stochastic basis is composed of a probability space $(\Omega, \mathbb{P})$ with filtration $(\mathscr{F}_t)_{t \in [0,T]}$—i.e. a non decreasing family of sigma algebras evolving in time. The noise term involves independent Wiener process components $(B_t^i)_{i=1,\dots,d}$ defined on an orthogonal basis $(e_n)_{n \in \mathbb{N}}$ of H as

$$(B_t^i)(x) = \sum_{n \in \mathbb{N}} \beta_t^n e_n^i(x), \tag{2.2}$$

where $(\beta_t^i)_{i \in \mathbb{N}}$ is a sequence of independent one dimensional standard Brownian motions. The spatial structure of the unresolved flow component is modelled by the correlation operator, σ_t, defined as an integral operator on H. Let a matrix kernel $\breve{\sigma} = (\breve{\sigma}_{ij})_{i,j=1,\dots,d}$ that is bounded in space and time, then define

$$\sigma_t f(x) = \int_{\mathscr{D}} \breve{\sigma}(x, y, t) f(y) \, dy, \quad f \in H, \quad x \in \mathscr{D}. \tag{2.3}$$

Here, we have assumed that this operator is deterministic. However, it is important to note that, if required, it could be defined as a random operator. An example of such a random correlation operator can be found in [20] where it is defined from the dynamic mode decomposition technique (DMD) [32] and a Girsanov transform. Girsanov transform enables in particular to introduce a drift term associated to a non-centred noise, through a change of probability measure. This procedure has proven particularly useful for noise calibration in data assimilation [10]. The composition of $\sigma_t[\bullet]$ and its adjoint operator $\sigma_t^*[\bullet]$ defines a compact self-adjoint positive operator, of which eigenfunctions and eigenvalues are denoted $\xi_n(\cdot, t)$ and $\lambda_n(t)$ respectively. These eigenvalues fulfil $\sum_{n \in \mathbb{N}} \lambda_n(t) < +\infty$ and decrease toward zero at infinity, and the eigenfunctions form an orthonormal basis of H. As such, the noise can be equivalently defined on this basis as the following spectral expression,

$$\sigma_t \, d\boldsymbol{B}_t(\boldsymbol{x}) = \sum_{n \in \mathbb{N}} \lambda_n^{1/2}(t) \boldsymbol{\xi}_n(\boldsymbol{x}, t) \, d\beta_t^n. \tag{2.4}$$

Consequently, the noise component is a H-valued Gaussian process of null mean and with bounded variance —that is $\mathbb{E}_{\mathbb{P}}[\int_0^t \sigma_s \, d\boldsymbol{B}_s] = \boldsymbol{0}, \forall t > 0$ and $\mathbb{E}_{\mathbb{P}}[\| \int_0^t \sigma_s \, d\boldsymbol{B}_s \|_H^2] < +\infty, \forall t > 0$—under the probability measure $\mathbb{P}$. Moreover, the auto-covariance at point $\boldsymbol{x} \in \mathscr{D}$ of the unresolved flow component at each instant $t \in [0, T]$ is given by the matrix kernel of the composite operator $\sigma\sigma^*$, and denoted by $\boldsymbol{a}(\boldsymbol{x}, t)$, namely

$$\boldsymbol{a}(\boldsymbol{x}, t) := \int_{\mathscr{D}} \breve{\sigma}(\boldsymbol{x}, \boldsymbol{y}, t) \breve{\sigma}^T(\boldsymbol{y}, \boldsymbol{x}, t) \, d\boldsymbol{y} = \sum_{n \in \mathbb{N}} \lambda_n(t) (\boldsymbol{\xi}_n \boldsymbol{\xi}_n^T)(\boldsymbol{x}, t). \tag{2.5}$$

The process $\int_0^t \boldsymbol{a}(\boldsymbol{x}, s) \, ds$ corresponds to the quadratic variation of $\int_0^t \sigma_s \, d\boldsymbol{B}_s(\boldsymbol{x})$ [2].

The stochastic integral defining the noise could have been defined in terms of a Stratonovich integral instead of an Itō integral [2]. For a deterministic correlation operator, $\boldsymbol{a}$ boils down to the local variance of the noise, due to the Itō isometry [26]. Hence, we referred to $\boldsymbol{a}$ as the variance tensor, although for a random correlation operator this denomination is misleading since it is a random process. Physically, the symmetric non-negative tensor $\boldsymbol{a}$ represents the friction coefficients of the unresolved fluid motions, and is physically homogeneous to a viscosity with unit m^2/s.

Consider an extensive random tracer Θ (e.g. temperature, salinity or buoyancy) transported by the stochastic flow, fulfilling the following conservation property along the trajectories: $\Theta(\boldsymbol{X}_{t+\delta t}, t + \delta t) = \Theta(\boldsymbol{X}_t, t)$ with δt an infinitesimal time variation. Thus, the evolution law of Θ is given by the following stochastic partial differential equation (SPDE),

$$\mathbb{D}_t \Theta = d_t \Theta + (\boldsymbol{v}^* \, dt + \sigma \, d\boldsymbol{B}_t) \cdot \nabla \Theta - \frac{1}{2} \nabla \cdot (\boldsymbol{a} \nabla \Theta) \, dt = 0, \tag{2.6}$$

where $\mathbb{D}_t$ is introduced as a stochastic transport operator and $d_t \Theta(\boldsymbol{x}) := \Theta(\boldsymbol{x}, t + \delta t) - \Theta(\boldsymbol{x}, t)$ stands for the forward time-increment of Θ at a fixed point $\boldsymbol{x} \in \mathscr{D}$. This operator $\mathbb{D}_t$ was derived in [22] using the generalized Itō formula (also called Itō-Wentzell formula in the literature) [18] and plays the role of a transport operator in a stochastic setting. Remarkably, it encompasses physically meaningful terms [2, 28]: the two first terms correspond to the classical terms of the material derivative, while the third term describes the tracer advection by the unresolved flow component. As shown in [2, 28], the resulting (non Gaussian) multiplicative noise $\sigma \, d\boldsymbol{B}_t \cdot \nabla \Theta$ continuously backscatters random energy to the system through the quadratic variation $1/2 (\nabla\Theta) \cdot (\boldsymbol{a}\nabla\Theta)$ of the random tracer. The last term in (2.6) represents the tracer diffusion due to the mixing of the unresolved scales. The energy loss by the diffusion term is exactly balanced with the energy brought by the noise. This pathwise balance (i.e for any realisation) leads to tracer energy conservation and highlights the parallel between the classical material derivative and the stochastic transport operator. This will be precised below.

Under specific noise and/or variance tensor definitions, the resulting diffusion [15, 22] can be connected to the additional eddy viscosity term introduced in many large-scale circulation models [27, 34]. As an additional feature of interest, this evolution law introduces an *effective* advection velocity v^* in (2.6) defined for an incompressible noise term as

$$v^* = v - \frac{1}{2}\nabla \cdot a. \tag{2.7}$$

This statistical eddy-induced velocity drift captures the action of inhomogeneity of the random field on the transported tracer and the possible divergence of the unresolved flow component. It is shown in [2] that the *turbophoresis* term, $v_s = 1/2\nabla \cdot a$, can be interpreted as a generalization of the Stokes drift associated to surface wave current and that it plays a key role in the triggering of secondary circulations such as the Langmuir circulation [6, 19]. Consequently, this velocity is termed Itō-Stokes drift (ISD) in [2]. Notice that, in the modified advection (2.7), the ISD cancels out for homogeneous random fields (since then the variance tensor is constant over space).

Many useful properties of the stochastic transport operator $\mathbb{D}_t$ have been explored by [28, 29]. In particular, if a random tracer is transported by the incompressible stochastic flow under suitable boundary conditions, then the pathwise p-th moment ($p \geq 1$) of the tracer is materially and integrally invariant, namely

$$\mathbb{D}_t\left(\frac{1}{p}\Theta^p\right) = 0, \quad d_t\left(\int_\mathscr{D} \frac{1}{p}\Theta^p \, dx\right) = 0. \tag{2.8}$$

It is worth noting that the transport Eq. (2.6) can be written in terms of Stratonovich integral [2], as

$$D_t\Theta = d_t\Theta + (v^* \, dt + \sigma \circ dB_t) \cdot \nabla \Theta = 0. \tag{2.9}$$

The Stratonovich integral has the advantage of fulfilling a "standard" chain rule, so that the notation D_t for the transport operator similar to the material derivative. Itō calculus introduces second order terms—such as the diffusion term—which becomes implicit in the Stratonovich integral. Because of this, Stratonovich noise is not anymore a martingale and is not of null expectation. Moreover, it is possible to move safely from one integral to the other under some regularity assumptions. In the following we will use the Stratonovich notation.

3   Non Linear Shallow Water Equations Under Location Uncertainty

In this section, we derive a stochastic representation of ocean waves in the near-shore zone, focusing on the derivation of models going from the Shallow water model to the Serre-Green-Nagdi model. Our stochastic derivation is similar to the scaling procedure described in [1]. Our derivation starts from the general 3D Euler equations in the LU setting, which read

$$
\mathrm{d}_t \boldsymbol{v} + \left(\boldsymbol{v} - \frac{1}{2}\, \boldsymbol{\nabla} \cdot \boldsymbol{a}\right) \cdot \boldsymbol{\nabla} \boldsymbol{v}\mathrm{d}t + (\boldsymbol{\sigma}\,\mathrm{od}\boldsymbol{B}_t \cdot \boldsymbol{\nabla})\boldsymbol{v} = -\frac{1}{\rho}\boldsymbol{\nabla}(P\mathrm{d}t + \mathrm{d}h_t^\sigma) - \boldsymbol{g}z\mathrm{d}t,
$$

$$(3.1)$$

$$
\boldsymbol{\nabla} \cdot \left(\boldsymbol{v} - \frac{1}{2}\, \boldsymbol{\nabla} \cdot \boldsymbol{a}\right) = 0,
\qquad\qquad\qquad (3.2)
$$

denoting $\boldsymbol{v} = (\boldsymbol{u}, w)$ the three-dimensional velocity decomposed in terms of horizontal, $\boldsymbol{u}$, and vertical, w, components. The pressure is denoted P, while ρ and $\boldsymbol{g}$ stand for the density and the gravitational vector directed along the vertical direction, respectively.

Shallow water conditions are characterised by a water depth h being much smaller than the wave length scale $L \sim 1/|\boldsymbol{k}|$, where $\boldsymbol{k}$ denotes the wave number. This condition is often expressed through the quantity $\beta = \boldsymbol{k}h_0 \ll 1$, which informs about the predominance of dispersion. In linear theory, the amplitude A of the wave is small and tends to zero when the characteristic scale tends to infinity. Another quantity is usually introduced to measure the non-linear effects: $\epsilon = A/h_0$. Shoaling processes start for $\beta \leq 1$ (i.e. when wavelength and depth have the same order) and ends when the waves break at $\epsilon \geq 1$ (an illustration is provided in Fig. 1, which was borrowed from [16]). Thus, shoaling requires asymptotic models with short wavelength and high wave amplitude. These adimensional numbers are used to build approximated solutions of the shallow water waves system, ranging from weakly nonlinear Boussinesq models (small amplitude regime, $\beta^2 \ll 1$ and $\epsilon = O(\beta^2)$) to the nonlinear Serre-Green-Naghdi system (large amplitude regime). As it will be shown in Sect. 3.1, within the modeling under uncertainty setting the Serre-Green-Naghdi model reads

$$
\overline{D}_t^H \eta = -h\left(\boldsymbol{\nabla}_H \cdot \left(\overline{\boldsymbol{u}} - \frac{1}{2}\Upsilon\epsilon\overline{\boldsymbol{u}}_s\right)\mathrm{d}t + \Upsilon\boldsymbol{\nabla}_H \cdot \overline{\boldsymbol{\sigma}\,\mathrm{od}\boldsymbol{B}_t^H}\right),
\qquad (3.3\mathrm{a})
$$

$$
\overline{D}_t^H \overline{\boldsymbol{u}} + \boldsymbol{\nabla}_H \eta\mathrm{d}t - \frac{1}{h}\epsilon\beta^2\boldsymbol{\nabla}_H\left(\frac{h^3}{3}(\mathrm{d}\overline{G})\right) = \mathcal{O}(\beta^4, \epsilon\beta^4),
\qquad (3.3\mathrm{b})
$$

with

$$
\mathrm{d}\overline{G}(\boldsymbol{x}, t) = \overline{D}_t^H(\boldsymbol{\nabla}_H \cdot \overline{\boldsymbol{u}}) - \epsilon\boldsymbol{\nabla}_H \cdot \left(\left(\overline{\boldsymbol{u}} - \frac{1}{2}\Upsilon\epsilon\overline{\boldsymbol{u}}_s\right)\mathrm{d}t + \Upsilon^{1/2}\overline{\boldsymbol{\sigma}\,\mathrm{od}\boldsymbol{B}_t^H}\right)\boldsymbol{\nabla}_H \cdot \overline{\boldsymbol{u}}.
\quad (3.4)
$$

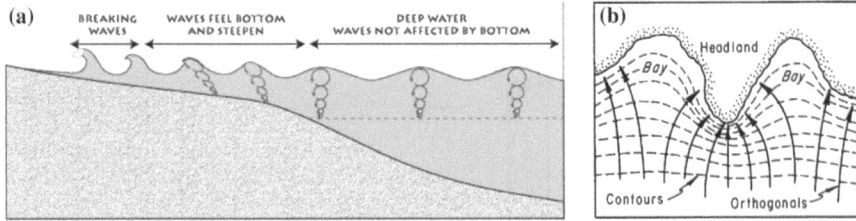

Fig. 1 Illustration of shoaling processes, borrowed from [16]. (**a**) Wave shoaling diagram. (**b**) Wave refraction diagram. Source: USACE coastal engineering Manual. As the waves approach the coasts, both the typical wavelength and the water depth decrease, the former much faster than the latter. Hence, β decreases and ϵ increases as the waves approach the coast. Thus, shoaling processes are characterised by moderately large values of the parameters ϵ and β—that is, away enough from the coast to avoid breaking waves, yet near enough for the bottom topography to influence the wave dynamics

The notation $D_t^H f$ stands for the Stratonovich transport operator with respect to vertically averaged stochastic flow components,

$$\overline{D}_t^H f = \mathrm{d}_t f + \epsilon \, \boldsymbol{\nabla}_H f \cdot \overline{\boldsymbol{u}}^* \, \mathrm{d}t + \Upsilon^{1/2} \epsilon \, \boldsymbol{\nabla}_H f \cdot \overline{\boldsymbol{\sigma} \circ \mathrm{d} \boldsymbol{B}_t^H}, \tag{3.5}$$

with the depth averaged horizontal velocity

$$\overline{\boldsymbol{u}}(x, t) = \frac{1}{h} \int_0^{h(x,t)} \boldsymbol{u}(z) \mathrm{d}z. \tag{3.6}$$

The last left-hand side term of (3.3b) can be understood as a pressure term associated to the vertical velocity component acceleration corrected by compressibility effects. This term is quite intuitively a function of the average horizontal velocity divergence. Overall, the system constitutes a stochastic version of the Serre-Green-Naghdi equations [12, 33]: compared to the original deterministic model, it includes an additional transport noise term and the contribution of the Itō-Stokes drift induced by the inhomogeneity of the small-scale fluctuations. A graphical illustration of the bottom topography, water depth and surface deformation is provided on Fig. 2.

3.1 Deriving the Stochastic Serre-Green-Naghdi System

In this section, we derive a stochastic representation of the Serre-Green-Naghdi model.

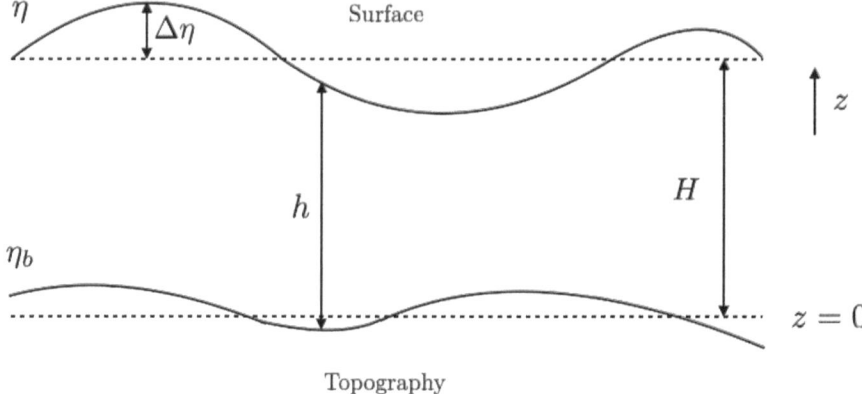

Fig. 2 Illustration of the bottom topography η_b, water depth h and surface deformation η

3.1.1 Evolution Equations

Scaling Relations

First, we proceed to the following adimentionalisation to define the asymptoptic models. Consider $x = L\tilde{x}$, $z = h_0\tilde{z}$, $v = \epsilon\, C_0\tilde{v}$ with $C_0 = \sqrt{gh_0}$ the characteristic long wave velocity. Also, time is scaled as $T \sim L/C_0$. The notation $\tilde{\bullet}$ stands for adimensioned variables. We will further assume the following incompressible assumption for the noise and the ISD,

$$\nabla \cdot \boldsymbol{\sigma}\,\text{od}\boldsymbol{B}_t = 0 \text{ and } \nabla \cdot \boldsymbol{v}_s = 0. \tag{3.7}$$

These assumptions on the noise lead to the global energy conservation—expressed as the sum of the potential and kinetic energies—of the shallow water system [5], as well as the energy conservation of transported scalar [2, 28]. The incompressibility condition on the flow boils down to the classical divergence-free condition

$$\nabla \cdot \boldsymbol{v} = 0. \tag{3.8}$$

This incompressibility condition leads to the scaling $w = \epsilon\beta C_0\tilde{w}$ for the vertical velocity. In addition, we scale the noise term and the variance tensor $\boldsymbol{a}$ by $\Upsilon^{1/2}$ and Υ, respectively. The incompressibility condition on the noise term writes

$$\boldsymbol{\sigma}\text{d}\boldsymbol{B}_t^h = \Upsilon^{1/2}L\widetilde{\boldsymbol{\sigma}\text{d}\boldsymbol{B}}_t^h, \tag{3.9}$$

with a similar scaling on the horizontal noise displacement. This provides a scaling on the vertical noise component as

$$\boldsymbol{\sigma}\text{od}\boldsymbol{B}_t^z = \Upsilon^{1/2}\epsilon\beta L\,\widetilde{\boldsymbol{\sigma}\text{od}\boldsymbol{B}}_t^z. \tag{3.10}$$

The variance tensor has the dimension of a viscosity in m^2/s. For a deterministic correlation tensor, $\boldsymbol{\sigma}$, it is related to the variance of the flow fluctuation, $\boldsymbol{v}' = (\boldsymbol{u}', w')^{\mathsf{T}}$ through

$$\boldsymbol{a} = \mathbb{E}(\boldsymbol{v}'\boldsymbol{v}'\mathsf{T})\tau, \tag{3.11}$$

where τ is a decorrelation time. Now denote $\boldsymbol{a}^{HH}$, $\boldsymbol{a}^{Hz}$ and a^{zz} the horizontal (2D) matrix variance tensor, the horizontal-vertical cross vector, and the vertical variance, respectively. Thus, we get the following scaling relations,

$$\boldsymbol{a}^{HH} = \Upsilon\epsilon^2 C_0 L\, \tilde{\boldsymbol{a}}^{HH} \tag{3.12a}$$

$$\boldsymbol{a}^{Hz} = \Upsilon\epsilon^2 \beta C_0 L\, \tilde{\boldsymbol{a}}^{Hz} \tag{3.12b}$$

$$a^{zz} = \Upsilon\epsilon^2 \beta^2 C_0 L\, \tilde{a}^{zz}. \tag{3.12c}$$

Note that these relations could have been inferred directly from the noise scalings and relation

$$\boldsymbol{a}\mathrm{d}t = \mathbb{E}(\boldsymbol{\sigma}\mathrm{d}\boldsymbol{B}_t(\boldsymbol{\sigma}\mathrm{d}\boldsymbol{B}_t)\mathsf{T}). \tag{3.13}$$

Additionally, the ISD component scales as

$$1/2\,\boldsymbol{\nabla}\cdot\boldsymbol{a} = (\boldsymbol{u}_s, w_s) = \Upsilon\epsilon^2 C_0\big(\nabla_{\tilde{H}}\cdot\tilde{\boldsymbol{a}}^{HH} + \partial_{\tilde{z}}\tilde{\boldsymbol{a}}^{Hz}, \beta\nabla_{\tilde{H}}\cdot\tilde{\boldsymbol{a}}^{Hz} + \beta\partial_{\tilde{z}}\tilde{a}^{zz}\big), \tag{3.14}$$

where $\boldsymbol{\nabla}_H$ refers to the gradient with respect to the horizontal coordinates. As in the classical setting, we will assume that the large-scale component of the flow is irrotational and thus

$$\partial_{\tilde{y}}\tilde{u} = \partial_{\tilde{x}}\tilde{v} \text{ and } \partial_{\tilde{z}}\tilde{\boldsymbol{u}} = \beta^2 \nabla_{\tilde{H}}\cdot\tilde{w}, \tag{3.15}$$

while the adimensional continuity equation is

$$\nabla_{\tilde{H}}\tilde{\boldsymbol{u}} = -\partial_{\tilde{z}}\tilde{w}. \tag{3.16}$$

In the long wave assumption, the pressure is scaled by the static pressure as follows

$$P \sim \rho g h_0 \text{ and thus } P = \rho g h_0 \tilde{P}. \tag{3.17}$$

The same scaling is assumed for the turbulent pressure, with the additional noise variance scaling $\Upsilon^{1/2}$,

$$\mathrm{d}h_t^\sigma = \Upsilon^{1/2}\rho g h_0 \mathrm{d}\tilde{h}_t^\sigma. \tag{3.18}$$

For simplicity, from now on we will drop the tilde accentuation for the adimensional variables in the following.

Incompressibility Condition

Integrating over the z axis, the incompressibility condition (3.2) gives

$$\int_0^{h(x,t)} \nabla_H \cdot (u - \frac{1}{2}\Upsilon \epsilon\, u_s) dz = -\left(w\big(h(x)\big) - \frac{1}{2}\Upsilon \epsilon\, w_s\big(h(x)\big)\right). \qquad (3.19)$$

Using Leibniz formula and introducing the depth averaged horizontal effective velocity $\overline{u^*}$,

$$\overline{u^*}(x, t) = \frac{1}{h}\int_0^{h(x,t)} u^*(z) dz, \qquad (3.20)$$

we have

$$\nabla_H \cdot \int_0^{h(x,t)} u^* dz - u^*(h)\cdot\nabla_H h = \nabla_H \cdot (\overline{u^*}\, h) - u^*(h)\cdot\nabla_H h = -w^*(h). \qquad (3.21)$$

In the same way, we have for the noise term

$$\nabla_H \cdot (h\,\overline{\sigma\,\mathrm{d}B_t^H}) - \nabla_H \cdot \big(h\sigma\mathrm{d}B_t^H(h)\big) = -\sigma\mathrm{d}B_t^z(h). \qquad (3.22)$$

Adimensionned Momentum Equations

Gathering the previous scaled relations, we obtain the following equations for the horizontal and vertical velocity,

$$\epsilon\mathrm{d}_t u + \epsilon^2\nabla_H\cdot\left((u - \frac{1}{2}\Upsilon\epsilon u_s)u\right)\mathrm{d}t + \epsilon^2\partial_z\left((w - \frac{1}{2}\Upsilon\epsilon w_s)u\right)\mathrm{d}t + \Upsilon^{1/2}\epsilon^2\nabla\cdot\left(\sigma\,\mathrm{od}B_t^H\, u\right)$$
$$+ \Upsilon^{1/2}\epsilon^2\partial_z\left(\sigma\,\mathrm{od}B_t^z\, u\right) = -\nabla_H P\mathrm{d}t - \Upsilon^{1/2}\partial_x \mathrm{d}h_t^\sigma, \qquad (3.23)$$

and

$$\epsilon\beta^2\mathrm{d}_t w + \epsilon^2\beta^2\left((u - \frac{1}{2}\Upsilon\epsilon u_s)\cdot\nabla_H w\right) + \epsilon^2\beta^2\left((w - \frac{1}{2}\Upsilon\epsilon w_s)\partial_z w\right)\mathrm{d}t$$
$$+ \epsilon^2\beta^2\Upsilon^{1/2}(\sigma\,\mathrm{od}B_t^H\cdot\nabla_H w)$$
$$+ \epsilon^2\beta^2\Upsilon^{1/2}(\sigma\,\mathrm{od}B_t^z\partial_z w) = -(\partial_z P + 1)\mathrm{d}t - \Upsilon^{1/2}\partial_z \mathrm{d}h_t^\sigma. \qquad (3.24)$$

The vertical momentum equation can be written in a more compact form, that is

$$\epsilon\beta^2 D_t w = -\partial_z \mathrm{d}P - 1\mathrm{d}t, \qquad (3.25)$$

where P represents the total pressure, sum of the finite variation and martingale pressures. In this formula, 1 stands for the rescaled gravity term.

3.1.2 Boundary Conditions

Bottom Boundary Conditions

For the boundary conditions we will assume that $w(x, z, t) = \sigma \text{o} d B_t^z = 0$ on the bottom $z = 0$. Due to this last condition we have also $a^{Hz} = a^{zz} = 0$ on the bottom. The constraint of the vertical ISD from (3.14) implies that $\partial_z a^{zz}(x, 0, t) = 0$ on the bottom. As a consequence,

$$\nabla \cdot a(x, 0, t) = \Upsilon \epsilon^2 C_0 \left(\nabla_H \cdot a^{HH}(x, 0, t) + \partial_z a^{Hz}(x, 0, t), 0 \right).$$

The divergence-free constraint of the ISD leads to following condition on the bottom,

$$\nabla_H \cdot \nabla_H \cdot a^{HH}(x, 0, t) = 0. \tag{3.26}$$

We will also consider that the noise and the ISD have the same characteristics as the large scale velocity near the bottom: due to the large-scale component being irrotational, the following hold in the vicinity of the bottom

$$\partial_y \sigma \text{o} d B_t^H = \partial_x \sigma \text{o} d B_t^H, \quad \beta^2 \nabla_H \sigma \text{o} d B_t^z = \partial_z \sigma \text{o} d B_t^H \text{ at } z = \delta z, \tag{3.27}$$

and

$$\partial_x v_s = \partial_y u_s, \quad \beta^2 \nabla_H w_s = \partial_z u_s \text{ at } z = \delta z. \tag{3.28}$$

Free Surface Boundary Conditions

At the free surface $z = h(x, t)$, we denote by η the surface elevation variation and consider the scaling

$$z = h(x, t) = h_0 \left(1 + \epsilon \eta(x, t) \right),$$

which leads to

$$\tilde{z} = \frac{1}{h_0} h(x, t) = \left(1 + \epsilon \eta(x, t) \right). \tag{3.29}$$

The evolution of the *dimensional* surface elevation is given by

$$(w - w_s) dt + \sigma \text{o} d B_t^z = D_t \eta, \tag{3.30}$$

where an effective vertical velocity driving the dynamics of η appears on the left-hand side. This velocity is composed of the vertical components of the velocity, the ISD, and an additive noise variable linked with rapid vertical oscillations. As usual in long wave approximations, we assume that the whole pressure is constant at the interface, and for sake of simplicity we consider that this constant is null, that is

$$dP_t(\boldsymbol{x}, h(\boldsymbol{x}, t), t) = 0. \tag{3.31}$$

The adimensional form of the free surface evolution (dropping the tilde accentuation) reads,

$$\left(w(h) - \frac{1}{2}\Upsilon\epsilon w_s(h)\right)dt + \Upsilon^{1/2}\,\boldsymbol{\sigma}\circ d\boldsymbol{B}_t^z(h)$$

$$= d_t\eta + \epsilon\boldsymbol{\nabla}_H\eta\cdot(\boldsymbol{u}(h) - \frac{1}{2}\Upsilon\epsilon\boldsymbol{u}_s(h))\,dt + \Upsilon^{1/2}\epsilon\,\boldsymbol{\nabla}_H\eta\cdot\boldsymbol{\sigma}\circ d\boldsymbol{B}_t^H(h). \tag{3.32}$$

Considering the averaged continuity Eq. (3.21) we obtain

$$d_t\eta + \epsilon\,\boldsymbol{\nabla}_H\eta\cdot(\overline{\boldsymbol{u}} - \frac{1}{2}\Upsilon\epsilon\overline{\boldsymbol{u}}_s)\,dt + \Upsilon^{1/2}\epsilon\,\boldsymbol{\nabla}_H\eta\cdot\overline{\boldsymbol{\sigma}\circ d\boldsymbol{B}_t^H}$$

$$= -h\Big(\boldsymbol{\nabla}_H\,\cdot\,(\overline{\boldsymbol{u}} - \frac{1}{2}\Upsilon\epsilon\overline{\boldsymbol{u}}_s)dt + \Upsilon^{1/2}\boldsymbol{\nabla}_H\,\cdot\,\overline{\boldsymbol{\sigma}\circ d\boldsymbol{B}_t^h}\Big), \tag{3.33}$$

which can be written more compactly as

$$\overline{D}_t^H\eta = -h\Big(\boldsymbol{\nabla}_H\,\cdot\,(\overline{\boldsymbol{u}} - \frac{1}{2}\Upsilon\epsilon\overline{\boldsymbol{u}}_s)dt + \Upsilon^{1/2}\boldsymbol{\nabla}_H\,\cdot\,\overline{\boldsymbol{\sigma}\circ d\boldsymbol{B}_t^H}\Big). \tag{3.34}$$

Notice that we have introduced an operator $\overline{D}_t^H$, that involves only depth average velocity components. Moreover, remark that for homogeneous noise (i.e. with no statistical dependence on space location), the variance tensor is constant over space and the ISD cancels out. In such a case, the surface elevation dynamics simplifies in the more intuitive form,

$$d_t\eta + \epsilon\,\boldsymbol{\nabla}_H\eta\cdot\overline{\boldsymbol{u}}\,dt + \Upsilon^{1/2}\epsilon\,\boldsymbol{\nabla}_H\eta\cdot\overline{\boldsymbol{\sigma}\circ d\boldsymbol{B}_t^H} = -h\Big(\boldsymbol{\nabla}_H\cdot\overline{\boldsymbol{u}}dt + \Upsilon^{1/2}\boldsymbol{\nabla}_H\cdot\overline{\boldsymbol{\sigma}\circ d\boldsymbol{B}_t^H}\Big). \tag{3.35}$$

Additionally, in the general case, the integrated ISD can be written in terms of horizontal quantities,

$$h\overline{\boldsymbol{u}}_s = \Big(\int_0^{h(x,t)}\boldsymbol{\nabla}_H\,\cdot\,\boldsymbol{a}^{HH} + \boldsymbol{a}^{Hz}(h)\Big) = \boldsymbol{\nabla}_H\,\cdot\,(h\overline{\boldsymbol{a}}^H) - \boldsymbol{a}^{HH}(h)\boldsymbol{\nabla}_H h + \boldsymbol{a}^{Hz}(h)$$

$$= \boldsymbol{\nabla}_H\,\cdot\,(h\overline{\boldsymbol{a}}^H) - \boldsymbol{a}^{HH}(h)\boldsymbol{\nabla}_H h + \langle\boldsymbol{\sigma}\circ d\boldsymbol{B}_t^z(h), \boldsymbol{\sigma}\circ d\boldsymbol{B}_t^H(h)\rangle$$

$$= \boldsymbol{\nabla}_H\,\cdot\,(h\overline{\boldsymbol{a}}^H) - \boldsymbol{a}^{HH}(h)\boldsymbol{\nabla}_H h - \frac{1}{2}h\boldsymbol{\nabla}_H\,\cdot\,\boldsymbol{a}^{HH}(h). \tag{3.36}$$

3.1.3 The Wave Model

Depth Averaged Horizontal Momentum Equation

By taking the average of the horizontal momentum equation, and using Leibniz rules in time and space for stochastic processes (see Appendix 1), the continuity relation and the elevation evolution together with the velocity boundary conditions, we obtain

$$\epsilon h \, d_t \overline{u} + \epsilon^2 h \left(\overline{u} - \frac{1}{2} \Upsilon \epsilon \, \overline{u}_s \right) \cdot \nabla_H) \overline{u} \, dt + \Upsilon^{1/2} \epsilon^2 h \left(\sigma \circ d B_t^H \cdot \nabla_H \right) \overline{u}$$

$$- h \epsilon^2 \nabla_H \cdot \int_0^{h(x,t)} \left((u - \frac{1}{2} \Upsilon \epsilon u_s) u \, dt + \Upsilon^{1/2} (\sigma \circ d B_t^H u) - (\overline{u} - \frac{1}{2} \Upsilon \epsilon \overline{u}_s) \overline{u} \, dt \right.$$

$$\left. + \Upsilon^{1/2} (\overline{\sigma \circ d B_t^H \overline{u}}) \right) dz = - \int_0^h \nabla_H d P \, dz. \tag{3.37}$$

Now we express below the depth-averaged pressure term involved in the right-hand side of this equation.

Dynamic Pressure Contribution

Decomposing $dP = p \, dt + dh_t^\sigma$, we obtain from Leibniz formula

$$\int_0^{h(x,t)} \nabla_H (dP) dz = \nabla_H (h \overline{dP}) - dP(h) \nabla_H h = \nabla_H (h \overline{dP}). \tag{3.38}$$

Moreover, remind that dP is given by integrating the vertical momentum equation (3.25), i.e

$$- \partial_z \, dP = dt + \epsilon \beta^2 D_t w. \tag{3.39}$$

Thus, integrating (3.38) vertically from depth z up to the surface, one has

$$- p(x, z, t) dt - dh_t^\sigma(x, z, t) = (z - h) dt - \epsilon \beta^2 \int_z^h D_t w(x, z', t) dz', \tag{3.40}$$

which implies

$$- h (\overline{p} dt + \overline{dh_t^\sigma}) = -\frac{1}{2} h^2 dt - \epsilon \beta^2 \int_0^h \left(\int_z^h D_t w(x, z', t) dz' \right) dz. \tag{3.41}$$

In the previous relation, the second right-hand side term can be simplified through integration by part as follows

$$-\epsilon \beta^2 \int_0^h \left(\int_z^h D_t w \, dz' \right) dz = -\epsilon \beta^2 \int_0^h z \, \partial_z \left(\int_0^z D_t w \, dz' - \int_0^h D_t w \, dz' \right) dz$$

$$+ \left[z \left(\int_0^z D_t w dz' - \int_0^h D_t w dz' \right) \right]_0^h = -\epsilon \beta^2 \int_0^h z D_t w dz.$$

Then, the horizontal momentum equation reads

$$d_t \overline{\boldsymbol{u}} + \epsilon \left((\overline{\boldsymbol{u}} - \frac{1}{2} \Upsilon \epsilon \overline{\boldsymbol{u}}_s) \cdot \nabla_H \right) \overline{\boldsymbol{u}} dt + \Upsilon^{1/2} \epsilon (\boldsymbol{\sigma} \operatorname{od} \boldsymbol{B}_t^H \cdot \nabla_H) \overline{b \boldsymbol{u}}$$

$$+ \nabla_H \eta dt + \frac{1}{h} \epsilon \beta^2 \nabla_H \int_0^h z D_t w dz$$

$$= -\frac{\epsilon}{h} \nabla_H \cdot \int_0^h \left[(\boldsymbol{u} - \frac{1}{2} \Upsilon \epsilon \boldsymbol{u}_s) \boldsymbol{u} - (\overline{\boldsymbol{u}} - \frac{1}{2} \Upsilon \epsilon \overline{\boldsymbol{u}}_s) \overline{\boldsymbol{u}} \right] dz$$

$$- \frac{\epsilon}{h} \Upsilon^{1/2} \nabla_H \cdot \int_0^h \left[(\boldsymbol{\sigma} \operatorname{od} \boldsymbol{B}_t^H \boldsymbol{u}) - (\overline{\boldsymbol{\sigma} \operatorname{od} \boldsymbol{B}_t^H \overline{\boldsymbol{u}}}) \right] dz. \qquad (3.42)$$

So far, no approximation has been introduced in the derivation of this equation. In order to compute the different terms of this averaged equation and build simplified systems, we introduce now some asymptotic approximations. In particular, the integral terms on the right-hand side of (3.42) involve both $\boldsymbol{u}$ and its vertical average value $\overline{\boldsymbol{u}}$. This means the system is not closed since one cannot deduce $\boldsymbol{u}$ from $\overline{\boldsymbol{u}}$ solely. Therefore, one needs to investigate reliable approximations of these integral terms, as well as of the vertical acceleration term on the left-hand side of (3.42).

Expansion of the Velocities at the Bottom

Since the velocity potential is harmonic and irrotational, expanding the velocity component through a Taylor expansion at $z = 0$ gives

$$\boldsymbol{u}(x, z, t) = \boldsymbol{u}(x, 0, t) + \partial_z \boldsymbol{u}(x, 0, t) z + \frac{1}{2} \partial_{zz}^2 \boldsymbol{u}(x, 0, t) z^2$$

$$+ \frac{1}{6} \partial_{zzz}^3 \boldsymbol{u}(x, 0, t) z^3 + \mathcal{O}(\beta^4 z^4)$$

$$= \boldsymbol{u}(x, 0, t) + \beta \underbrace{\nabla_H w(x, 0, t)}_{=0} z + \frac{1}{2} \beta \partial_z \nabla_H w(x, 0, t) z^2$$

$$+ \frac{1}{6} \beta \partial_{zz}^2 \nabla_H w(x, 0, t) z^3 + \mathcal{O}(\beta^4 z^4)$$

$$= \boldsymbol{u}(x, 0, t) - \frac{1}{2} \beta^2 \nabla_H \nabla_H \cdot \boldsymbol{u}(x, 0, t) z^2$$

$$- \frac{1}{6} \beta^3 \underbrace{\Delta_H \nabla_H w(x, 0, t)}_{=0} z^3 + \mathcal{O}(\beta^4 z^4). \qquad (3.43)$$

Notice that, due to both harmonicity and irrotationality conditions, the expansion involves only even orders terms. Hence, since z at most of the order of h_0, we obtain

$$u(x, z, t) = u(x, 0, t) - \frac{1}{2}\beta^2 \nabla_H \nabla_H \cdot u(x, 0, t)z^2 + \mathcal{O}(\beta^4). \tag{3.44}$$

Averaging this equation, the bottom horizontal velocity $u^b(x, t) = u(x, 0, t)$ is expressed as

$$u^b(x, t) = \overline{u}(x, t) + \frac{1}{6}\beta^2 h^2 \nabla_H \nabla_H \cdot u^b(x, t) + \mathcal{O}(\beta^4). \tag{3.45}$$

Using the following expansion for u^b in (3.45),

$$u^b = \overline{u}(x, t) + \epsilon x' \cdot \nabla_H \overline{u}, \tag{3.46}$$

we obtain

$$u(x, z, t) = \overline{u}(x, t) + \frac{1}{6}h^2\beta^2 \nabla_H \nabla_H \cdot \overline{u}(x, t) - \frac{1}{2}\beta^2 \nabla_H \nabla_H \cdot \overline{u}(x, t)z^2 + \mathcal{O}(\beta^4, \epsilon\beta^4). \tag{3.47}$$

Remark that this expansion brings an additional error term of order $\mathcal{O}(\epsilon\beta^4)$.

For the ISD, proceeding in the exact same way, and assuming it is irrotational at the bottom, we obtain as well

$$u_s(x, z, t) = u_s(x, 0, t) + \partial_z u_s(x, 0, t)z + \frac{1}{2}\partial_{z^2}^2 u_s(x, 0, t)z^2$$

$$+ \frac{1}{6}\partial_{z^3}^3 u_s(x, 0, t)z^3 + \mathcal{O}(\beta^4 z^4)$$

$$= u_s(x, 0, t) - \frac{1}{2}\beta^2 \nabla_H \nabla_H \cdot u_s(x, 0, t)z^2 + \mathcal{O}(\beta^4), \tag{3.48}$$

where we used irrotationality at the bottom and incompressibility of the ISD. Averaging along depth, and replacing in the above expression the bottom ISD, we get

$$u_s(x, z, t) = \overline{u}_s(x, t) + \mathcal{O}(\beta^2, \epsilon\beta^2). \tag{3.49}$$

Furthermore, for the vertical velocity component w we have

$$w(x, y, t) = \partial_z w(x, 0, t)z + \frac{1}{2}\partial_{z^2}^2 w(x, 0, t)z^2 + \frac{1}{6}\partial_{z^3}^3 w(x, 0, t)z^3 + \mathcal{O}(\beta^4 z^4),$$

$$= -\nabla_H \cdot u(x, 0, t)z - \beta^2 \frac{1}{2} \underbrace{\Delta_H w(x, 0, t)}_{=0} z^2$$

$$+ \beta^2 \frac{1}{6}\Delta_H \nabla_H \cdot u(x, 0, t)z^3 + \mathcal{O}(\beta^4 z^4),$$

$$= -\nabla_H \cdot u(x, 0, t)z + \beta^2 \frac{1}{6}\Delta_H \nabla_H \cdot u(x, 0, t)z^3 + \mathscr{O}(\beta^4 z^4).$$

(3.50)

Making use of the expression of the bottom velocity, we get

$$w(x, y, t) = -\nabla_H \cdot \bar{u}(x, t)z + \mathscr{O}(\beta^2).$$

(3.51)

Similarly, for the vertical component of the ISD, one has

$$w_s(x, y, t) = \partial_z w_s(x, 0, t)z + \frac{1}{2}\partial_{z^2}^2 w_s(x, 0, t)z^2 + \frac{1}{6}\partial_{z^3}^3 w_s(x, 0, t)z^3 + \mathscr{O}(\beta^4 z^4),$$

$$= -\nabla_H \cdot u_s(x, 0, t)z + \beta^2 \frac{1}{6}\Delta_H \nabla_H \cdot u_s(x, 0, t)z^3 + \mathscr{O}(\beta^4 z^4),$$

(3.52)

then

$$w_s(x, y, t) = -\nabla_H \cdot \bar{u}_s(x, t)z + \mathscr{O}(\beta^2).$$

(3.53)

For the noise term, following the same procedure, the horizontal component is expressed as

$$\sigma \circ dB_t^H(x, z, t) = \overline{\sigma \circ dB_t^H}(x, t) + \frac{1}{6}\beta^2 h^2 \nabla_H \nabla_H \cdot \overline{\sigma \circ dB_t^H}(x, t)$$

$$- \frac{1}{2}\beta^2 \nabla_H \nabla_H \cdot \overline{\sigma \circ dB_t^H}(x, t)z^2 + \mathscr{O}(\beta^4, \epsilon\beta^4).$$

(3.54)

and the vertical one as

$$\sigma \circ dB_t^z(x, y, t) = -\nabla_H \cdot \overline{\sigma \circ dB_t^H}(x, t)z + \mathscr{O}(\beta^2).$$

(3.55)

From these formulae, we can approximate the variance tensor components a^H, a^{Hz} and a^{zz}:

$$a^H dt = \mathbb{E}(\sigma dB_t^H \otimes \sigma dB_t^H) = \mathbb{E}(\overline{\sigma dB_t^H} \otimes \overline{\sigma dB_t^H}) + \mathscr{O}(\beta^2, \epsilon\beta^2)$$

$$= \bar{a}^H dt + \mathscr{O}(\beta^2, \epsilon\beta^2),$$

(3.56a)

$$a^{Hz} dt = \mathbb{E}(\sigma dB_t^H \sigma dB_t^z) = -\beta\mathbb{E}(\overline{\sigma dB_t^H} \nabla_H \cdot \overline{\sigma dB_t^H})z + \mathscr{O}(\beta^2, \epsilon\beta^2)$$

$$= -\frac{1}{2}z \nabla_H \cdot \bar{a}^H dt + \mathscr{O}(\beta^2, \epsilon\beta^2),$$

(3.56b)

$$a^{zz}dt = \mathbb{E}(\sigma dB_t^z \, \sigma dB_t^z) = \mathbb{E}(\nabla_H \cdot \overline{\sigma dB_t^H} \, \nabla_H \cdot \overline{\sigma \operatorname{od}B_t^H}) \, z^2 + \mathscr{O}(\beta^4, \epsilon\beta^4)$$

$$= z^2 \, \overline{a}^{\operatorname{div}} dt + \mathscr{O}(\beta^4), \tag{3.56c}$$

With the above approximations, we observe that all the quadratic integrals scale as

$$\int_0^h \left[(u - \frac{1}{2}\Upsilon \epsilon u_s)u - (\overline{u} - \frac{1}{2}\Upsilon \epsilon \overline{u}_s)\overline{u} \right] dz \sim \mathscr{O}(\beta^4, \epsilon\beta^4), \tag{3.57}$$

$$\int_0^h \left[(\sigma \operatorname{od}B_t^H u) - (\overline{\sigma \operatorname{od}B_t^H \overline{u}}) \right] dz \sim \mathscr{O}(\beta^4, \epsilon\beta^4) \tag{3.58}$$

Besides, the ISD (3.36) reads now

$$u_s = \frac{1}{h}\left(\nabla_H \cdot (h\overline{a}^H) - a^{HH}(h)\nabla_H h - \frac{1}{2}h\nabla_H \cdot a^{HH}(h)\right)$$

$$= \frac{1}{2}(\nabla_H \cdot \overline{a}^H) + \mathscr{O}(\beta^2, \epsilon\beta^2), \tag{3.59}$$

$$w_s = \nabla_H \cdot a^{Hz} + \partial_z a^{zz}$$

$$= -z(\frac{1}{2}\nabla_H \cdot \nabla_H \cdot \overline{a}^H - 2\overline{a}^{\operatorname{div}}) + \mathscr{O}(\beta^2, \epsilon\beta^2). \tag{3.60}$$

Using the approximation of the vertical velocity (3.51), the vertical acceleration relation (3.25) reads

$$D_t w = -z\left(d_t \nabla_H \cdot \overline{u} + \epsilon((\overline{u} - \frac{1}{2}\Upsilon \epsilon \overline{u}_s) \cdot \nabla_H)\nabla_H \cdot \overline{u}dt - \epsilon(\nabla_H \cdot (\overline{u} - \frac{1}{2}\Upsilon \epsilon \overline{u}_s)\nabla_H \cdot \overline{u})dt + \right.$$

$$\left. \epsilon\Upsilon^{1/2}(\overline{\sigma \operatorname{od}B_t^H} \cdot \nabla_H)\nabla_H \cdot \overline{u} - \epsilon\Upsilon^{1/2}(\nabla_H \cdot \overline{\sigma \operatorname{od}B_t^H}\nabla_H \cdot \overline{u})\right) + \mathscr{O}(\beta^2, \epsilon\beta^2). \tag{3.61}$$

Eventually, we find the evolution equations for the surface elevation and the depth averaged velocity,

$$\overline{D}_t^H \eta = -h\left(\nabla_H \cdot (\overline{u} - \frac{1}{2}\Upsilon \epsilon \overline{u}_s)dt + \Upsilon \nabla_H \cdot \overline{\sigma \operatorname{od}B_t^H}\right), \tag{3.62a}$$

$$d_t \overline{u} + \epsilon((\overline{u} - \frac{1}{2}\Upsilon \epsilon \overline{u}_s) \cdot \nabla_H)\overline{u}dt + \Upsilon^{1/2}\epsilon(\overline{\sigma \operatorname{od}B_t^H} \cdot \nabla_H)\overline{u}$$

$$+ \nabla_H \eta dt - \frac{1}{h}\epsilon\beta^2\nabla_H\left(\frac{h^3}{3}(d\overline{G})\right) = \mathscr{O}(\beta^4, \epsilon\beta^4) \tag{3.62b}$$

with

$$
\begin{aligned}
d\overline{G}(x,t) = \Big(& d_t \nabla_H \cdot \overline{u} + \epsilon\big((\overline{u} - \tfrac{1}{2}\Upsilon\epsilon\overline{u}_s)\cdot\nabla_H\big)\nabla_H \cdot \overline{u}dt \\
& - \epsilon\big(\nabla_H \cdot (\overline{u} - \tfrac{1}{2}\Upsilon\epsilon\overline{u}_s)\nabla_H \cdot \overline{u}\big)dt \\
& + \epsilon\Upsilon^{1/2}(\overline{\sigma \circ dB_t^H}\cdot\nabla_H)\nabla_H \cdot \overline{u}dt - \epsilon\Upsilon^{1/2}(\nabla_H \cdot \overline{\sigma \circ dB_t^H}\nabla_H \cdot \overline{u})\Big),
\end{aligned}
$$
$$(3.63)$$

which can be more compactly written as

$$
d\overline{G}(x,t) = \overline{D}_t^H (\nabla_H \cdot \overline{u}) - \epsilon\nabla_H \cdot \big((\overline{u} - \tfrac{1}{2}\Upsilon\epsilon\overline{u}_s)dt + \Upsilon^{1/2}\overline{\sigma \circ dB_t^H}\big)\nabla_H \cdot \overline{u}. \qquad (3.64)
$$

This last expression can readily be understood as the acceleration of the average horizontal velocity divergence corrected by compressibility effects. This system constitutes a stochastic version of the Serre-Green-Naghdi equations [12, 33]. Compared to the original *deterministic* model, additional noise terms are involved. Those terms correspond to small scale advection and are accompanied with a corresponding ISD correction term in the large scale advection.

Let us now exhibit some simplified models that arise by neglecting higher order terms.

3.2   Shallow Water Waves Approximated Models

In this section we present two stochastic representations of classical approximations of the Serre-Green-Naghdi equations, namely the Shallow Water and the Boussinesq wave models. In addition, we briefly mention a stochastic version of the Kordeveg-De Vries equation. Through the section, we will make use of the Stokes number (also called Ursel number), defined as $S = \epsilon/\beta^2$.

Shallow Water (or Saint-Venant) Long Waves Approximation

For long waves regimes such as tidal waves, we have $\beta \ll \epsilon = A/h \ll 1$, which corresponds to a Stokes number $S \gg 1$. Neglecting in system (3.62) the terms of order higher than ϵ gives

$$
\overline{D}_t^H \eta = -h\big(\nabla_H \cdot (\overline{u} - \tfrac{1}{2}\Upsilon\epsilon\overline{u}_s)dt + \Upsilon^{1/2}\nabla_H \cdot \overline{\sigma \circ dB_t^H}\big), \qquad (3.65a)
$$

$$
d_t\overline{u} + \epsilon\big((\overline{u} - \tfrac{1}{2}\Upsilon\epsilon\overline{u}_s)\cdot\nabla_H\big)\overline{u}dt + \Upsilon^{1/2}\epsilon(\overline{\sigma \circ dB_t^H}\cdot\nabla_H)\overline{u} + \nabla_H\eta dt = 0. \qquad (3.65b)
$$

This system corresponds to a stochastic version of the 2D Shallow water model [5]. We note that the noise term is kept assuming Υ is of order 1 or higher. The diffusion term is in balance with the energy brought by the noise and must be kept to ensure energy conservation. For lower noise amplitude, the system boils down to the classical deterministic system. Notice that this system results from neglecting the vertical acceleration, which corresponds to the usual hydrostatic assumption. However, differently from the deterministic (linear) shallow water system, this corresponding linear stochastic system admits dispersive waves as solutions due to the noise term [23].

Boussinesq Approximation

Assuming that $S \approx 1$ and $\beta \ll 1$, we retain only the terms of order ϵ and β^2 in the system (3.62). Thus, we obtain a stochastic interpretation of Boussinesq wave equation. This system reads

$$D_t^H \eta = -h\big(\nabla_H \cdot (\overline{\boldsymbol{u}} - \tfrac{1}{2}\Upsilon\epsilon\overline{\boldsymbol{u}}_s)\mathrm{d}t + \Upsilon^{1/2}\nabla_H \cdot \overline{\boldsymbol{\sigma}\mathrm{od}\boldsymbol{B}_t^H}\big), \tag{3.66a}$$

$$\mathrm{d}_t\overline{\boldsymbol{u}} + \epsilon\big((\overline{\boldsymbol{u}} - \tfrac{1}{2}\Upsilon\epsilon\overline{\boldsymbol{u}}_s)\cdot\nabla_H\big)\overline{\boldsymbol{u}}\mathrm{d}t + \Upsilon^{1/2}\epsilon(\overline{\boldsymbol{\sigma}\mathrm{od}\boldsymbol{B}_t^H}\cdot\nabla_H)\overline{\boldsymbol{u}} + \nabla_H\eta\mathrm{d}t$$

$$- \epsilon\beta^2\big(\frac{h^2}{3}\nabla_H\nabla_H \cdot \mathrm{d}_t\overline{\boldsymbol{u}}\big) = \mathcal{O}(\beta^4, \epsilon\beta^4) \tag{3.66b}$$

Remark that an additional dispersive term appears compared to the previous system. From the LU Boussinesq system, one can also derive a stochastic version of the Kordeveg-De Vries (KdV) equation, as developed in Appendix 3.

3.3 Discussion on Conserved Quantities

In this section, we discuss the conservation of the following quantities: mass, momentum and (mechanical) energy. We consider a bounded horizontal domain S_H and assume that the noise term $\overline{\boldsymbol{\sigma}\mathrm{od}\boldsymbol{B}_t^H}$ is zero on the boundary ∂S_H. In particular, $\overline{\boldsymbol{\sigma}\mathrm{od}\boldsymbol{B}_t^H}$ is periodic.

Mass

We regard mass conservation first. Notice that the total mass m fulfils $m = \int_{S_H}\int_0^h \rho \, \mathrm{d}z \, \mathrm{d}S_H$, where S_H is the horizontal domain—which can be 1D or 2D depending on the considered problem—and $h = (1 + \epsilon\eta)$. Assuming that $\rho = \rho_0$ is constant, we get $m = \rho_0 h(|S_H| + \epsilon \int_{S_H} \eta \, \mathrm{d}S_H)$. Hence, $\mathrm{d}_t m \propto \mathrm{d}_t\big(\int_{S_H} \eta \, \mathrm{d}S_H\big)$.

Moreover, in the three models studied in this chapter, the evolution equation of the surface elevation η remains unchanged and reads

$$d_t \eta + \epsilon (\overline{u}^* \cdot \nabla_H) \eta \, dt + \Upsilon^{1/2} \epsilon \, \overline{(\sigma \, od B_t^H \cdot \nabla_H)} \eta = -h \left(\nabla_H \cdot \overline{u}^* dt + \Upsilon^{1/2} \nabla_H \cdot \overline{\sigma \, od B_t^H} \right),$$
$$(3.67)$$

that is to say, in conservative form,

$$d_t \eta + \nabla_H \cdot (\overline{u}^* h) dt + \Upsilon^{1/2} \nabla_H \cdot \overline{(\sigma \, od B_t^H h)} = 0. \qquad (3.68)$$

Thus, integrating over the horizontal domain S_H and using the divergence theorem, Eq. (3.68) yields

$$d_t \left(\int_{S_H} \eta \, dS_H \right) + \int_{\partial S_H} \left(\overline{u}^* h \, dt + \Upsilon^{1/2} \overline{\sigma \, od B_t^H h} \right) \cdot dn_{S_H} = 0 \qquad (3.69)$$

Therefore, under suitable boundary conditions—which are periodic with a 1D domain S_H in our study—the horizontal integral of the surface elevation $\int_{S_H} \eta \, dS_H$ is conserved, and consequently so is the total mass.

Momentum

To investigate momentum conservation, we define the total momentum $\mathbf{p} = \int_{S_H} \int_0^h \rho \overline{u} \, dz \, dS_H = \rho_0 \int_{S_H} h \overline{u} \, dS_H$. Then, we derive its evolution equation: starting from the LU Serre-Green-Naghdi model, we get

$$d_t (h\overline{u}) = h d_t \overline{u} + \overline{u} d_t h = h d_t \overline{u} + \epsilon \overline{u} d_t \eta$$

$$= - \epsilon (h \overline{u}^* \cdot \nabla_H) \overline{u} dt - \Upsilon^{1/2} \epsilon \overline{(h \sigma \, od B_t^H \cdot \nabla_H)} \overline{u} - h \nabla_H \eta dt + \epsilon \beta^2 \nabla_H \left(\frac{h^3}{3} (d\overline{G}) \right)$$

$$- \epsilon \overline{u} \left(\nabla_H \cdot (\overline{u}^* h) dt + \Upsilon^{1/2} \nabla_H \cdot \overline{(\sigma \, od B_t^H h)} \right)$$

$$= - \epsilon \nabla_H \cdot (h \overline{u} \otimes \overline{u}^*) dt - \epsilon \Upsilon^{1/2} \nabla_H \cdot \overline{(h \overline{u} \otimes \sigma \, od B_t^H)}$$

$$\underbrace{- \frac{1}{2\epsilon} \nabla_H h^2}_{= \nabla_H \left(\eta + \frac{\epsilon \eta^2}{2} \right)} dt + \epsilon \beta^2 \nabla_H \left(\frac{h^3}{3} (d\overline{G}) \right). \qquad (3.70)$$

Similarly as for the mass, for $i \in \{x, y\}$ in 2D, and for $i = x$ in 1D, Eq. (3.70) yields

$$d_t \left(\int_{S_H} h \overline{u}_i \, dS_H \right) + \epsilon \int_{\partial S_H} \left(h \overline{u}_i \overline{u}^* \, dt + \Upsilon^{1/2} h \overline{u}_i \overline{\sigma \, od B_t^H} \right) \cdot dn_{S_H}$$

$$+ \int_{\partial S_H} \left((\eta + \frac{\epsilon \eta^2}{2}) dt - \frac{\epsilon \beta^2 h^3}{3} d\overline{G} \right) d(\mathbf{n}_{S_H})_i = 0, \qquad (3.71)$$

using the gradient and the divergence theorems. Remind that $d\overline{G}$ is defined as

$$d\overline{G}(x,t) = \overline{D}_t^H (\nabla_H \cdot \overline{u}) - \epsilon \nabla_H \cdot \left(\overline{u}^* dt + \Upsilon^{1/2} \overline{\sigma \circ dB_t^H}\right) \nabla_H \cdot \overline{u}. \tag{3.72}$$

Again, under suitable boundary conditions—i.e periodic in our work—it is immediate that

$$\epsilon \int_{\partial S_H} \left(h\overline{u}_i \overline{u}^* \, dt + \Upsilon^{1/2} h\overline{u}_i \overline{\sigma \circ dB_t^H}\right) \cdot dn_{S_H} + \int_{\partial S_H} (\eta + \frac{\epsilon \eta^2}{2}) dt \, d(n_{S_H})_i = 0. \tag{3.73}$$

Consequently, one has

$$d_t \left(\int_{S_H} h\overline{u}_i \, dS_H \right) = \frac{\epsilon \beta^2}{3} \int_{\partial S_H} h^3 d\overline{G} \, d(n_{S_H})_i. \tag{3.74}$$

The water height h being periodic by assumption, it is enough to show that $d\overline{G}$ is periodic. In the LU Saint-Venant model, the RHS of Eq. (3.74) is completely neglected, which is equivalent to assuming $d\overline{G} = 0$. In such case, momentum conservation is immediate. In the LU Boussinesq model, $d\overline{G}$ is approximated as

$$d\overline{G} = d_t(\nabla_H \cdot \overline{u}) =: d\overline{G}_B.$$

Since $\overline{u}$ is periodic, $\nabla_H \cdot \overline{u}$ also is, as long as the 1st order space derivatives of $\overline{u}$ are well defined. Then $d_t(\nabla_H \cdot \overline{u})$ is periodic as well—as long as this term is well-defined—which proves momentum conservation. Regarding the LU Serre-Green-Naghdi model, we use the "full" Eq. (3.72) on $d\overline{G}$, namely

$$d\overline{G} = d\overline{G}_{SGN} := \overline{D}_t^H (\nabla_H \cdot \overline{u}) - \epsilon \nabla_H \cdot \left(\overline{u}^* dt + \Upsilon^{1/2} \overline{\sigma \circ dB_t^H}\right) \nabla_H \cdot \overline{u}$$

$$= d\overline{G}_B + \epsilon (\overline{u}^* \cdot \nabla_H)(\nabla_H \cdot \overline{u}) \, dt + \Upsilon^{1/2} \epsilon \, (\overline{\sigma \circ dB_t^H} \cdot \nabla_H)(\nabla_H \cdot \overline{u})$$

$$- \epsilon \nabla_H \cdot \left(\overline{u}^* dt + \Upsilon^{1/2} \overline{\sigma \circ dB_t^H}\right) \nabla_H \cdot \overline{u}.$$

By similar arguments, the new terms on the RHS are periodic since the 2nd order space derivatives of $\overline{u}$ are, as long as they are well-defined. Hence, $d\overline{G}_{SGN}$ is periodic as well, that is momentum is conserved.

Energy

For shallow water models—in particular, the LU Saint-Venant model—the total energy E_{SW} is defined as follows (using dimensioned velocities $\overline{u}$ and water height h),

$$E_{SW} = \int_{S_H} \int_0^h \frac{1}{2}\rho_0 \|\overline{u}\|^2 \, dz \, dS_H + \int_{S_H} \int_0^h \rho_0 g z \, dz \, dS_h$$

$$= \frac{\rho_0}{2} \int_{S_H} h\|\overline{u}\|^2 dS_H + \frac{\rho_0 g}{2} \int_{S_H} h^2 dS_H, \tag{3.75}$$

where the two terms on the RHS respectively correspond to the kinetic and potential energies. Scaling $\overline{u}$ and h as before, we find the equation on the following rescaled energy equation [8, 36]

$$E_{SW} = \frac{\epsilon^2}{2}\int_{S_H} h\|\overline{u}\|^2 dS_H + \frac{1}{2}\int_{S_H} h^2 dS_H = \frac{\epsilon^2}{2}(h\overline{u},\overline{u})_{L^2(S_H,\mathbb{R}^2)}$$

$$+ \frac{1}{2}\|h\|_{L^2(S_H,\mathbb{R})}^2 := E_c + E_p, \tag{3.76}$$

denoting E_c and E_p the scaled total kinetic and potential energies. We also denote $e_c = \frac{\epsilon^2}{2}h\|\overline{u}\|^2$ and $e_p = \frac{1}{2}h^2$. Now, we derive the evolution equation of this energy in the LU Saint-Venant model, using Einstein's notation on i,

$$d_t e_c = \frac{\epsilon^2}{2}\overline{u}\cdot d_t(h\overline{u}) + \frac{\epsilon^2}{2}h\overline{u}\cdot d_t\overline{u}$$

$$= -\frac{\epsilon^3}{2}\sum_{j\in J}\overline{u}_i\partial_j(h\overline{u}_i\overline{u}_j^*)dt - \frac{\epsilon^3\Upsilon^{1/2}}{2}\sum_{j\in J}\overline{u}_i\partial_j(h\overline{u}_i(\overline{\sigma\,\text{od}B_t^H})_j) - \frac{\epsilon}{2}h\overline{u}_i\partial_i h\,dt)$$

$$- \frac{\epsilon^3}{2}\sum_{j\in J}h\overline{u}_i\overline{u}_j^*\partial_j\overline{u}_i dt - \frac{\epsilon^3\Upsilon^{1/2}}{2}\sum_{j\in J}h\overline{u}_i(\overline{\sigma\,\text{od}B_t^H})_j\partial_j\overline{u}_i - \frac{\epsilon}{2}h\overline{u}_i\partial_i h\,dt)$$

$$= -\frac{\epsilon^3}{2}\nabla_H\cdot(h\overline{u}_i^2\overline{u}^*)dt - \frac{\epsilon^3}{2}\Upsilon^{1/2}\nabla_H\cdot(h\overline{u}_i^2(\overline{\sigma\,\text{od}B_t^H})_j) - \epsilon h\overline{u}_i\partial_i h\,dt),$$

and

$$d_t e_p = h\cdot d_t h = -\epsilon h\partial_i(h\overline{u}_i^*)dt - \epsilon\Upsilon^{1/2}h\partial_i(h(\overline{\sigma\,\text{od}B_t^H})_i),$$

where $J = \{x\}$ is the problem is 1D and $J = \{x,y\}$ if it is 2D. Thus, the quantity $e = e_c + e_p$ fulfils

$$\nabla\cdot_t e = -\frac{\epsilon^3}{2}\nabla_H\cdot(h\|\overline{u}\|^2\overline{u}^*)dt - \frac{\epsilon^3}{2}\Upsilon^{1/2}\nabla_H\cdot(h\|\overline{u}\|^2\overline{\sigma\,\text{od}B_t^H}) - \epsilon\nabla_H\cdot(h^2\overline{u})dt \tag{3.77}$$

$$+ \frac{\epsilon\Upsilon}{4}h\nabla_H\cdot(h\overline{u}_s)\,dt - \epsilon\Upsilon^{1/2}h\nabla_H\cdot(h\overline{\sigma\,\text{od}B_t^H}).$$

Integrating over the domain S_H, using the divergence theorem and periodic boundary conditions, we get by integration by parts the evolution equation of the total Saint-Venant energy E_{SV},

$$d_t E_{SW} = \int_{S_H}\left[-\frac{\epsilon\Upsilon}{8}\overline{u}_s\cdot\nabla_H h^2\,dt + \frac{\epsilon}{2}\Upsilon^{1/2}\overline{\sigma\,\text{od}B_t^H}\cdot\nabla_H h^2\right]dS_H \tag{3.78}$$

$$= \int_{S_H} \left[\frac{\epsilon \Upsilon}{8} h^2 \nabla_H \cdot \overline{u}_s \, dt - \frac{\epsilon}{2} \Upsilon^{1/2} h^2 \nabla_H \cdot \overline{\sigma \, odB_t^H} \right] dS_H.$$

Consequently, for the LU Saint-Venant model—that is assuming $d\overline{G} = 0$—energy conservation is enforced by choosing the noise term σodB_t such that $\nabla_H \cdot \overline{\sigma odB_t^H} = \nabla_H \cdot \overline{u}_s = 0$. We denote this assumption (**DF-BHN-ISD**), standing for "divergence free barotropic horizontal noise and Itō-Stokes drift". However, in 1D problems, this condition does not make much physical sense since it is equivalent to considering a constant horizontal noise over space. For this reason, energy conservation is not ensured in our numerical simulations, since they were performed with more general noises which do not fulfil this assumption. Nevertheless, we anticipate that performing 2D test simulations of this stochastic Saint-Venant equation with non trivial divergence free noise would lead to numerical energy conservation results.

For the LU Serre-Green-Naghdi model, the energy is rather defined as in [8, 24],

$$E_{SGN} = \frac{\epsilon^2}{2} (h\overline{u}, \overline{u})_{L^2(S_H, \mathbb{R}^2)} + \frac{1}{2} \|h\|^2_{L^2(S_H, \mathbb{R})}$$

$$+ \frac{\epsilon^3 \beta^2}{6} (h^3 \nabla_H \cdot \overline{u}, \nabla_H \cdot \overline{u})_{L^2(S_H, \mathbb{R})}. \tag{3.79}$$

Using the same notations e_c and e_p as before, and defining $e_{pv} = \frac{1}{6} h^3 (\nabla_H \cdot \overline{u})^2$ and $e = e_c + e_p + e_{pv}$, one has similarly

$$d_t E_{SGN} = \int_{S_H} \left[-\frac{\epsilon^3 \beta^2}{3} h^3 (\nabla_H \cdot \overline{u}) d\overline{G} + \frac{\epsilon^3 \beta^2}{2} h^2 d_t h (\nabla_H \cdot \overline{u})^2 \right.$$

$$\left. + \frac{\epsilon^3 \beta^2}{3} h^3 (\nabla_H \cdot \overline{u}) d_t (\nabla_H \cdot \overline{u}) \right] dS_H, \tag{3.80}$$

using that $\nabla_H \cdot \overline{\sigma odB_t^H} = \nabla_H \cdot \overline{u}_s = 0$. Now, computing the first term in the integrand yields

$$-\frac{\epsilon^3 \beta^2}{3} h^3 (\nabla_H \cdot \overline{u}) d\overline{G} = -\frac{\epsilon^4 \beta^2}{3} h^3 (\nabla_H \cdot \overline{u}) (\overline{u}^* dt + \Upsilon^{1/2} \overline{\sigma odB_t^H}) \cdot \nabla_H (\nabla_H \cdot \overline{u})$$

$$-\frac{\epsilon^3 \beta^2}{3} h^3 (\nabla_H \cdot \overline{u}) d_t (\nabla_H \cdot \overline{u}) + \frac{\epsilon^4 \beta^2}{3} h^3 (\nabla_H \cdot \overline{u})^2 \nabla_H \cdot (\overline{u}^* dt + \Upsilon^{1/2} \overline{\sigma odB_t^H}), \tag{3.81}$$

that is

$$-\frac{\epsilon^3 \beta^2}{3} h^3 (\nabla_H \cdot \overline{u}) d\overline{G} + \frac{\epsilon^3 \beta^2}{3} h^3 (\nabla_H \cdot \overline{u}) d_t (\nabla_H \cdot \overline{u})$$

$$= -\frac{\epsilon^4 \beta^2}{6} h^3 (\overline{u}^* dt + \Upsilon^{1/2}\overline{\sigma \, \mathrm{o} dB_t^H}) \cdot \nabla_H (\nabla_H \cdot \overline{u})^2 \qquad (3.82)$$

$$+ \frac{\epsilon^4 \beta^2}{3} h^3 (\nabla_H \cdot \overline{u})^2 \nabla_H \cdot (\overline{u}^* \, dt + \Upsilon^{1/2}\overline{\sigma \, \mathrm{o} dB_t^H}).$$

In addition, the second term in the integrand is

$$\frac{\epsilon^3 \beta^2}{2} h^2 \mathrm{d}_t h (\nabla_H \cdot \overline{u})^2 = -\frac{\epsilon^4 \beta^2}{2} h^2 (\nabla_H \cdot \overline{u})^2 \nabla_H \cdot (h\overline{u}^* dt + \Upsilon^{1/2} h \overline{\sigma \, \mathrm{o} dB_t^H})$$

$$= -\frac{\epsilon^4 \beta^2}{2} h^3 (\nabla_H \cdot \overline{u})^2 \nabla_H \cdot (\overline{u}^* dt + \Upsilon^{1/2}\overline{\sigma \, \mathrm{o} dB_t^H})$$

$$-\frac{\epsilon^4 \beta^2}{6} (\nabla_H \cdot \overline{u})^2 (\overline{u}^* dt + \Upsilon^{1/2}\overline{\sigma \, \mathrm{o} dB_t^H}) \cdot \nabla_H h^3. \qquad (3.83)$$

Consequently,

$$\mathrm{d}_t E_{SGN} = \frac{\epsilon^4 \beta^2}{6} \int_{S_H} \left[h^3 (\overline{u}^* dt + \Upsilon^{1/2}\overline{\sigma \, \mathrm{o} dB_t^H}) \cdot \nabla_H (\nabla_H \cdot \overline{u})^2 \right] dS_H$$

$$+ \frac{\epsilon^4 \beta^2}{6} \int_{S_H} \left[h^3 (\nabla_H \cdot \overline{u})^2 \nabla_H \cdot (\overline{u}^* \, dt + \Upsilon^{1/2}\overline{\sigma \, \mathrm{o} dB_t^H}) \right] dS_H$$

$$+ \frac{\epsilon^4 \beta^2}{6} \int_{S_H} \left[(\nabla_H \cdot \overline{u})^2 (\overline{u}^* dt + \Upsilon^{1/2}\overline{\sigma \, \mathrm{o} dB_t^H}) \cdot \nabla_H h^3 \right] dS_H$$

$$= \frac{\epsilon^4 \beta^2}{6} \int_{S_H} \nabla_H \cdot \left[h^3 (\nabla_H \cdot \overline{u})^2 (\overline{u}^* dt + \Upsilon^{1/2}\overline{\sigma \, \mathrm{o} dB_t^H}) \right] dS_H = 0,$$

$$(3.84)$$

using the divergence theorem and the periodic boundary conditions again. Notice that no assumptions were made in addition to the one for the Saint-Venant model, that is (DF-BHN-ISD). As before, this assumption leads to a space constant noise in the 1D case, therefore it is anticipated that the energy is not conserved in our simulations. Moreover, the previous calculations show that (DF-BHN-ISD) *is not enough to enforce energy conservation* in the LU Boussinesq model for both the energies E_{SW} and E_{SGN}, which is coherent with the deterministic Boussinesq model.

4   Numerical Simulations

In this section, we present some numerical simulations we made to test the three models derived. The Julia code that we produced is based on the work of Vincent

Duchêne and Pierre Navarro, who proposed a variety of wave models implementations in the deterministic setting—see the documentation in the following
link: http://waterwavesmodels.github.io/WaterWaves1D.jl/dev/ [9]. These models
are essentially based on pseudo-spectral resolution methods, which justifies the
use of periodic boundary conditions. We adapted their numerical framework to the
stochastic case, introducing implementations of the noise terms and the ISDs for
this purpose.

Regarding the purely stochastic aspects, we consider noises with wave spatial
structure. This is justified by the shape of solutions found in [23]: considering a
constant noise a first, the authors showed that the system admits progressive wave
solutions. Then, they extend the analysis to systems where the noise is itself a
progressive wave. In the end, the 1D noise we consider is the following,

$$\sigma(x)dW_t = A\cos(kx)d\beta_t^1 + A\sin(kx)d\beta_t^2, \qquad (4.1)$$

where A denotes the amplitude of the noise and $d\beta_t^1$, $d\beta_t^2$ denote Brownian motions.
Using a Box-Muller argument, this shape is equivalent to

$$\sigma(x)dW_t = A\cos(kx + \phi_t)d\beta_t, \qquad (4.2)$$

where ϕ_t is a uniformly distributed random phase on $(-\pi, \pi)$, such that for all
$t \neq s$, ϕ_t and ϕ_s are independent. Additionally, $d\beta_t$ is a Brownian motion. In our
simulations, the noise wave number is set to $k = 2\pi/100$, and the noise amplitude
may take the following values: $A = 0.001$, $A = 0.005$ or $A = 0.01$. Notice that
the dimensioned noise scales like $A\epsilon\sqrt{gh_0}$ as a consequence. The value of wave
number k is chosen to be at least one order of magnitude smaller than the typical
wave number of the deterministic wave. This is because our simulations showed
that noise terms with too small space scale oscillations lead to numerical instability,
and enables us to further discuss the presence of an additive noise term in the water
elevation dynamics. The values of amplitude where chosen to be much smaller than
the typical height of the wave. Numerically, we have observed that $A = 0.001$ yields
slight perturbations of the deterministic waves—that is typical realisations of each
LU model is similar to its deterministic counterpart—while $A = 0.01$ induces a
more "noisy" dynamics—that is typical realisations are essentially noise driven. In
addition, we chose $A = 0.005$ as an enlightening intermediate case.

Our tests are based on computing the evolution of the deformation surface η, with
a "heap of water" type initial condition. Namely, the initial surface deformation is
set to be $\eta(x, t = 0) = \exp(-x^4)$, while the initial velocity is set to $u(x, t = 0) = 0$.
All of our tests were performed on a numerical 1D tank $[-L, L]$ with $L = 50$, which
is discretised with $N = 2^{11}$ spatial points. The timestep is chosen to be $dt = 0.005s$,
and we assume periodic boundary conditions. To enforce these conditions on the
noise terms as well, we multiply them the function $s_\alpha(x) = \exp\left(\frac{1}{\alpha^2}\left(1 - \frac{1}{1-(x/L)^2}\right)\right)$,
with $\alpha = 10$, in order to make them vanish on the boundary. The initial condition
on η and the profile of s_α are given on Fig. 3.

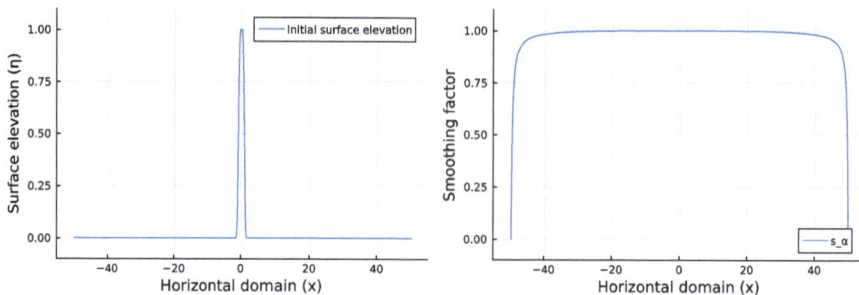

Fig. 3 Initial surface deformation (left) and boundary conditions enforcement function s_α (right, $\alpha = 10$)

Moreover, we propose two sets of parameters for testing our models

- $\beta = 0.01$ and $\epsilon = 0.1$ ($\mathscr{P}_1$), so that the scaling conditions associated with the Shallow Water model hold (and in particular, they hold for the Boussinesq model as well). Therefore, the three models should give qualitatively similar results.
- $\beta = 0.1$ and $\epsilon = 0.1$ ($\mathscr{P}_2$), so that the scaling conditions of the Shallow Water and the Boussinesq models do not hold. Thus, these models should give qualitatively distinct—yet not dramatically different—results.

Additionally, we will compare our stochastic models to the deterministic one dimensional Water Waves model.

Furthermore, two options are available for simulating our models, depending on whether we consider a "true" Stratonovitch noise or if we rewrite it in Itō form. In the former case, we may adopt a stochastic Euler-Heun approach as developed in [38], where the computation of the stochastic diffusion becomes implicit. In the latter case however, one needs to give an analytical expression for the stochastic diffusion. Even though one can compute them in the LU Saint-Venant and the LU Boussinesq systems, this correction term becomes extremely complex in the LU Serre-Green-Naghdi model due to the term $\overline{dG}$. Therefore, we rather adopted the first approach using the stochastic Euler-Heun method.

Moreover, for numerical stability reasons, we used a stochastic version of the order 4 Runge-Kutta algorithm (RK4), rather than the (simpler) Euler-Maruyama algorithm. For instance, the (deterministic) Saint-Venant equations are known to be dramatically more stable when solved with the order 4 Runge-Kutta rather than the Euler method. In summary, the solving algorithm we used is essentially the following: we treat the bounded variation term as in the classical RK4 method, and the martingale term as in the stochastic Euler-Heun method. Such approach has been studied in more details in [11, 13, 25].

Furthermore, since water elevation equations on η are the same in the three models we study—regarding for example Eq. (3.67)—one may notice the presence of the term

$$\Upsilon^{1/2} h \nabla_H \cdot \overline{(\sigma \circ d\boldsymbol{B}_t^H)} = \Upsilon^{1/2} \nabla_H \cdot \overline{(\sigma \circ d\boldsymbol{B}_t^H)} + \epsilon \Upsilon^{1/2} \eta \nabla_H \cdot \overline{(\sigma \circ d\boldsymbol{B}_t^H)}.$$

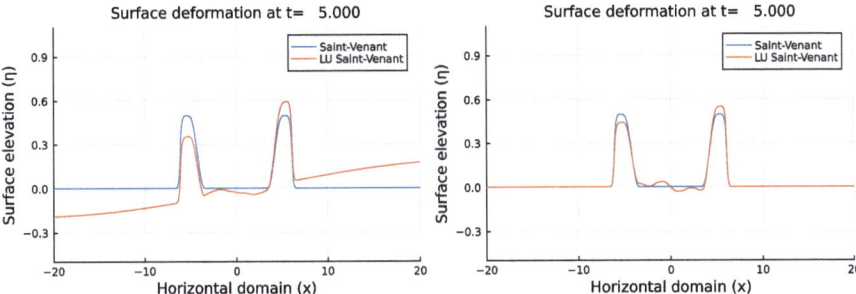

Fig. 4 Realisations of the LU Saint-Venant model, with additive noise (left) and without (right). We only plot the solution over the domain $[-20, 20]$, at $t = 5s$. Parameter set: ($\mathscr{P}_1$)—Wave number: $k = 2\pi/100$—Amplitude: 0.001

This shows the existence of an additive noise term $\Upsilon^{1/2}\nabla_H \cdot \overline{(\boldsymbol{\sigma} \mathrm{od}\boldsymbol{B}_t^H)}$ in the water elevation dynamics. Its effect is illustrated by Fig. 4: due to this term, the surface elevation is *not flat* "away from the wave", which is a strong difference compared to the deterministic setting. To facilitate the comparison between our models and their deterministic counterparts, we chose to *disregard* this additive noise term in our simulations. This can be interpreted as a filtering of the lower wave numbers—i.e. large scale dynamics.

4.1 Qualitative Analysis on the Effect of the Noise

In this subsection, we give some qualitative insights about the effect of the noise on the dynamics of the system. For this purpose, we compare each deterministic solution to a solution of the associated LU interpretation, disregarding the effect of the additive noise term previously mentioned.

4.1.1 Parameters ($\mathscr{P}_1$)

Considering the parameters ($\mathscr{P}_1$), we compared the deterministic Saint-Venant, Boussinesq and Serre-Green-Naghdi models (blue curves) to realisations of their associated LU models (orange curves), using the 1D Water Waves as a reference (green curves). The noise we chose has the shape of a stationary wave, with wave number $k = 2\pi/10$. Figure 5 show realisations of these models at $t = 5s$, for different noise amplitudes.

We observe that deterministic solutions are close to the Water Waves solution. This is expected regarding the scaling of $\beta = 0.01 \ll 1$. Moreover, the LU systems seem to converge to their respectively associated deterministic solution for a vanishing noise, thus giving consistency to the LU interpretation of the wave

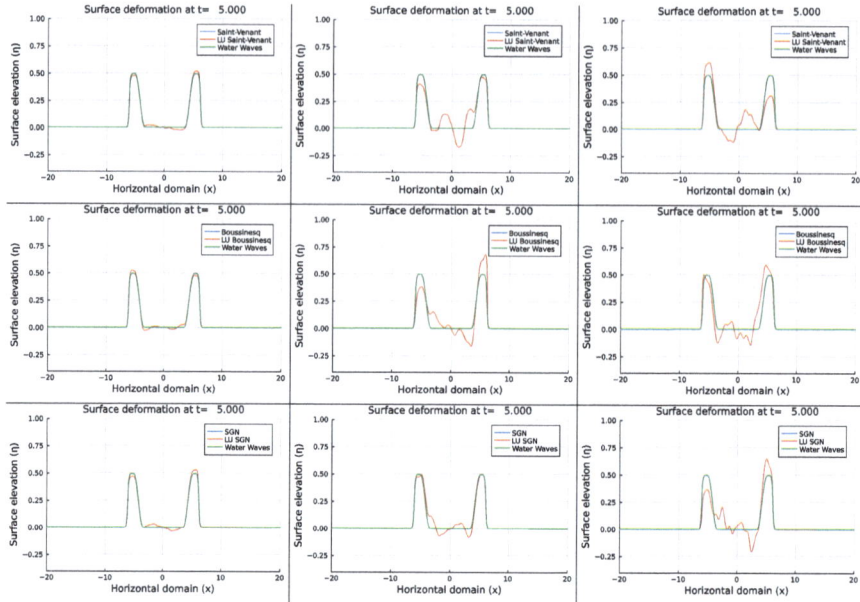

Fig. 5 Comparison between the surface deformation of the deterministic Saint-Venant (1st row), Boussinesq (2nd row) and Serre-Green-Naghdi (3rd row) models and realisations of their LU interpretations. Parameter set: $(\mathscr{P}_1)$—Wave number: $k = 2\pi/100$—Amplitude, from left to right: 0.001, 0.005 and 0.01

equations. In both cases, the noise tends to break the spatial symmetry of the wave. This stems from the shape of the noise, which is a sum of a symmetric function— $A\cos(kx)d\beta_t^1$—and an antisymmetric function—$A\sin(kx)d\beta_t^2$.

4.1.2 Parameters $(\mathscr{P}_2)$

The symmetry breaking mentioned in the previous subsection is also observed in Fig. 6, where we use parameters $(\mathscr{P}_2)$. However, in this case, the parameter $\beta = 0.1$ does not match the validity conditions of the Saint-Venant equations since $\beta \sim \epsilon$. Therefore, the Water Waves solution is expected to differ from the Saint-Venant ones, in both deterministic and LU forms. This is observed in Fig. 6, indeed. In addition, the LU Boussinesq and LU Serre-Green-Naghdi appear to give an interesting variability to their deterministic versions. However, the LU Serre-Green-Naghdi model yields numerical instabilities when choosing larger values of β—e.g. $\beta = 1$. This is due to a periodicity default caused by the symmetry breaking of the wave, which does not exist in the deterministic setting. To tackle this issue, we think of investigating numerical models in conservative form—that is computing the water elevation/momentum variables $(\eta, h\overline{u})$ rather than the water elevation/momentum variables $(\eta, \overline{u})$. For similar reasons, our

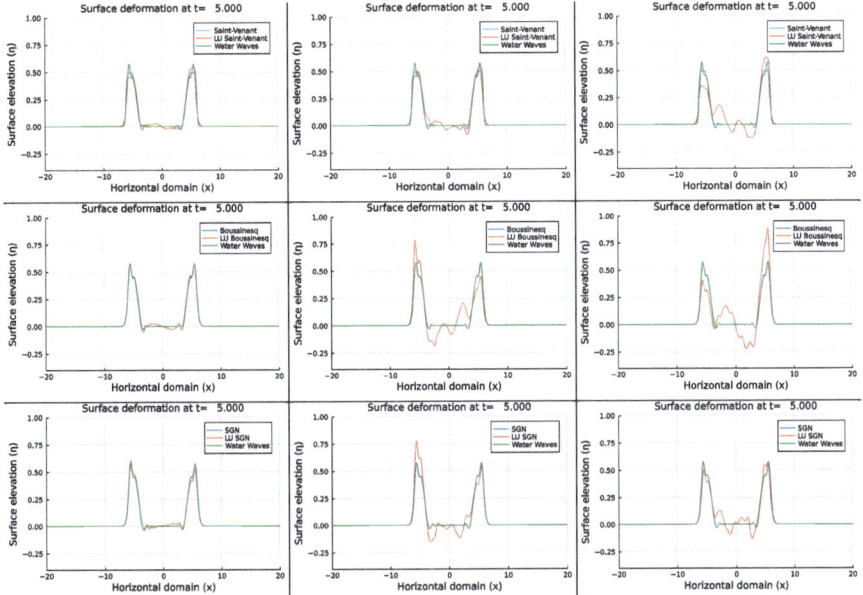

Fig. 6 Comparison between the surface deformation of the deterministic Saint-Venant (1st row), Boussinesq (2nd row) and Serre-Green-Naghdi (3rd row) models and their LU interpretations. Parameter set: ($\mathscr{P}_2$)—Wave number: $k = 2\pi/100$—Amplitude, from left to right: 0.001, 0.005 and 0.01

implementation does not enjoy conservation of physical quantities discussed earlier. Although it is relatively simple to translate our LU Saint-Venant algorithm into conservative form—which then conserves mass, momentum and energy up to machine accuracy—our LU Boussinesq and LU Serre-Green-Naghdi algorithms are more challenging to adapt due to the presence of the term $\overline{dG}$. We expect such numerical method to be more stable than the currently used one, and to allow investigating how multi-scale location uncertainty affects the waves dynamics. This is subject to further work.

4.2 Numerical Estimation of the Noise-Induced Spreading

In this section, we analyse the first and second order statistics of each LU wave model in the setting described above, at time $t = 5s$. Again, we will study these models in the sets of parameters ($\mathscr{P}_1$) for the LU Saint-Venant model, and ($\mathscr{P}_2$) for the LU Boussinesq and Serre-Green-Naghdi models. The LU models statistics we analyse are computed with 130 realisations of each stochastic model.

A spreading arises from LU Saint-Venant model, which appears to grow linearly with the amplitude of the noise —see Fig. 7 . Also, is concentrated in the "upstream"

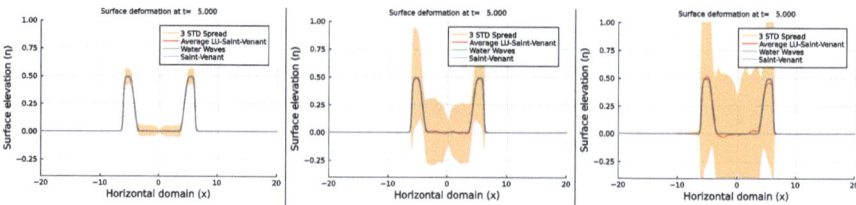

Fig. 7 First and second order statistics of the LU Saint-Venant model at $t = 5s$, compared to the associated deterministic model and the water waves one. In addition, an evaluation of the spreading—defined here as 3 times the empirical standard deviation—is given for different values of noise amplitude (orange area). We only plot the solution over the domain $[-20, 20]$. Parameter set: $(\mathscr{P}_1)$—Wave number: $k = 2\pi/100$—Amplitude, from left to right: 0.001, 0.005 and 0.01

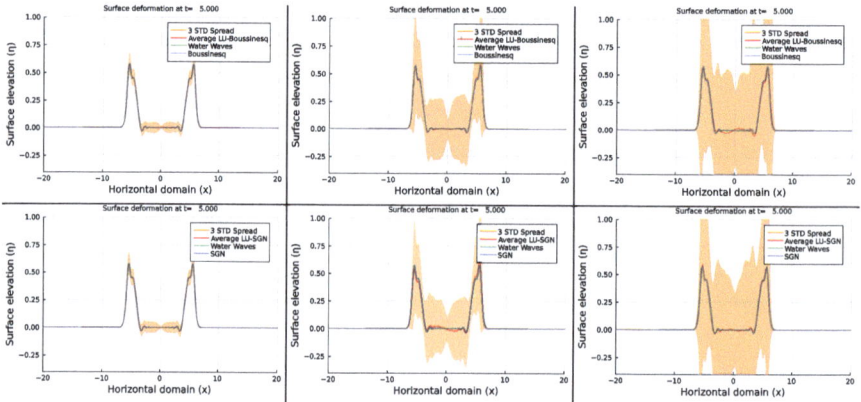

Fig. 8 First and second order statistics of the LU Boussinesq (1st row) and LU Serre-Green-Naghdi (2nd row) models at $t = 5s$, compared to the associated deterministic models and the water waves one. In addition, an evaluation of the spreading—defined here as 3 times the empirical standard deviation—is given for different values of noise amplitude (orange area). We only plot the solution over the domain $[-20, 20]$. Parameter set: $(\mathscr{P}_2)$—Wave number: $k = 2\pi/100$—Amplitude, from left to right: 0.001, 0.005 and 0.01

of the wave, which is expected since it would be physically irrelevant for the wave to affect a region where it has not passed. In addition, the means and the standard deviations appear to be space symmetric—up to statistical error—which suggests the distribution is also space symmetric. Moreover, there exists a significant spreading at the peak of the wave for strong enough noise amplitudes.

Regarding the LU Boussinesq and Serre-Green-Naghdi models, the same remarks on the peak spreading and the symmetry of the means and standard deviations apply Fig. 8. Again, the spreading is concentrated in the "upstream" of the wave, specifically at its peaks and troughs. In particular, the maximum height value of the LU models varies depending on the stochastic realisation. Such observation gives lines of approach for building stochastic models by selecting the noise σ to data, using for instance calibration or data assimilation techniques.

Although the Serre-Green-Naghdi and the Boussinesq models give similar results for the parameters ($\mathscr{P}_2$), we expect that much more differences would be observed between the two models with $\beta = 1$ and $\epsilon = 0.1$. As mentioned before, wave models written in conservative form should be more stable and allow to perform such tests in this configuration.

5 Conclusion and Discussion

In this work, we investigated the stochastic representation of several shallow water coastal wave models within the LU framework. These stochastic models maintain the same conservation properties formally and thus exhibit physical consistency with their deterministic counterparts. We demonstrated numerically that they induce a pathwise symmetry breaking, accompanied by a restoration of this symmetry in law, which is a property that should be expected in representing turbulent effects. For low noise amplitudes, the studied models have shown to converge toward the deterministic solutions. For large amplitudes, we need to transition to a numerical scheme in conservation form. This will be the subject of a future study.

Appendix 1: Leibniz Formula

We give in the following the expression of the Leibniz integral rule for a stochastic process. We want to evaluate the derivative of an integral of the form

$$\int_{a(x,t)}^{b(x,t)} f(x, z, t)\,\mathrm{d}z,$$

with respect to the space, x, and time, t, and where $a(x, z, t)$, $b(x, z, t)$ and $f(x, z, t)$ are continuous C^2-semimartingale. For the space variable, no time derivative is at play and the usual formula stands:

$$\frac{\partial}{\partial x} \int_{a(x,t)}^{b(x,t)} f(x, z, t)\,\mathrm{d}z = \int_{a(x,t)}^{b(x,t)} \partial_x f(x, z, t)\,\mathrm{d}z + \partial_x b\, f(x, b(x, t), t)$$
$$- \partial_x a\, f(x, a(x, t), t).$$

Differentiation with respect to time, involves now a stochastic integration. Understanding the stochastic integral in the Stratonovich setting, we have,

$$\mathrm{d}_t \int_{a(x,t)}^{b(x,t)} f(x, z, t)\,\mathrm{d}z = \int_{a(x,t)}^{b(x,t)} \mathrm{d}_t f(x, z, t)\,\mathrm{d}z + \mathrm{d}_t b\, f(x, b(x, t), t)$$

$$- \mathrm{d}_t a \, f(x, b(x, t), t).$$

The key argument of proof consists in defining the functions $F(x, y, t) = \int_{z_1}^{y} f(x, z, t)\mathrm{d}z$ and $G(x, a, b, t) = \int_{a(x,t)}^{b(x,t)} f(x, z, t)\mathrm{d}z$, then link them with a functional relation, that is

$$G(x, a, b, t) = \int_{0}^{b(x,t)} f(x, z, t)\mathrm{d}z - \int_{0}^{a(x,t)} f(x, z, t)\mathrm{d}z$$

$$= F(x, b(x, t), t) - F(x, a(x, t), t).$$

Now we introduce the following chain rule for Stratonovich calculus,

Theorem A.1.1 (Generalized Itō's Formula—Stratonovich Form) *Let* $\theta(x, t)$, $x \in \Omega$ *be a continuous C^3-process and a continuous C^2-semimartingale, let X_t be a continuous semimartingale with values in Ω. Then, the following formula is satisfied:*

$$\mathrm{d}\theta(X_t, t) = \mathrm{d}_t \theta(X_t, t) + \frac{\partial \theta}{\partial x_i}(X_t, t) \circ \mathrm{d}X_t^i. \qquad (A.1.1)$$

Upon applying this on $G(x, a, b, t)$, we obtain

$$\mathrm{d}_t G = \mathrm{d}_t F(x, b(x, t), t) + \partial_b F(x, b(x, t), t)\mathrm{d}_t b$$

$$- \mathrm{d}_t F(x, a(x, t), t) - \partial_b F(x, b(x, t), t)\mathrm{d}_t a$$

$$= \int_{a(x,t)}^{b(x,t)} \mathrm{d}_t f(x, z, t)\mathrm{d}z + \mathrm{d}_t b \, f(x, b(x, t), t) - \mathrm{d}_t a \, f(x, b(x, t), t),$$

To interpret the stochastic integral in the Itō setting, we would need to adapt the previous chain rule as follows,

Theorem A.1.2 (Generalized Itō's Formula—Itō Form) *Let* $\theta(x, t)$, $x \in \Omega$ *be a continuous C^2-process and a continuous C^1-semimartingale, let X_t be a continuous semimartingale with values in Ω. Then, $\theta(X_t, t)$ is a continuous semimartingale satisfying*

$$\mathrm{d}\theta(X_t, t) = \mathrm{d}_t \theta(X_t, t) + \frac{\partial \theta}{\partial x_i}(X_t, t)\mathrm{d}X_t^i + \frac{1}{2}\frac{\partial^2 \theta}{\partial x_i \partial x_j}(X_t, t)\mathrm{d}\langle X^i, X^j \rangle \Big|_t$$

$$+ \mathrm{d}\Big\langle \frac{\partial \theta}{\partial x_i}(X, \cdot), X^i \Big\rangle \Big|_t. \qquad (A.1.2)$$

Applying it to the process $G(x, a, b, t)$, we obtain

$$d_t G = \int_{a(x,t)}^{b(x,t)} d_t f(x, z, t) dz + d_t b\, f(x, b(x, t), t) - d_t a\, f(x, b(x, t), t) +$$

$$d\langle f(x, b(x, t), t), b\rangle_t dt - d\langle f(x, a(x, t), t), a\rangle_t dt +$$

$$\frac{1}{2} d\langle b, b\rangle_t \partial_{b^2}^2 F(x, b(x, t), t) - \frac{1}{2} d\langle a, a\rangle_t \partial_{a^2}^2 F(x, a(x, t), t) dt. \qquad \text{(A.1.3)}$$

This latter includes several additional quadratic variation terms, and is more cumbersome to use in formal developments. However, this expression is necessary to access to the mathematical expectation.

Appendix 2: Quadratic Covariation

In stochastic calculus, the quadratic covariation (or cross-variance) of two processes X and Y is defined as

$$\langle X, Y\rangle_t = \lim_{n \to 0} \sum_{i=1}^{p_n} (X_i^n - X_{i-1}^n)(Y_i^n - Y_{i-1}^n), \qquad \text{(A.2.1)}$$

where $0 = t_0^n < t_1^n < \cdots < t_{p_n}^n = t$ is a partition of the interval $[0, t]$. The previous limit, if it exists, is defined in the sense of convergence in probability.

Let X and Y two continuous semimartingales, defined as $X_t = X_0 + A_t + M_t$, $Y_t = Y_0 + B_t + N_t$, M, N being martingales and A, B finite variation processes. In this context, their quadratic covariation (A.2.1) exists, and is given by

$$\langle X, Y\rangle_t = \langle M, N\rangle_t. \qquad \text{(A.2.2)}$$

In particular, the quadratic variation of a standard Brownian motion B (as a martingale) is $\langle B\rangle_t := \langle B, B\rangle_t = t$, by definition.

Quadratic covariations play an important role in stochastic calculus, as they arise in Itō's lemma, which can be interpreted as a stochastic chain rule. In particular, these terms are involved in the Itō integration by parts formula,

$$d_t(XY) = X d_t Y + Y d_t X + d\langle X, Y\rangle_t, \qquad \text{(A.2.3)}$$

and in Itō's isometry, expressing the covariance of two Itō integrals: let f and g two predictable processes such that $\int_0^t f^2 d\langle M, M\rangle_s$ and $\int_0^t g^2 d\langle N, N\rangle_s$ are finite, then

$$\mathbb{E}\left[\left(\int_0^t f dM_s\right)\left(\int_0^t g dN_s\right)\right] = \mathbb{E}\left[\int_0^t f g\, d\langle M, N\rangle_s\right]. \qquad \text{(A.2.4)}$$

Appendix 3: LU Kordeveg–De Vries Equation

To derive the LU interpretation of the Kordeveg–De Vries equation (KdV), we will adapt a standard procedure to the stochastic case [14]. For simplification purpose, we will assume homogeneity in the transverse direction and consider in the following a 1D version of the Boussinesq model. Assuming that the bottom is flat, the 1D Boussinesq model reads

$$D_t^H \eta = -h\big(\partial_x(\overline{u} - \tfrac{1}{2}\Upsilon\epsilon\overline{u}_s)dt + \Upsilon^{1/2}\partial_x\overline{\boldsymbol{\sigma}\text{od}\boldsymbol{B}_t^x}\big), \qquad (A.3.1a)$$

$$\begin{aligned}
d_t\overline{u} + \epsilon\big(\overline{u} - \tfrac{1}{2}\Upsilon\epsilon\overline{u}_s\big)\partial_x\overline{u}dt \\
+ \Upsilon^{1/2}\epsilon\overline{\boldsymbol{\sigma}\text{d}\boldsymbol{B}_t^x}\partial_x\overline{u} + \partial_x\eta dt - \epsilon\beta^2\big(\frac{h^2}{3}\partial_{xx}^2 d_t\overline{u}\big) = \mathcal{O}(\beta^4, \epsilon\beta^4)
\end{aligned} \qquad (A.3.1b)$$

The solutions we consider for the above Boussinesq system are assumed to be waves, and we apply the following change of variable: $\xi = x - t + \varphi_t$, where $\varphi_t = f(B_t)$ is a random phase that does not depends on x. The wave shape is also assumed to change on a large temporal scale $\tau = \epsilon t$. The depth averaged horizontal velocity $u(\xi, \tau, \varphi)$ and the surface elevation $\eta(\xi, \tau, \varphi)$ are assumed to be smooth functions. The noise $\boldsymbol{\sigma}\text{od}\boldsymbol{B}_t^x(\xi, \tau)$ is assumed to be homogeneous, and thus is associated to a constant variance tensor and a zero ISD. In this new formalism the surface elevation equation reads

$$\begin{aligned}
- \partial_\xi\eta \, dt + \epsilon\partial_\tau\eta \, dt + \partial_\varphi\eta f'(B_t)dB_t + \epsilon u\partial_\xi\eta \, dt + \Upsilon^{1/2}\epsilon \, \boldsymbol{\sigma}\text{o} \, d\boldsymbol{B}_t^x\partial_\xi\eta \\
+ h\big(\partial_\xi u \, dt + \Upsilon^{1/2}\partial_\xi \, \boldsymbol{\sigma}\text{o} \, d\boldsymbol{B}_t^x\big) = 0.
\end{aligned} \qquad (A.3.2)$$

After converting this equation in Itō form, terms of finite variations (ie "dt" terms) and martingale terms (i.e. "dB_t" terms) can be rigorously separated by the Biechteller-Delacherie theorem. For the martingale terms we have

$$\partial_\varphi\eta \, f'(B_t)dB_t = -\Upsilon^{1/2}\partial_\xi(h\boldsymbol{\sigma}d\boldsymbol{B}_t^x), \qquad (A.3.3)$$

and for the finite variation terms

$$- \partial_\xi\eta + \epsilon\partial_\tau\eta + \epsilon u\partial_\xi\eta - \frac{1}{2}\Upsilon\epsilon^2 a^H\partial_{\xi\xi}\eta + h\partial_\xi u = 0. \qquad (A.3.4)$$

Moreover, for the velocity equation we have

$$- \partial_\xi u \, dt + \epsilon \partial_\tau u \, dt + \partial_\varphi u f'(B_t) dB_t + \epsilon u \partial_\xi u \, dt + \Upsilon^{1/2} \epsilon \, \boldsymbol{\sigma} \circ dB_t^x \partial_\xi u$$

$$- \frac{1}{2} \Upsilon \epsilon^2 \bar{a}^{hh} \partial_{\xi\xi}^2 u \, dt + \partial_\xi \eta \, dt \tag{A.3.5}$$

$$- \frac{1}{h} \epsilon \beta^2 \left(\frac{h^3}{3} \partial_{\xi\xi}^2 (-\partial_\xi u + \epsilon \partial_\tau u) \right) dt = \mathcal{O}(\beta^4, \epsilon \beta^4).$$

Then, the martingale terms yield

$$\partial_\varphi u f'(B_t) dB_t = -\Upsilon^{1/2} \epsilon \, \boldsymbol{\sigma} \circ dB_t^x \partial_\xi u \tag{A.3.6}$$

and for the finite variation terms

$$- \partial_\xi u + \epsilon \partial_\tau u + \epsilon u \partial_\xi u - \frac{1}{2} \Upsilon \epsilon^2 a^H \partial_{\xi\xi} u + \partial_\xi \eta + \frac{1}{h} \epsilon \beta^2 \left(\frac{h^3}{3} (\partial_{\xi\xi\xi}^3 u - \epsilon \partial_{\xi\xi\tau}^3 u) \right) = 0. \tag{A.3.7}$$

Expanding u and η in terms of the small parameter ϵ as $q = q_0 + \epsilon q_1 + \cdots$, and identifying the equations term by term of corresponding order, we get for the zero order terms,

$$\partial_\xi \eta_0 = \partial_\xi u_0 \text{ and hence } \eta_0 = u_0.$$

At order ϵ, for a noise of magnitude up to $\Upsilon \sim \mathcal{O}(1)$, we obtain the system

$$- \partial_\xi u_1 + \partial_\tau \eta_0 + \eta_0 \partial_\xi \eta_0 + \partial_\xi \eta_1 + \frac{1}{3} \partial_{\xi\xi\xi}^3 \eta_0 = 0,$$

$$- \partial_\xi \eta_1 + \partial_\tau \eta_0 + 2\eta_0 \partial_\xi \eta_0 + \partial_\xi u_1 = 0,$$

from which one gets immediately the classical KdV equation with random phase

$$\partial_\tau \eta_0 + \frac{3}{2} \eta_0 \partial_\xi \eta_0 + \frac{1}{6} \partial_{\xi\xi\xi}^3 \eta_0 = 0. \tag{A.3.8}$$

This equation has the structure of a Burger equation with an additional dispersive term. A modified KdV equation is obtained by considering a stronger noise of amplitude $\Upsilon \sim 1/\epsilon$. In that case the second order terms must be kept in the Taylor development of terms at ϵ order. Eventually, one obtains the dissipative KdV equation (with random phase)

$$\partial_\tau \eta_0 + \frac{3}{2} \eta_0 \partial_\xi \eta_0 - \frac{1}{2} a^H \partial_{\xi\xi} \eta_0 + \frac{1}{6} \partial_{\xi\xi\xi}^3 \eta_0 = 0. \tag{A.3.9}$$

The random phase of both KdV equation is determined by (A.3.3) and (A.3.6). From the latter, we notice immediately that $f'(B)$ and ϵ must share the same order for the solution not to be trivial. At order ϵ from (A.3.6), one has

$$\partial_\varphi \eta_0 \, f'(B_t) \mathrm{d}B_t = -\Upsilon^{1/2} \partial_\xi \eta_0 \sigma \mathrm{d}\mathbf{B}_t^{0,x}.$$

A simple way to choose f is to impose $f(B_t) = -k\Upsilon^{1/2}\epsilon k^\sigma B_t$ and $\sigma = k^\sigma$, where k^σ is associated to the waves form

$$\eta_0 = h_0 e^{i(kx - \omega t - \epsilon \Upsilon^{1/2} k k^\sigma B_t)}. \tag{A.3.10}$$

This solution corresponds to stochastic linear waves solutions as derived in [7]. Note that the derivations above have been done using the strong hypothesis of u and η being smooth functions of time—more precisely of finite variation. Looking now for more general stochastic solutions, one considers the following variables

$$u(\xi, \tau), \quad \sigma \mathrm{od}\mathbf{B}_t^H(\xi, \tau), \quad \eta(\xi, \tau), \tag{A.3.11}$$

which are not differentiable with respect to τ (i.e. they are semi-martingale stochastic processes). Making the same assumptions and deriving the equations in Stratonovich form in the same way as previously, we obtain two coupled SPDE's,

$$-\partial_\xi \eta \mathrm{d}t + \epsilon \mathrm{d}_\tau \eta + \epsilon u \partial_\xi \eta \, \mathrm{d}t + \Upsilon^{1/2}\epsilon \, \boldsymbol{\sigma} \circ \mathrm{d}\mathbf{B}_t \partial_\xi \eta$$

$$+ h\big(\partial_\xi u \, \mathrm{d}t + \Upsilon^{1/2}\partial_\xi \, \boldsymbol{\sigma} \circ \mathrm{d}\mathbf{B}_t^x\big) = 0, \tag{A.3.12}$$

and

$$-\partial_\xi u \mathrm{d}t + \epsilon \mathrm{d}_\tau u + \epsilon u \partial_\xi u \, \mathrm{d}t + \Upsilon^{1/2}\epsilon \, \boldsymbol{\sigma} \circ \mathrm{d}\mathbf{B}_t^x \partial_\xi u$$

$$+ \partial_\xi \eta \, \mathrm{d}t - \frac{1}{h}\epsilon\beta^2\Big(\frac{h^3}{3}\partial_{\xi\xi}^2(-\partial_\xi u + \epsilon\partial_\tau u)\Big)\mathrm{d}t = \mathcal{O}(\beta^4, \epsilon\beta^4). \tag{A.3.13}$$

At zeroth order the system reads

$$\partial_\xi \eta_0 = \partial_\xi u_0, \text{ hence } \eta_0 = u_0 \text{ together with } \partial_\xi \, \boldsymbol{\sigma} \circ \mathrm{d}\mathbf{B}_t^{0,x} = 0.$$

At order ϵ, one has

$$-\partial_\xi u_1 \mathrm{d}t + \mathrm{d}_\tau \eta_0 + \eta_0\partial_\xi \eta_0 \mathrm{d}t + \Upsilon^{1/2} \, \boldsymbol{\sigma} \circ \mathrm{d}\mathbf{B}_t^{0,x}\partial_\xi \eta_0 + \partial_\xi \eta_1 \mathrm{d}t + \frac{1}{3}\partial_{\xi\xi\xi}^3\eta_0\mathrm{d}t = 0,$$

$$-\partial_\xi \eta_1 \mathrm{d}t + \mathrm{d}_\tau \eta_0 + \Upsilon^{1/2} \, \boldsymbol{\sigma} \circ \mathrm{d}\mathbf{B}_t^{0,x}\partial_\xi \eta_0 + 2\eta_0\partial_\xi \eta_0\mathrm{d}t + \partial_\xi u_1 \mathrm{d}_t$$

$$+ \Upsilon^{1/2}\partial_\xi \, \boldsymbol{\sigma} \circ \mathrm{d}\mathbf{B}_t^{1,x} = 0.$$

Assuming that the noise terms $\boldsymbol{\sigma} \mathrm{od}\mathbf{B}_t^{1,x}$ does not depend on space, yet making no assumption on $\boldsymbol{\sigma} \mathrm{od}\mathbf{B}_t^{0,x}$, one obtains a stochastic KdV equation with transport noise,

$$d_\tau \eta_0 + \frac{3}{2}\eta_0 \partial_\xi \eta_0 dt + \Upsilon^{1/2} \boldsymbol{\sigma} \circ d\boldsymbol{B}_t^{0,x} \partial_\xi \eta_0 + \frac{1}{6}\partial_{\xi\xi\xi}^3 \eta_0 dt = 0. \tag{A.3.14}$$

Relaxing the spatially constant noise assumption on $\boldsymbol{\sigma} \circ d\boldsymbol{B}_t^{1,x}$ would add an additive stochastic forcing $\frac{1}{2}\Upsilon^{1/2}\partial_\xi \boldsymbol{\sigma} \circ d\boldsymbol{B}_t^{1,x}$. In such a case, we would get a stochastic KdV equation forced by an additive white noise of the form studied in [4], and for which existence and unicity of solution have been shown in the Sobolev space $H^1(\mathbb{R})$.

Without this additive forcing term, we face a simpler system that boils down to the deterministic one. As a matter of fact, proceeding to the change of variable suggested by Wadati [37], one has

$$X = \xi - \frac{2}{3}\Upsilon^{1/2}\int_0^t \boldsymbol{\sigma} \circ d\boldsymbol{B}_s^{0,x}, \tag{A.3.15}$$

with $\boldsymbol{\sigma}$ being constant over time and space. Now reparametrising $\eta(\xi, \tau)$ as $\eta'(X, t)$, we obtain the unperturbed KdV equation

$$d_t \eta_0' + \frac{3}{2}\eta_0' \partial_X \eta_0' dt + \frac{1}{6}\partial_{XXX}^3 \eta_0' dt = 0, \tag{A.3.16}$$

using the chain rules $\partial_\xi \eta = \partial_X \eta' \partial_\xi X = \partial_X \eta'$ and $d_\tau \eta = \partial_X \eta' d_t X + d_t \eta'$. Such an equation admits a solitary travelling wave solution given by

$$\eta_0'(X, t) = A \operatorname{sech}^2\left(\sqrt{\frac{3}{4}A}(X - (1 + \frac{1}{2}A)t)\right). \tag{A.3.17}$$

Note that considering a solution with a random phase would lead to the presence of a new additive noise term $\partial_\varphi \eta_0 \, \epsilon f'(B_t)dB_t$. Plus, the noise $\boldsymbol{\sigma} \circ d\boldsymbol{B}_t^{0,x}$ may not be constant in space anymore.

References

1. E. Barthélemy. "Nonlinear Shallow Water Theories for Coastal Waves". In: *Surveys in Geophysics* 25.3 (2004), pp. 315–337. DOI: 10.1007/s10712-003-1281-7. URL: https://doi.org/10.1007/s10712-003-1281-7.
2. W. Bauer et al. "Deciphering the Role of Small-Scale Inhomogeneity on Geophysical Flow Structuration: A Stochastic Approach". In: *Journal of Physical Oceanography* 50.4 (1Apr. 2020), pp. 983–1003. DOI: 10.1175/JPO-D-19-0164.1. URL: https://journals.ametsoc.org/view/journals/phoc/50/4/jpo-d-19-0164.1.xml.
3. P. S. Berloff and J. C. McWilliams. "Material Transport in Oceanic Gyres. Part II: Hierarchy of Stochastic Models". In: *Journal of Physical Oceanography* 32.3 (2002), pp. 797–830. DOI: 10.1175/1520-0485(2002)032<0797:MTIOGP>2.0.CO;2. URL: https://journals.ametsoc.org/view/journals/phoc/32/3/1520-0485_2002_032_0797_mtiogp_2.0.co_2.xml.
4. A. de Bouard and A. Debussche. "On the Stochastic Korteweg de Vries Equation". In: *Journal of Functional Analysis* 154 (1998), pp. 215–251.

5. R. Brecht et al. "Rotating shallow water flow under location uncertainty with a structure-preserving discretization". In: *Journal of advances in modelling earth systems* 13.12 (2021).

6. A. Craik and S. Leibovich. "Rational model for Langmuir circulations". In: *J. Fluid Mech.* 73 (1976), pp. 401–426.

7. E. Dinvay and E. Mémin. "Hamiltonian formulation of the stochastic surface wave problem". In: *Proceedings of the Royal Society A: Mathematical, Physical and Engineering Sciences* 478–2265 (2022), p. 20220050. DOI: 10.1098/rspa.2022.0050. URL: https://royalsocietypublishing.org/doi/abs/10.1098/rspa.2022.0050.

8. D. G. Dritschel and M. R. Jalali. "On the regularity of the Green–Naghdi equations for a rotating shallow fluid layer". In: *Journal of Fluid Mechanics* 865 (2019), pp. 100–136. DOI: 10.1017/jfm.2019.47.

9. V. Duchêne and P. Navarro. *WaterWaves1D.jl*. June 2022. URL: https://perso.univ-rennes1.fr/vincent.duchene/post/waterwaves1d/ (visited on 03/21/2024).

10. B. Dufée, E. Mémin, and D. Crisan. "Stochastic parametrization: An alternative to inflation in ensemble Kalman filters". In: *Quarterly Journal of the Royal Meteorological Society* 148.744 (2022/08/29 2022), pp. 1075–1091. DOI: https://doi.org/10.1002/qj.4247. URL: https://doi.org/10.1002/qj.4247.

11. C. Fiorini et al. "A Two-Step Numerical Scheme in Time for Surface Quasi Geostrophic Equations Under Location Uncertainty". In: *Stochastic Transport in Upper Ocean Dynamics*. Ed. by B. Chapron et al. Cham: Springer International Publishing, 2023, pp. 57–67. ISBN: 978-3-031-18988-3.

12. A. E. Green and P. M. Naghdi. "A derivation of equations for wave propagation in water of variable depth". In: *Journal of Fluid Mechanics* 78.2 (1976), pp. 237–246.

13. F. Gugole and C. L. E. Franzke. "Numerical Development and Evaluation of an Energy Conserving Conceptual Stochastic Climate Model". In: *Mathematics of Climate and Weather Forecasting* 5.1 (2019), pp. 45–64. DOI: doi:10.1515/mcwf-2019-0004. URL: https://doi.org/10.1515/mcwf-2019-0004.

14. R. S. Johnson. *A Modern Introduction to the Mathematical Theory of Water Waves*. Cambridge Univ. Press, 1997.

15. S. Kadri Harouna and E. Mémin. "Stochastic representation of the Reynolds transport theorem: revisiting large-scale modeling". In: *Computers & Fluids* 156 (Aug. 2017), pp. 456–469. DOI: 10.1016/j.compfluid.2017.08.017. URL: https://hal.inria.fr/hal-01394780.

16. R. Kozlovsky and J. Grobman. "The blue garden: coastal infrastructure as ecologically enhanced wave-scapes". In: *Landscape Research* (Dec. 2016). DOI: 10.1080/01426397.2016.1260702.

17. R. Kraichnan. "The structure of isotropic turbulence at very high Reynolds numbers". In: *J. of Fluids Mech.* 5 (1959), pp. 477–543.

18. H. Kunita. *Stochastic flows and stochastic differential equations*. Cambridge University Press, 1990.

19. S. Leibovich. "On wave-current interaction theories of Langmuir circulations". In: *J. Fluid Mech.* 99.4 (1980), pp. 715–724.

20. L. Li, E. Mémin, and G. Tissot. "Stochastic parameterization with dynamic mode decomposition". In: *STUOD Proceedings*. Springer Verlag, 2022.

21. L. Li et al. "Stochastic Data-Driven Parameterization of Unresolved Eddy Effects in a Baroclinic Quasi-Geostrophic Model". In: *Journal of Advances in Modeling Earth Systems* 15.2 (2023), e2022MS003297.

22. E. Mémin. "Fluid flow dynamics under location uncertainty". In: *Geophys. & Astro. Fluid Dyn.* 108.2 (2014), pp. 119–146. DOI: 10.1080/03091929.2013.836190.

23. E. Mémin et al. "Linear Wave Solutions of a Stochastic Shallow Water Model Check for updates". In: *Stochastic Transport in Upper Ocean Dynamics II: STUOD 2022 Workshop, London, UK, September 26–29*. Vol. 11. Springer Nature. 2023, p. 223.

24. J. Miles and R. Salmon. "Weakly dispersive nonlinear gravity waves". In: *Journal of Fluid Mechanics* 157 (1985), pp. 519–531.

25. G. Pavliotis and A. Stuart. *Multiscale methods: averaging and homogenization*. Springer Science & Business Media, 2008.
26. G. D. Prato and J. Zabczyk. *Stochastic equations in infinite dimensions*. Cambridge University Press, 1992.
27. M. H. Redi. "Oceanic Isopycnal Mixing by Coordinate Rotation". In: *Journal of Physical Oceanography* 12.10 (1982), pp. 1154–1158.
28. V. Resseguier, E. Mémin, and B. Chapron. "Geophysical flows under location uncertainty, Part I Random transport and general models". In: *Geophys. & Astro. Fluid Dyn.* 111.3 (2017), pp. 149–176.
29. V. Resseguier et al. "New trends in ensemble forecast strategy: uncertainty quantification for coarsegrid computational fluid dynamics". In: *Archives of Computational Methods in Engineering* (2020), pp. 1886–1784.
30. J. M. Restrepo. "Wave Breaking Dissipation in the Wave-Driven Ocean Circulation". In: *Journal of Physical Oceanography* 37.7 (2007), pp. 1749–1763. DOI: 10.1175/JPO3099.1. URL: https://journals.ametsoc.org/view/journals/phoc/37/7/jpo3099.1.xml.
31. J. M. Restrepo et al. "Multiscale Momentum Flux and Diffusion due to Whitecapping in Wave–Current Interactions". In: *Journal of Physical Oceanography* 41.5 (2011), pp. 837–856. DOI: 10.1175/2010JPO4298.1. URL: https://journals.ametsoc.org/view/journals/phoc/41/5/2010jpo4298.1.xml.
32. P. Schmid. "Dynamic mode decomposition of numerical and experimental data". In: *J. Fluid Mech.* 656 (2010), pp. 5–28.
33. F. Serre. "Contribution à l'étude des écoulements permanents et variables dans les canaux". In: *Houille Blanche* 8 (1953), pp. 374–388.
34. J. Smagorinsky. "General circulation experiments with the primitive equation: I. The basic experiment". In: *Monthly Weather Review* 91 (1963), pp. 99–165.
35. F. L. Tucciarone, E. Mémin, and L. Li. "Primitive Equations Under Location Uncertainty: Analytical Description and Model Development". In: *Stochastic Transport in Upper Ocean Dynamics*. Ed. by B. Chapron et al. Cham: Springer International Publishing, 2023, pp. 287–300. ISBN: 978-3-031-18988-3.
36. G. Vallis. *Atmospheric and Oceanic Fluid Dynamics*. Cambridge University Press, 2017.
37. M. Wadati. "Stochastic Korteweg-de Vries Equation". In: *Journal of the Physical Society of Japan* 52.8 (1983), pp. 2642–2648. DOI: 10.1143/JPSJ.52.2642. eprint: https://doi.org/10.1143/JPSJ.52.2642. URL: https://doi.org/10.1143/JPSJ.52.2642.
38. A. Xiao and X. Tang. "High strong order stochastic Runge-Kutta methods for Stratonovich stochastic differential equations with scalar noise". In: *Numerical Algorithms* 72 (2016), pp. 259–296.

The Effects of Unresolved Scales on Analogue Forecasting Ensembles

Paul Platzer and Bertrand Chapron

1 Introduction

Analogues are nearest neighbors of a system's state according to a given similarity measure. The term "analogue" originally refers to the idea that certain geophysical systems are likely to undergo similar states at distant times, a notion close to the recurrence theorem of Poincaré [1] which states that certain dynamical systems will almost surely return infinitesimally close to any initial condition in the attractor. The term "analogue" was coined by Lorenz [2] who built a proxy of atmospheric predictability from the time-growth of the Euclidean distance between any observed geopotential height field and its closest neighbour (analogue) in other years of record. Since then, analogues have been used in other contexts and can be more broadly understood as similar observations of a system's state occurring at distant times either within a given long trajectory of the system, or within independent simulations of the same system.

More specifically, analogues are found in a database of events often called a "catalogue", such as a long series of observations [3], a reanalysis [4], past deterministic forecasts [5], one long numerical simulation [6], or ensemble simulations [7, 8]. One key advantage of analogue methods is that they are fast and therefore allow a cheap generation of ensembles. As far as forecasting is concerned, analogues have been used for post-processing of deterministic forecasts with "the Analogue Ensemble (AnEn)" [9], in weather generators with the "Analogue Weather Generator (AnaWEGE)" [10], in data assimilation with "the Analog Data Assimilation (AnDA)" [11], and finally for the forecast of dynamical systems with "Kernel Analog Forecasting (KAF)" [12].

P. Platzer (✉) · B. Chapron
Laboratoire d'Océanographie Physique et Spatiale (LOPS), Ifremer, Plouzané, France
e-mail: paul.platzer@ifremer.fr

© The Author(s) 2025   223
B. Chapron et al. (eds.), *Stochastic Transport in Upper Ocean Dynamics III*,
Mathematics of Planet Earth 13, https://doi.org/10.1007/978-3-031-70660-8_10

There are several mechanisms driving uncertainties of analogue forecasts. First, analogue forecasting generally assumes that the distance between the analogues and the target[1] image is small, however these distances are never zero [13]. These non-trivial distances introduce forecasting errors. Depending on the nature of the system under study and on the target image, this initial dissimilarity may increase or decrease with time. In a chaotic system, the mismatch in initial conditions grows according to scaling laws given by Lyapunov exponents [14]. This type of deterministic analogue forecasting error was studied thoroughly [15–17].

Second, for real systems one does not have access to the full state of the system, and the similarity measure used to find analogues is bound to be under-representative of the system's complexity. That is, analogues and corresponding distances are computed on observables of the system, not in the system's true state-space. In particular, fine-scale structures of atmospheric or ocean flows are generally unknown, and these may impact the time-evolution of the system. Therefore, even if the analogues were at distance zero to the target initial large-scale image according to the chosen similarity measure, their time-evolution would likely differ from the one of the target. In the present study we focus on this second source of uncertainty, referred to here as the effect of unresolved spatial scales on analogue forecasting ensembles.

In real-life case studies, these two sources of uncertainty are non-linearly combined and hard to disentangle. Here, we use numerical simulations to separate the effect of unresolved spatial scales from the effect of initial analogue-to-target distance. We base our analysis on numerical simulations of a stochastic version [18] of the well-known three-variable chaotic Lorenz system [19]. In this set of stochastic ordinary differential equation, the effects of unresolved spatial scales on the low-order model of Lorenz are rigorously derived from a reformulated material derivative, and expressed as stochastic multiplicative noise terms. This gives a physically consistent yet numerically affordable multi-scale system which can easily be used to understand the effects of unresolved spatial scales on analog forecasting ensembles.

Section 2 details the numerical simulations of the stochastic Lorenz system under location uncertainty as well as the different ensemble forecasting procedures and statistical tests considered in this work. Section 3 outlines numerical experiments and their results. Section 4 draws conclusion and gives perspectives for future work.

---

[1] The "target", also called "query" in machine-learning terminology, is the (vector of) value(s) of the observed variable(s) of which analogues are sought for. In the example of this article, the target is the three-dimensional vector of values of the coordinates of the stochastic Lorenz at a given time t.

2 Methods

2.1 Lorenz System Under Location Uncertainty

Flow modelling under location uncertainty introduces errors through the assumption that Lagrangian fluid particle displacement results from a smooth deterministic component and another component which is uncorrelated in time but correlated in space [20, 21]. This framework allows to derive the Lorenz system under location uncertainty [18] (hereafter "stochastic Lorenz") as a stochastic representation of the classical Lorenz-63 model [19] (hereafter "deterministic Lorenz") describing incompressible flow undergoing Rayleigh–Bénard convection in a simplified 2D atmospheric convection scenario.

Traditionally, large-scale geophysical flow representations rely on Reynolds decomposition and eddy viscosity models, which dissipate energy without considering local backscattered energy or inhomogeneous turbulence. In contrast, the stochastic approach from which we borrow decomposes the flow into a large-scale, smooth component and a small-scale, fast oscillating velocity component modeled as a random field. It is derived rigorously from physical conservation principles and offers efficient exploration of the attractor at low computational cost [18]. The derivation of the stochastic Lorenz follows closely the original derivation of the deterministic Lorenz, except that the location uncertainty framework allows to express the effects of truncated higher-order spatial modes (i.e., small spatial scales) on the dominant large-scale modes through multiplicative stochastic terms. Therefore, the noise terms in the model represent the effects of small spatial scales on the three large-scale variables, and will be loosely referred to as "small-scale variables" in the following.

Stochastic differential equations for this system are as follows:

$$\frac{dx}{dt} = Pra\,(y - x) - \frac{2}{\Upsilon}x, \tag{1}$$

$$dy = \left((\rho - z)x - \left(1 + \frac{2}{\Upsilon}\right)y\right)dt + \frac{\rho - z}{\Upsilon^{1/2}}dB_t, \tag{2}$$

$$dz = \left(xy - \left(b + \frac{4}{\Upsilon}\right)z\right)dt + \frac{y}{\Upsilon^{1/2}}dB_t. \tag{3}$$

It comprises a deterministic differential equation for the velocity variable x and two coupled stochastic differential equations associated with the small-scale temperature fluctuations, written in Itô formulation. Scalar parameters include the ones from the deterministic Lorenz model, which are here set to the usual chaotic values $Pra = 10$, $\rho = 28$, $b = 8/3$, and one parameter that controls noise amplitude, set to $\Upsilon = 10$ (this value is labelled "strong noise" by [18]). The noise terms dB_t are Wiener processes, which can be viewed heuristically as independent, identically distributed, centered normal random variables with variance dt, where dt is the time-increment of the numerical scheme.

Here, this system is numerically simulated with Euler–Maruyama integration for the stochastic differential equations with time step 10^{-5} as in [18].

In the following, we will use a parallel between the non-dimensional time of the stochastic Lorenz system and the typical time of evolution of atmospheric synoptic circulation. We will consider, as in [11], that 0.08 non-dimensional time intervals of the stochastic Lorenz system correspond to 6 hours of atmospheric time. This choice is close to the one of [22] who set non-dimensional time intervals of length-1 of the Lorenz system to correspond to 5 days. This parallel is a rough estimation for comparison purposes and should be interpreted with care.

2.2 Ensemble Forecasts

We place ourselves in the context of the forecast of a "true" trajectory, which corresponds to the numerical integration of the stochastic Lorenz system ((1)–(3)) with time step 10^{-5}, from a given initial condition $\mathbf{x}_t = (x_t, y_t, z_t)$ on the large-scale variables, and with a given sequence of noises $\{dB_t, \ldots, dB_{t+h}\}$ from time t to time $t + h$, where h is the forecast horizon.

- True trajectory :

Large-scale variables (initial condition) = $\mathbf{x}_t$,

Small-scale variables = $\{dB_t, \ldots, dB_{t+h}\}$.

To forecast this true trajectory using analogues, we assume that one is given a "catalogue" which is here a sub-sampling of a long numerical resolution of the stochastic Lorenz system ((1)–(3)), generated with one particular realization of the noises. Each catalogue has a set of elements $\{\mathbf{x}_{t'}, t' \in \mathcal{C}\}$ in which analogues are sought for, and a corresponding set of elements $\{\mathbf{x}_{t'+h}, t' \in \mathcal{C}\}$ in which the successors of the analogues can be found. To generate the different catalogues used in this study, fixed-size random subsamples were drawn from the same long numerical solution of ((1)–(3)) from time $t' = 0$ to $t' = 7008$ in non-dimensional time-unit. The notation t' is here to emphasis that the time of the catalogue is different from the time of the true trajectory (typically, analogues are sought for in a catalogue of past observations). As the catalogues are random subsamples, the time-difference between different analogues are irregular, but each analogue $\mathbf{x}_{t'}$ is associated to its successor $\mathbf{x}_{t'+h}$ with the same horizon h.

We employ the simplest type of analogue ensemble forecasting where the forecast is given by an ensemble of K successors $\mathbf{x}_{t'_i+h}$ of analogues $\mathbf{x}_{t'_i}$ of the true initial point $\mathbf{x}_t$, and each successor is given the same weight $\frac{1}{K}$. In all numerical experiments, the number of analogues is set to 50. Analogues are computed as K-

nearest neighbours, in terms of Euclidean distance, of the large-scale variables' initial value $\mathbf{x}_t$, which are assumed to be observed, while the noises $\mathrm{d}B_t$ are unknown. Other similarity measures could be envisaged (see e.g., [23–25]), but this is beyond the scope of this study. The analogue ensemble forecast is then made of the set successors $\left\{ \mathbf{x}_{t'_1+h} \ \ldots, \ \mathbf{x}_{t'_K+h} \right\}$. Therefore, although $\mathbf{x}_{t'_i}$ should be close to $\mathbf{x}_t$, the sequence of noises $\left\{ \mathrm{d}B_{t'_i}, \ldots, \mathrm{d}B_{t'_i+h} \right\}$ is completely different (independent) from $\{ \mathrm{d}B_t, \ldots, \mathrm{d}B_{t+h} \}$.

- Analogue ensemble forecast :

Large-scale variables (initial conditions) $= \left[\mathbf{x}_{t'_1}, \ldots, \mathbf{x}_{t'_K} \right]$

Small-scale variables $= \left[\left\{ \mathrm{d}B_{t'_1}, \ldots, \mathrm{d}B_{t'_1+h} \right\}, \ldots, \left\{ \mathrm{d}B_{t'_K}, \ldots, \mathrm{d}B_{t'_K+h} \right\} \right]$

All catalogues are sub-samples of a long trajectory from time $t' = 0$ to $t' = 7008$ in non-dimensional time-unit, which corresponds to 60 years in atmospheric time. Note that our stochastic Lorenz system does not include seasonality, while in practice one must search for analogues in historical data within a seasonality condition, typically ± 2 months, and therefore our 60-years "season-less" catalogue will allow us to emulate the amount of data that one can access with ~ 180 years of data. Sub-catalogues of different sizes will be drawn from this large initial catalogue to asses the effect of catalogue size on analogue ensemble. For each catalogue size, different catalogues can be drawn from the original large catalogue, allowing to assess the uncertainties in analogue ensemble forecasts even when the catalogue size is known.

To understand the contribution of different types of uncertainty to the analogue ensemble forecast, we also compute two other ensemble forecasts.

First, we build what we refer to as the "small-scale ensemble forecast". This ensemble is given the same size as the analogue ensemble : $K = 50$ members. Every member is initiated from the same true initial condition on the large-scale variables $\mathbf{x}_t$ (i.e., for horizon $h = 0$ the ensemble is a Dirac-delta function centered on $\mathbf{x}_t$). Each member is propagated with numerical simulations of the stochastic Lorenz and randomly generated noises $\left\{ \mathrm{d}B'_t, \ldots, \mathrm{d}B'_{t+h} \right\}$ that are independent, different realizations of the same process as the true noises $\{ \mathrm{d}B_t, \ldots, \mathrm{d}B_{t+h} \}$. Each member i corresponds to a particular realization $\left\{ \mathrm{d}B'^{(i)}_t, \ldots, \mathrm{d}B'^{(i)}_{t+h} \right\}$ of the noises. Members are given equal weights $\frac{1}{K}$.

- Small-scale ensemble forecast :

Large-scale variables (initial conditions) $= [\mathbf{x}_t, \ldots, \mathbf{x}_t]$

Small-scale variables $= \left[\left\{ \mathrm{d}B'^{(1)}_t, \ldots, \mathrm{d}B'^{(1)}_{t+h} \right\}, \ldots, \left\{ \mathrm{d}B'^{(K)}_t, \ldots, \mathrm{d}B'^{(K)}_{t+h} \right\} \right]$

This ensemble represents the effect of uncertainty associated with the unresolved small spatial scales modelled by the noises. Since analogue forecasting is based only on the knowledge of resolved large-spatial-scale variables, one desired property of analogue forecasting ensemble is that it resembles the small-scale ensemble.

Second, we build what we call the "large-scale ensemble forecast". We compute ensemble forecasts from numerical simulations of the stochastic Lorenz starting at the analogues' positions in large-scale variables space $\left[\mathbf{x}_{t'_1}, \ldots, \mathbf{x}_{t'_K}\right]$, but using the same noise $\{dB_t, \ldots, dB_{t+h}\}$ as the "ground truth" trajectory. Therefore, while each member of the small-scale ensemble differs by the realization of the noises, on the contrary each member of the large-scale ensemble differs by the initial condition which corresponds to the initial value of the analogue ensemble. Members are again given equal weights $\frac{1}{K}$.

- Large-scale ensemble forecast :

Large-scale variables (initial conditions) = $\left[\mathbf{x}_{t'_1}, \ldots, \mathbf{x}_{t'_K}\right]$

Small-scale variables = $[\{dB_t, \ldots, dB_{t+h}\}, \ldots, \{dB_t, \ldots, dB_{t+h}\}]$

This large-scale ensemble allows to assess what the analogues would have evolved to if they were forced by the same small-scale as the ground truth trajectory. This ensemble allows to isolate the effect of mismatch in large-scale initial condition, also called "analogue-to-target distance" [13].

The properties of the different ensembles with respect to the ground truth and the analogues are summarized in Table 1 and Fig. 1. Figure 2 shows a schematic example of ensemble forecast trajectories in large-scale and small-scale space compared to the ground-truth. This figure is only illustrative and was not generated using the real stochastic Lorenz system. It highlights that the large-scale ensemble shares large-scale initial conditions with the analogue ensemble, while the small-scale ensemble shares large-scale initial condition with the ground truth. Conversely, the large-scale ensemble shares small-scale trajectory with the ground truth, while the analogue ensemble and the small-scale ensemble have different, independent small-scale trajectories.

Note that this procedure would be hardly feasible with a real physical model where spatial and temporal scales are not separated but intertwined. The stochastic Lorenz model not only allows to perform numerical simulations at a low cost, but

Table 1 Properties of the different forecast ensembles considered

	Initial condition on resolved large-scale variables (x, y, z)	Trajectory of unresolved small-scale variables dB_t
Analogue ensemble	Past large-scale analogues of ground-truth	Past realizations associated with large-scale analogues of ground truth
Small-scale ensemble	Same as ground-truth ("present")	Random ("that could have been")
Large-scale ensemble	Same as analogues ("past")	Same as ground truth ("present")

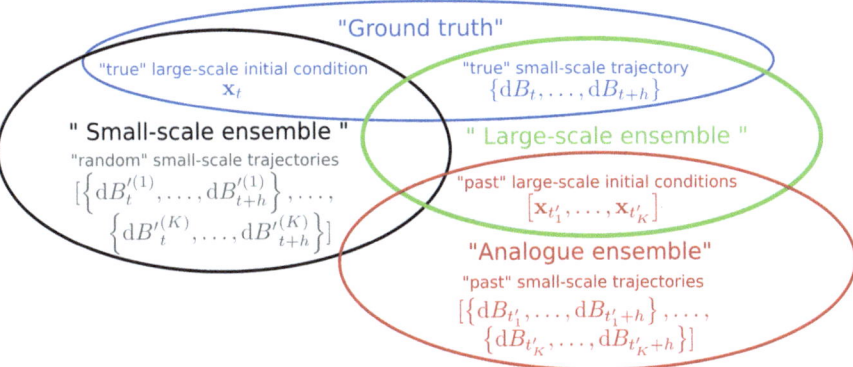

Fig. 1 Schematic of the shared properties of the different forecast ensembles considered and the ground truth trajectory

Fig. 2 Schematic of the trajectories of the different forecast ensembles considered and the ground truth trajectory, with only two ensemble members ($K = 2$) for readability purposes

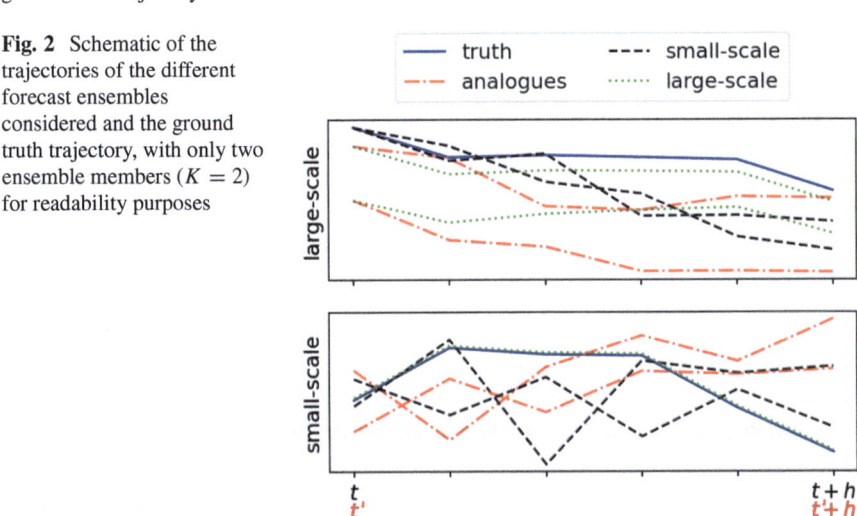

also facilitates the separation between the effects of small-scale and large-scale components of the flow.

The ground truth to which these ensemble forecasts is compared is made of one long trajectory of 1168 non-dimensional time steps which corresponds to 10 years in atmospheric time-scale. The noises dB_t associated with this ground-truth trajectory are saved in memory to allow for the generation of the large-scale ensemble forecasts.

2.3   Statistical Tests to Compare Ensembles

In Sect. 3.2, we will make use of three statistical tests to assess the potential differences between two given forecast ensembles. The tests will be performed on each large-scale variable separately, either x, y or z.

The first test is called Welch's t-test [26], and is used to test the null hypothesis that the two ensembles' background distributions have the same mean value without necessarily having the same variance. It is based on the assumption that the two samples are normally distributed.

The second test is called Levene's test [27] and is used to test the null hypothesis that the two ensembles' background distributions have the same variance. This test does not make assumption of normality.

The third test is called the Kolmogorov-Smirnov test [28] and is used to test the null hypothesis that the two ensembles' background distributions are equal. It does not make assumption of normality, and can also be used for complex distributions such as multi-modal distributions, for which the notions of mean and variance are inoperative.

For all three tests, we used built-in functions from Python's SciPy package, namely "ttest_ind", "levene" and "kstest". We use the p-value as an indicator, compared to the reference values 0.01, 0.05 and 0.1. p-values below 0.05 indicate that the distributions are different (high confidence if below 0.01), while p-values above 0.05 indicate that the distributions are equal (high confidence if above 0.1).

3   Numerical Experiments

3.1   Small-scale and Large-scale Variability

Figure 3 shows an example of the three different forecast ensembles outlined in the previous section, along with the ground truth trajectory, for a given initial condition inside the attractor, with forecast ensembles of size $K = 20$ (we use a smaller value for visualization purposes only). The forecast horizon of 0.1 corresponds to between 7 and 8 hours in atmospheric time scale. The large-scale ensemble is less dispersed as the two other ensembles, and is close to the ground truth. The fact that this ensemble appears to shrink with time for small forecast horizon is a consequence of the dissipative nature of the Lorenz system. The analogue and small-scale ensembles are close to each other at horizon $h = 0.1$, which is a desired property of analogue ensemble forecasting because it is based only on the knowledge of large-scale variables x, y, z, and it is not informed by the values of the small-scale variables (here, the noises). However, for smaller time horizon one can witness that the small-scale ensemble tends to the initial position, which is not the case of the analogue and large-scale ensembles. This undesired property of analogue forecasting errors for very short forecast-horizon was described previously in [17].

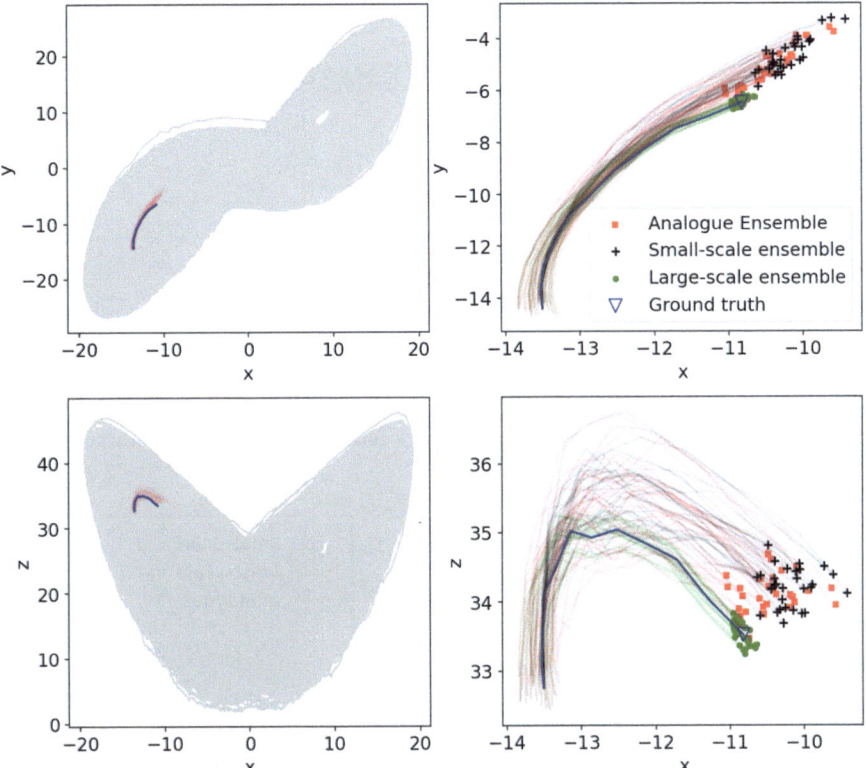

Fig. 3 Showing an example of three types of ensemble forecasts: small-scale (black lines and crosses), large-scale (green lines and circles) and analogue forecasting (red lines and squares) ensembles, versus ground truth trajectory (blue line and empty triangle). Final positions are given by the symbols. The forecast horizon is 0.1. We show only 20 members out of 50 for readability purposes

Focusing on ensemble spread, we now define the generalized variance (GV) as the determinant of the ensemble covariances, noted $|\Sigma_s|$ (small-scale), $|\Sigma_l|$ (large-scale), and $|\Sigma_a|$ (analogues).

Depending on the initial point, the situation may differ from Fig. 3. To test this assertion, Fig. 4 shows the ratio $(|\Sigma_s|/|\Sigma_l|)^{1/2}$ in log-scale, depending on the position of the initial point in the attractor, using 1000 points from the 10-years ground truth trajectory. Figure 4 shows that the relative spread of the large-scale and small-scale ensembles strongly depends on the values of the large-scale variables at the initial point. In particular and at this forecast horizon, small-scale variability dominates for values of $z < 20$, and large-scale variability dominates at the wing's borders for $z > 25$ and $-5 < y < 5$. Investigating the variability of $|\Sigma_s|$ and $|\Sigma_l|$ with initial point position shows that what drives high values of the ratio $|\Sigma_s|/|\Sigma_l|$ for low values of z is $|\Sigma_s|$, with a strong influence of small-scale noises in this area (not shown), while the smaller values of the ratio $|\Sigma_s|/|\Sigma_l|$ at the wings' borders are

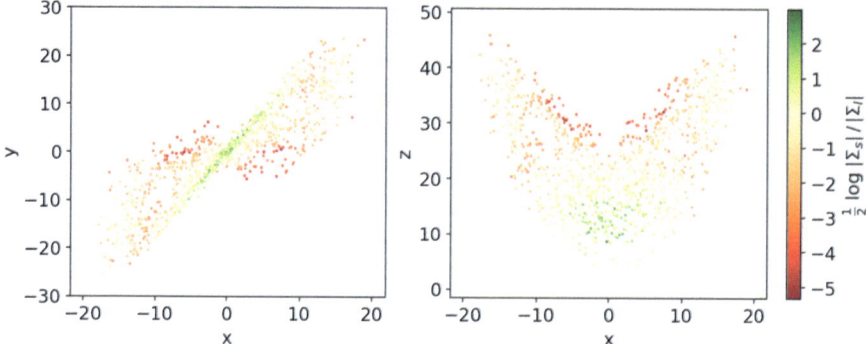

Fig. 4 Showing the ratio of small-scale to large-scale generalized standard-deviation ensemble variability (square-root of the covariance matrix's determinant), for forecast horizon 0.06 (equivalent to 4.5 hours in atmospheric time-scale), as a function of the initial point's position, in logarithmic scale

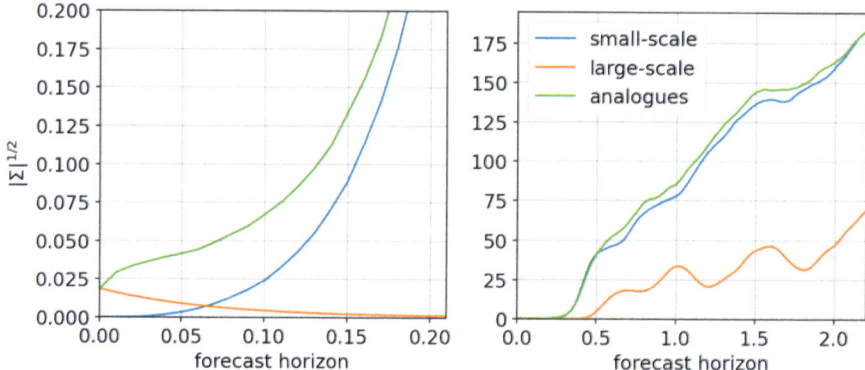

Fig. 5 Showing the attractor-averaged ensemble variability (given by the square root of the covariance matrix's determinant), against forecast horizon, for small-scale, large-scale and analogue ensemble forecasts. Left: zoom on small forecast horizon (<0.2). Right: forecast horizon up to 2.24

due to a larger values of $|\Sigma_l|$ which are caused by a smaller density of points (not shown). This experiment shows that analogue forecasting ensemble spread may be influenced both by the varying intensity of perturbation by unresolved spatial scales and by the varying density of points in the attractor (and therefore of analogue-to-target distances), both of which highly depend on the position in the attractor.

As shown in the particular case of Fig. 3, the ration between these ensemble GV depends not only the initial large-scale position but also on forecast horizon. The square root of GVs (which can be viewed as a generalized standard-deviation) $|\Sigma_s|^{1/2}$, $|\Sigma_l|^{1/2}$ and $|\Sigma_a|^{1/2}$, are averaged over 1000 points in the attractor (randomly drawn from the 10-year ground truth trajectory), and plotted against forecast horizon in Fig. 5. By construction, the small-scale ensemble GV tends to zero for small forecast horizon. Large-scale GV decreases as a function of time for small time-horizons because the Lorenz system is dissipative, however it remains non-zero. The analogue and large-scale ensemble GV are equal at horizon-zero by construction, and then the analogue ensemble GV grows similarly to small-scale ensemble GV. This shows that the analogue ensemble variability is highly linked to the small-scale ensemble variability, although it naturally overestimates the latter for short forecast-horizon because of the non-zero distance between the analogues and the initial point. Therefore, the only way that analogues can adequately represent the small-scale driven variability is to decrease analogue-to-target distance by increasing the catalog size.

3.2 Convergence of Analogues Towards the Small-scale Ensemble

In Figs. 6 and 7, two examples of convergence of the distribution of analogues to the small-scale ensemble distribution are shown, at fixed forecast horizon 0.1 (7–8 hours). The analogue forecast ensembles converge the small-scale ensembles in the limit of large catalogue size. In addition to showing quantiles of distribution as a function of catalogue size, we also show p-values for three statistical tests where the null hypothesis is that the analogue and small-scale ensembles are drawn from the same background continuous probability distribution. The catalog size needed for convergence depends on the point of reference and on the chosen variable. For instance it is fastest for variable z and Fig. 6, and slowest for variable z and Fig. 7. This is because in the second example the initial point is in an area of smaller attractor density. In Fig. 6, the average of the distribution converges faster than the standard deviation, while in the second example all three statistical tests indicate a convergence for approximately the same catalogue size. When the initial point is in an area of large attractor density, the analogue ensemble is likely to have a precise mean even for small catalogue size, although it will necessarily overestimate the variance of the small-scale ensemble. On the contrary, for small catalogue sizes and in areas of small attractor density such as the tail of the butterfly's wings, the analogue average is biases towards the attractor's average.

The features highlighted in the particular cases of Figs. 6 and 7 allow to better understand the general law given in Fig. 8. In the latter, we test the convergence of the analogue ensemble to the small-scale ensemble (for horizon 0.13, $\sim$10 hours, and horizon 0.43, $\sim$32 hours) according to the three statistical tests. The p-values are stored for 10 catalogues for each catalogue size and for 1000 points in the attractor taken from the ground truth trajectory, and quantiles are extracted for each catalogue size. The observed behaviour for the top plots (a-c, horizon 0.13, $\sim$10 hours) resembles the one of Fig. 6, which is the case where forecast horizon is small, and the attractor density is high at the initial point. This shows that for short forecast horizon the analogue ensemble demands a high catalogue size to adequately represent the standard-deviation while the average can be well represented even with small catalogue size. However, this must be tempered by the fact that the

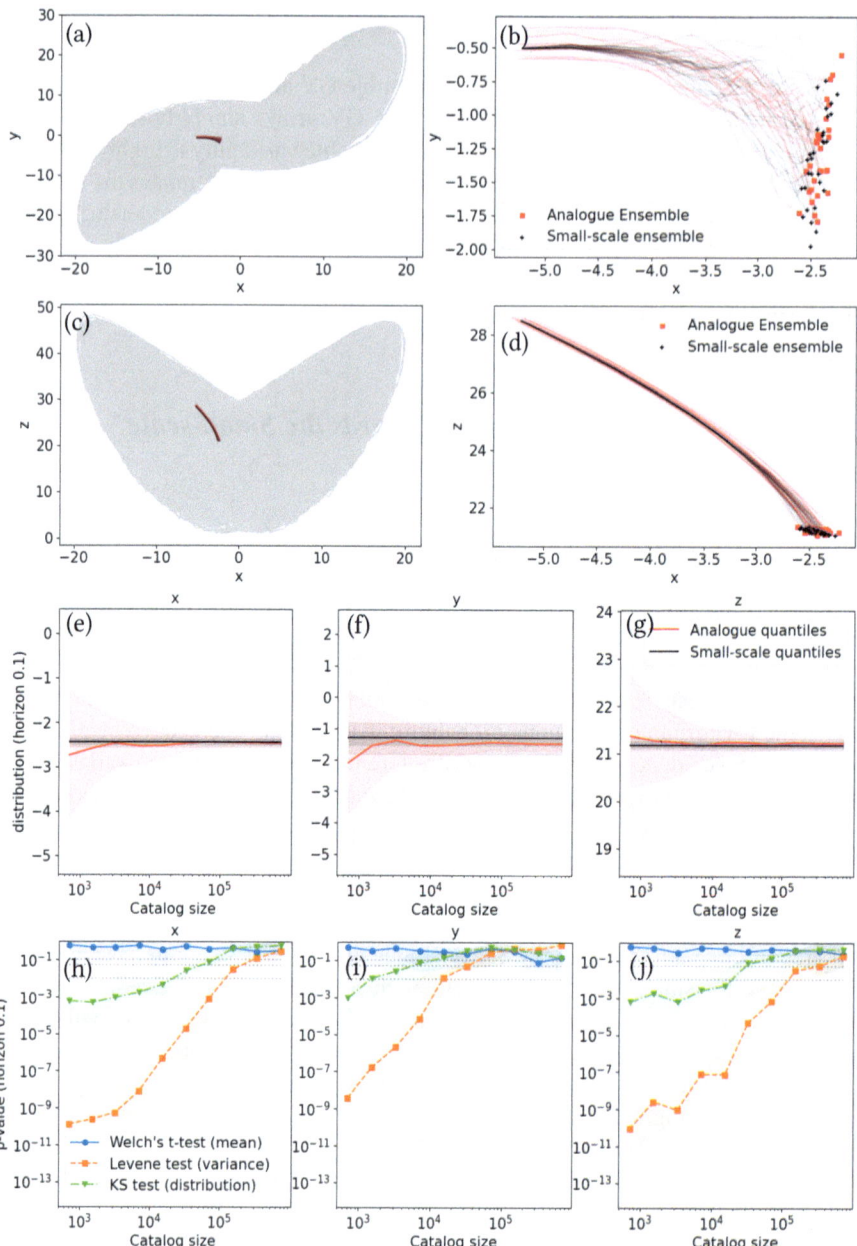

Fig. 6 One example of analogue ensemble forecast versus small-scale perturbed forecast, at forecast horizon 0.1. Top (**a**)–(**d**): forecast trajectories, unzoomed (left) and zoomed (right). Middle (**e**)–(**g**): analogue ensemble distribution (red) as a function of catalogue size, and small-scale ensemble distribution (black) shown for comparison, for each variable (x: (a), y: (b), z: (c)), showing quantiles 0.05-0.25-0.5-0.75-0.95. Bottom (**h**)–(**j**): p-value for Welch's t-test (full blue line and circles), Levene's test (dashed orange line and squares), and the Kolmogorov-Smirnov test (dash-dotted green line with triangles), versus catalogue size, where the null hypothesis is that the analogue ensemble and the small-scale ensemble are drawn from the same distribution, for each variable (x: (a), y: (b), z: (c)), showing quantiles 0.05-0.25-0.5-0.75-0.95 for different draws of catalogues for each catalogue size. Reference p-values of 0.01-0.05-0.1 are shown (dotted gray lines)

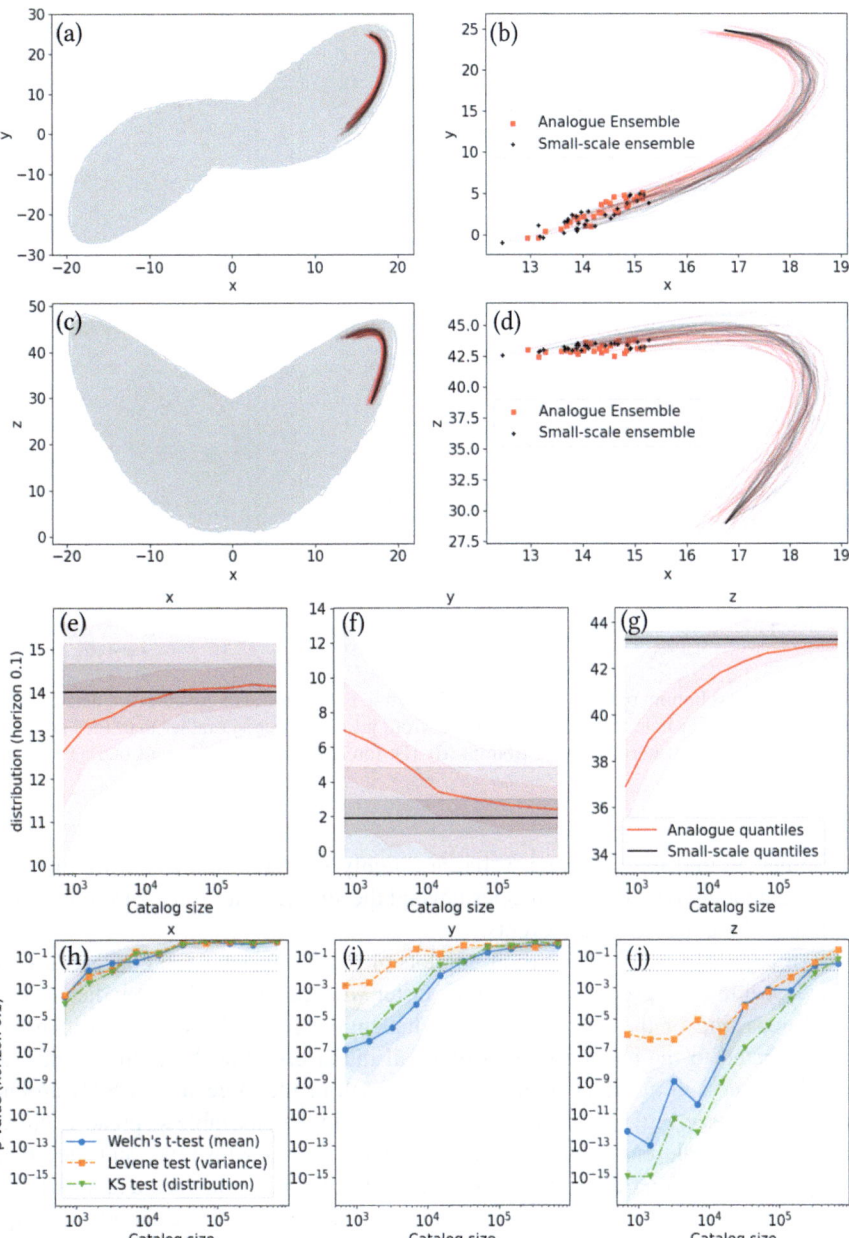

Fig. 7 Same as Fig. 6 but for a different initial condition

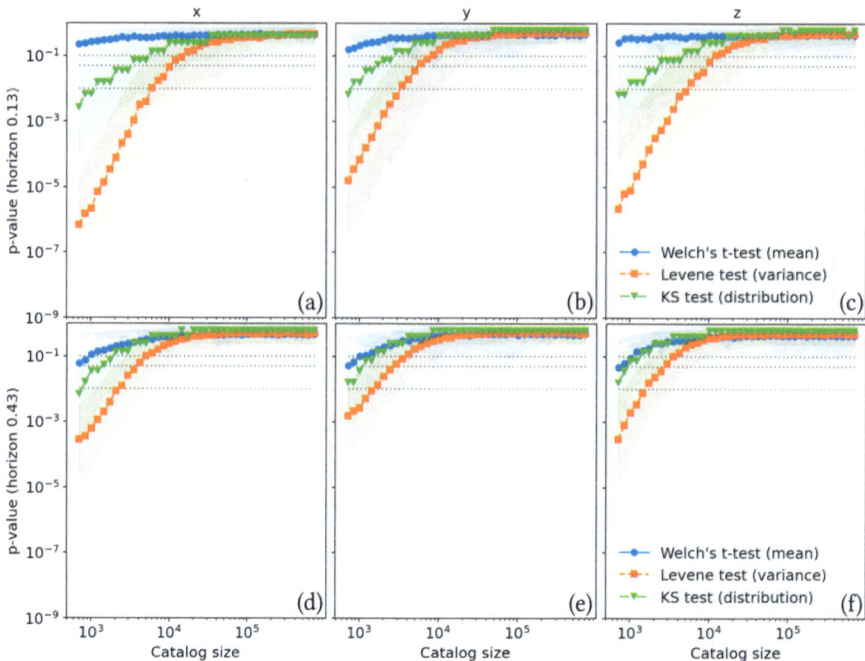

Fig. 8 Same as bottom plots of Fig. 6h–j and Fig. 7h–j, but quantiles of p-values are derived from statistics over the whole attractor (1000 points from a long trajectory) rather than for one point. Top (**a**)–(**c**): forecast horizon=0.13. Bottom (**d**)–(**f**): forecast horizon=0.43. Left (a,d): x. Middle (b,e): y. Right (c,f): z

quantiles around the median p-value are highly scattered, indicating the need to adapt analogue forecasting strategies beyond the simple brute-force ensemble when dealing with extreme or rare events.

For larger forecast horizon (bottom plots of Fig. 8), Welch's test converges more slowly while the Levene test converges faster with catalogue size. It indicates that the forecasting of the average is more challenging for such horizons, which can be attributed to the chaotic growth of small initial errors at these time scales (as can be witnessed in the beginning of the growth of the large-scale GSTD around horizon 0.5, see Fig. 5). On the contrary, the analogue ensemble's spread is closer to the small-scale ensemble spread, as the latter grows with time, and the effect of analogue-to-target distance starts to fade at these horizons (see again in Fig. 5 how the small-scale and analogue ensemble GSTD get closer to each other at large forecast horizon).

Building from the results of Fig. 8, we define a minimum catalogue size for each statistical test and large-scale variable, as the catalogue size for which a given quantile of p-value exceeds the threshold value of 0.1, where quantiles of p-value are calculated over 1000 points on the attractor and for 10 catalogues of equal size. To make this computation, we consider 40 different catalogue sizes

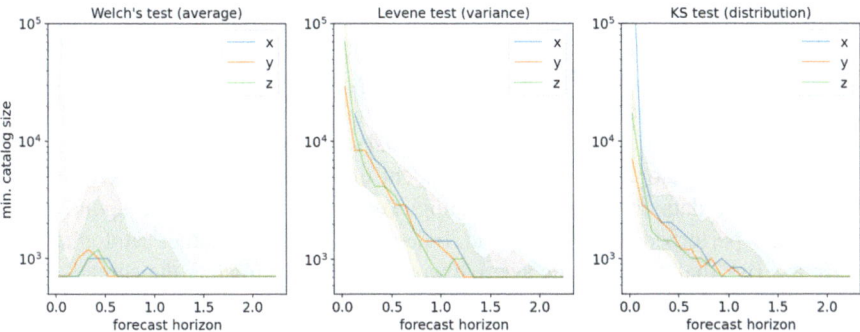

Fig. 9 Minimum catalogue size so that the quantile (0.3,0.5,0.7) of p-value over the attractor (1000 points from a long trajectory) exceeds the threshold value of 0.1, versus forecast horizon, for three tests (Welch: left, Levene: middle, Kolmogorov-Smirnov: right) and variables (x: blue, y: orange, z: green)

ranging from 700 to 700,000 independent elements, equally-spaced on a log-scale. Results for each variable (x, y, z) and for three quantiles of p-value (0.3,0.5,0.7) are shown in Fig. 9. To interpret these minimum catalog sizes, we make the parallel with atmospheric circulation data. We assume that one can find 60 independent circulation patterns as "analogue candidates" per year of data, if one restricts the search to a "season" of 4 months and if two circulation patterns are uncorrelated after 2 days on average.

The minimum catalogue size for Welch's t-test is highest around forecast horizon 0.4, for which a catalogue size of 10^3 is needed for the median p-value to exceed 0.1. Using our interpretation in terms of atmospheric data, a catalog size of 10^3 elements amounts to 20 years of data, which is reasonable with respect to nowadays's typical length of observational records. The minimum catalogue size for Welch's t-test then decreases with time horizon, which would not be the case for a deterministic system. Here, since the unresolved small-scale induce inevitable uncertainty in the forecast, the mismatch in initial condition due to the finite catalogue size has a lesser influence on the analogue ensemble average for larger forecast horizon. We consider horizons up to 2.24 which is equivalent to 1 week in atmospheric time-scale. For such large forecast horizons it appears that one does not require perfect knowledge of the large-scale initial condition in order to evaluate the average forecast under small-scale perturbation, which is due to the fact that the uncertainty due to the unknown small-scale features has more influence on the forecast than the initial large-scale perturbation due to the analogue-to-target distances. This particular experiment shows that, for chaotic multi-scale systems, analogue forecasting may actually be more successful in predicting the average of the stochastic large-scale system, than to predict the deterministic multi-scale system altogether.

The curves for the Levene and Kolmogorov-Smirnov tests are very similar to each other, but very different from the curves of Welch's t-test, indicating that what the analogue ensemble struggles to estimate is not the average but the spread of the small-scale distribution. As we have seen, for small horizons and small catalogue size the analogue ensemble overestimates the ensemble spread attributed to the unresolved small-scales. According to Fig. 9, this effect cannot be neglected until horizon 1.2, which corresponds to ~4 days in atmospheric time-scale. Taking this analogy further, for a 1-day (horizon 0.4) analogue ensemble forecast one would need a catalogue of 6000 independent elements, which would correspond to 120 years of data, in order for the analogue ensemble spread to correspond to the small-scale perturbed ensemble spread. Indeed, if one uses a smaller catalogue size, there is a risk that although the average would be well estimated, the spread would not, which is an unwanted property.

This result urges for the development of new ways of estimating the forecast spread based on the analogue ensemble, beyond using directly the spread of the analogue ensemble. Although optimizing the value of K might help in reducing the analogue ensemble spread, it will not be able to reduce the non-zero analogue-to-target distance due to finite catalogue size. Indeed, analogue-to-target distances depend on the analogue rank K as $(K/L)^{1/d}$ where d is the local dimension of the attractor and L is the catalogue size [13]. Therefore the distance between two analogues is lower-bounded by a constant proportional to $L^{-1/d}$, which gives a lower-bound for the analogue ensemble spread as well. Moreover, for moderately high-dimensional systems such as large-scale atmospheric circulation, the dimension ranges from 10 to 20 [29], and therefore analogue-to-target distances increase very slowly with the number of analogue K, therefore decreasing the number of analogues is likely not to reduce efficiently the analogue ensemble spread.

4   Conclusion and Perspectives

We have studied the effects of unresolved spatial scales on analogue forecast ensembles, using numerical simulations of a set of stochastic differential equations that emulate the effects of small-scales on the three large-scale variables of the chaotic Lorenz system in a physically consistent way. We have interpreted the behaviour of analogue ensemble forecasts based on two other ensembles, the large-scale ensemble which has the same noises (i.e. the same small-scale) as the ground truth but the same large-scale initial conditions as the analogues, and the small-scale ensemble which has the same large-scale initial condition as the ground truth but different noises (i.e. different small-scales). We have shown that the ratio of large-scale to small-scale ensemble spread strongly depends on the initial position in the attractor and on forecast horizon. In particular, the large-scale (respectively, small-scale) spread dominates for short-term (respectively, long-term) forecasts. The analogue ensemble spread is dominated by large-scale effects for short term forecasts and gradually tends to the small-scale forecast spread for larger time-horizons. The analogue ensemble converges in distribution to the small-scale ensemble in the limit of large catalogue size, which is a desired property of analogue ensemble forecasts. The rate of convergence depends on the position in the attractor and on the forecast horizon. The convergence of the analogue ensemble average to the small-scale ensemble average is rather fast, while the convergence of ensemble spread is slow. In particular, for short-term forecast of 1–2 days in atmospheric time-scale, our results in this idealized setup suggest that the catalogue size needed for the analogue ensemble to adequately represent the small-scale driven spread might exceed hundreds of years of data. In order for analogue ensembles to reliably estimate uncertainties associated to unresolved small-scales, one might benefit from the development of methodologies that go beyond the simple spread of the analogue ensemble.

This work should be extended to higher-dimensional physical systems with scale separation, and systematic tests of the adequacy between the analogue ensemble spread and the uncertainty of the analogue ensemble forecast with respect to the ground truth trajectory should be performed using atmospheric and ocean circulation data such as reanalysis products. We advocate for an adaption of uncertainty quantification from analogue ensembles using leave-one-out procedures, famous in the machine-learning community [30], which could also be used to produce unbiased analogue ensemble average forecasts in data-scarce areas of the attractor. Finally, proper tuning of the distance using metric-learning algorithms could help calibrate the analogue ensemble towards the desired distribution.

Acknowledgments This work was financially supported by the ERC project 856408-STUOD. We thank Pierre Dérian for providing code snippets on numerical simulations of the Lorenz System under location uncertainty. We thank two anonymous reviewers for helpful comments and suggestions.

References

1. H. Poincaré, "Sur le problème des trois corps et les équations de la dynamique," *Acta mathematica*, vol. 13, no. 1, pp. A3–A270, 1890.
2. E. N. Lorenz, "Atmospheric predictability as revealed by naturally occurring analogues," *Journal of Atmospheric Sciences*, vol. 26, no. 4, pp. 636–646, 1969.
3. R. Alexander, Z. Zhao, E. Székely, and D. Giannakis, "Kernel analog forecasting of tropical intraseasonal oscillations," *Journal of the Atmospheric Sciences*, vol. 74, no. 4, pp. 1321–1342, 2017.
4. P. Yiou and C. Déandréis, "Stochastic ensemble climate forecast with an analogue model," *Geoscientific Model Development*, vol. 12, no. 2, pp. 723–734, 2019.
5. S. Alessandrini, L. Delle Monache, S. Sperati, and G. Cervone, "An analog ensemble for short-term probabilistic solar power forecast," *Applied energy*, vol. 157, pp. 95–110, 2015.
6. D. Burov, D. Giannakis, K. Manohar, and A. Stuart, "Kernel analog forecasting: Multiscale test problems," *Multiscale Modeling & Simulation*, vol. 19, no. 2, pp. 1011–1040, 2021.
7. H. Ding, M. Newman, M. A. Alexander, and A. T. Wittenberg, "Diagnosing secular variations in retrospective enso seasonal forecast skill using cmip5 model-analogs," *Geophysical Research Letters*, vol. 46, no. 3, pp. 1721–1730, 2019.
8. Y. Zhen, P. Tandeo, S. Leroux, S. Metref, T. Penduff, and J. Le Sommer, "An adaptive optimal interpolation based on analog forecasting: application to ssh in the gulf of mexico," *Journal of Atmospheric and Oceanic Technology*, vol. 37, no. 9, pp. 1697–1711, 2020.

9. L. Delle Monache, F. A. Eckel, D. L. Rife, B. Nagarajan, and K. Searight, "Probabilistic weather prediction with an analog ensemble," *Monthly Weather Review*, vol. 141, no. 10, pp. 3498–3516, 2013.

10. P. Yiou, "Anawege: a weather generator based on analogues of atmospheric circulation," *Geoscientific Model Development*, vol. 7, no. 2, pp. 531–543, 2014.

11. R. Lguensat, P. Tandeo, P. Ailliot, M. Pulido, and R. Fablet, "The analog data assimilation," *Monthly Weather Review*, vol. 145, no. 10, pp. 4093–4107, 2017.

12. R. Alexander and D. Giannakis, "Operator-theoretic framework for forecasting nonlinear time series with kernel analog techniques," *Physica D: Nonlinear Phenomena*, vol. 409, p. 132520, 2020.

13. P. Platzer, P. Yiou, P. Naveau, J.-F. Filipot, M. Thiebaut, and P. Tandeo, "Probability distributions for analog-to-target distances," *Journal of the Atmospheric Sciences*, vol. 78, no. 10, pp. 3317–3335, 2021.

14. A. Wolf, J. B. Swift, H. L. Swinney, and J. A. Vastano, "Determining lyapunov exponents from a time series," *Physica D: nonlinear phenomena*, vol. 16, no. 3, pp. 285–317, 1985.

15. J. D. Farmer and J. J. Sidorowich, "Predicting chaotic time series," *Physical review letters*, vol. 59, no. 8, p. 845, 1987.

16. J. D. Farmer and J. J. Sidorowichl, "Exploiting chaos to predict the future and reduce noise," in *Evolution, learning and cognition*, pp. 277–330, World Scientific, 1988.

17. P. Platzer, P. Yiou, P. Naveau, P. Tandeo, J.-F. Filipot, P. Ailliot, and Y. Zhen, "Using local dynamics to explain analog forecasting of chaotic systems," *Journal of the Atmospheric Sciences*, vol. 78, no. 7, pp. 2117–2133, 2021.

18. B. Chapron, P. Dérian, E. Mémin, and V. Resseguier, "Large-scale flows under location uncertainty: a consistent stochastic framework," *Quarterly Journal of the Royal Meteorological Society*, vol. 144, no. 710, pp. 251–260, 2018.

19. E. N. Lorenz, "Deterministic nonperiodic flow," *Journal of the atmospheric sciences*, vol. 20, no. 2, pp. 130–141, 1963.

20. V. Resseguier, E. Mémin, and B. Chapron, "Geophysical flows under location uncertainty, part i random transport and general models," *Geophysical & Astrophysical Fluid Dynamics*, vol. 111, no. 3, pp. 149–176, 2017.

21. V. Resseguier, E. Mémin, and B. Chapron, "Geophysical flows under location uncertainty, part ii quasi-geostrophy and efficient ensemble spreading," *Geophysical & Astrophysical Fluid Dynamics*, vol. 111, no. 3, pp. 177–208, 2017.

22. T. N. Palmer, "Extended-range atmospheric prediction and the lorenz model," *Bulletin of the American Meteorological Society*, vol. 74, no. 1, pp. 49–66, 1993.

23. Z. Toth, "Intercomparison of circulation similarity measures," *Monthly weather review*, vol. 119, no. 1, pp. 55–64, 1991.

24. K. Fraedrich and B. Rückert, "Metric adaption for analog forecasting," *Physica A: Statistical Mechanics and its Applications*, vol. 253, no. 1–4, pp. 379–393, 1998.

25. C. Matulla, X. Zhang, X. Wang, J. Wang, E. Zorita, S. Wagner, and H. Von Storch, "Influence of similarity measures on the performance of the analog method for downscaling daily precipitation," *Climate Dynamics*, vol. 30, pp. 133–144, 2008.

26. B. L. Welch, "The generalization of 'student's' problem when several different population varlances are involved," *Biometrika*, vol. 34, no. 1–2, pp. 28–35, 1947.

27. H. Levene, "Robust tests for equality of variances," *Contributions to probability and statistics*, pp. 278–292, 1960.

28. J. W. Pratt, J. D. Gibbons, J. W. Pratt, and J. D. Gibbons, "Kolmogorov-smirnov two-sample tests," *Concepts of nonparametric theory*, pp. 318–344, 1981.

29. D. Faranda, G. Messori, and P. Yiou, "Dynamical proxies of north atlantic predictability and extremes," *Scientific reports*, vol. 7, no. 1, pp. 1–10, 2017.

30. D. W. Aha, "Lazy learning," in *Lazy learning*, pp. 7–10, Springer, 1997.

Particle-Based Algorithm for Stochastic Optimal Control

Sebastian Reich

1 Introduction

We consider controlled nonlinear diffusion processes of the form

$$\mathrm{d}X_t = b(X_t)\mathrm{d}t + G(X_t)u_t\mathrm{d}t + \sigma(X_t)\mathrm{d}B_t, \qquad X_0 = x_0. \tag{1}$$

Here $X_t \in \mathbb{R}^{d_x}$ denotes the random state variable at time $t \geq 0$ and $B_t \in \mathbb{R}^{d_b}$ standard d_b-dimensional Brownian motion. Furthermore, the functions $b(x) \in \mathbb{R}^{d_x}$, $G(x) \in \mathbb{R}^{d_x \times d_u}$, and $\sigma(x) \in \mathbb{R}^{d_x \times d_b}$ are all assumed to be given.

The cost to optimize via an appropriate choice of the time-dependent control $u_{0:T} = \{u_t\}_{t \in [0,T]}$ is given by

$$J_T(x_0, u_{0:T}) = \mathbb{E}\left[\int_0^T \left(c(X_t) + \frac{1}{2}u_t^\mathsf{T} R^{-1} u_t\right)\mathrm{d}t + f(X_T)\right]. \tag{2}$$

Expectation is taken with regard to the path measure generated by the stochastic differential equation (SDE) (1) conditioned on the initial $X_0 = x_0$ and a given control law $u_{0:T}$, which we assume to be state-dependent. Here

$$c(x) = \frac{1}{2}h(x)^\mathsf{T} S^{-1} h(x) \tag{3}$$

denotes the running cost and

S. Reich (✉)
Institut für Mathematik, Universität Potsdam, Potsdam, Germany
e-mail: sebastian.reich@uni-potsdam.de

© The Author(s) 2025
B. Chapron et al. (eds.), *Stochastic Transport in Upper Ocean Dynamics III*,
Mathematics of Planet Earth 13, https://doi.org/10.1007/978-3-031-70660-8_11

$$f(x) = \frac{1}{2}\xi(x)^T V^{-1}\xi(x) \tag{4}$$

the terminal cost. The matrices $R \in \mathbb{R}^{d_u \times d_u}$, $S \in \mathbb{R}^{d_h \times d_h}$, $V \in \mathbb{R}^{d_\xi \times d_\xi}$ as well as the functions $h(x) \in \mathbb{R}^{d_h}$, $\xi(x) \in \mathbb{R}^{d_\xi}$ are again all assumed to be given. See [19] for an introduction to diffusion processes and [5, 18] for an introduction to stochastic control.

Inspired by the success of the ensemble Kalman filter (EnKF) for very high-dimensional data assimilation problems, which relies on ensemble-based linear estimation, we aim in this chapter at finding control laws of the form

$$u_t(x) = RG(x)^T(A_t x + c_t) \tag{5}$$

to provide robust approximations to the optimal feedback control law denoted here by $u_t^*(x)$. While the optimal control law can be found via the associated Hamilton–Jacobi–Bellman (HJB) equation [5]; solving such a PDE is computationally demanding [10]. Popular alternatives include those based on forward-backward SDEs [6] in combination with machine learning techniques [9, 10]. Here we follow the work of [16] instead, which in turn has been inspired by [2, 17, 21], to reformulate the problem in terms of two McKean–Vlasov SDEs over state space $\mathbb{R}^{d_x}$. Those SDEs need to be solved in forward and reverse time, respectively, only once and are related to generative models using diffusion processes [23, 24].

While the work [16] is restricted to control problems for which the Cole–Hopf transformation linearizes the associated HJB equation, we consider a more general class of control problems for which the HJB equation cannot be linearized. See Remark 1 below. Furthermore, in order to obtain robust and easy to compute approximations, we employ the EnKF methodology to approximate the arising McKean–Vlasov interaction terms [4, 11]. A related EnKF-based approach to optimal control has been considered in [14] and is based on a direct approximation of the HJB equation via a McKean–Vlasov SDE. We use stabilization of an inverted pendulum position [18] to demonstrate the efficiency of our method. While the numerical experiments in [16] and [14] utilize ensemble sizes on the order of 10^3, our method has been implemented with an ensemble of size $M = d_x + 1 = 3$. We also propose a more general methodology which combines the EnKF methodology with diffusion map approximations for the arising grad-log density terms [7, 8]. This extension is useful whenever the underlying densities cannot be approximated well by Gaussian distributions. We will illustrate this aspect through a controlled nonlinear Langevin dynamics process.

The remainder of this chapter is structured as follows. The mathematical background on the HJB equation for stochastic optimal control problems is summarized in Sect. 2. We demonstrate in Sect. 3 how the value function defined by the HJB equation can be expressed as the ratio of two probability density functions. Here we extend previous work [16] to the wider class of optimal control problems defined by (1) and (2). Both the associated forward and reverse time evolution equations can be expressed in terms of McKean–Vlasov SDEs in the state variable, x. Before discussing numerical approximations of those McKean–Vlasov equations in Sect. 5, we demonstrate in Sect. 4 how our formulation is related to generative modeling using diffusion processes [23, 24]. We first discuss numerical approximations using EnKF-type methodologies [4, 11] in Sect. 5.1. Next we also employ diffusion maps [7, 8] in order to approximate grad-log density terms in Sect. 5.2. Numerical implementation details are discussed in Sect. 5.3, while numerical results for an inverted pendulum and nonlinear Langevin dynamics are presented in Sect. 6. Possible extension of the proposed methodology to infinite horizon optima control problems are discussed in Appendix.

2   Mathematical Problem Formulation

In this section, we recap the essential aspects of finding the optimal control law, $u_t^*(x)$, for controlled SDE (1) with cost (2). See [5, 18] for more detailed expositions.

The law, π_t, of the diffusion process, X_t, defined by the SDE (1) satisfies the Fokker–Planck equation

$$\partial_t \pi_t = -\nabla_x \cdot (\pi_t (b + Gu_t)) + \frac{1}{2} \nabla_x \cdot (\pi_t \Sigma) \tag{6a}$$

$$= -\nabla_x \cdot \left(\pi_t \left(b + Gu_t - \frac{1}{2} \nabla_x \cdot \Sigma - \frac{1}{2} \Sigma \nabla_x \log \pi_t \right) \right), \tag{6b}$$

where

$$\Sigma(x) = \sigma(x)\sigma(x)^{\mathrm{T}}. \tag{7}$$

We introduce the weighted norm $\| \cdot \|_R$ via

$$\|u\|_R^2 = u^{\mathrm{T}} R^{-1} u \tag{8}$$

and the Frobenius inner product $A : B = \mathrm{tr}\,(AB^{\mathrm{T}})$ of two $d_x \times d_x$ matrices A and B. The optimal feedback control is provided by

$$u_t^*(x) = -RG(x)^{\mathrm{T}} \nabla_x y_t^*(x), \tag{9}$$

where the optimal value function $y_t^*(x)$ satisfies the HJB equation

$$-\partial_t y_t^* = b \cdot \nabla_x y_t^* + \frac{1}{2} \Sigma : D_x^2 y_t^* + c + \min_u \left(Gu \cdot \nabla_x y_t^* + \frac{1}{2} \|u\|_R^2 \right) \tag{10a}$$

$$= b \cdot \nabla_x y_t^* + \frac{1}{2} \Sigma : D_x^2 y_t^* + c - \frac{1}{2} \|RG^{\mathrm{T}} \nabla_x y_t^*\|_R^2 \tag{10b}$$

with terminal condition $y_T^* = f$. Here $D_x^2 y_t^*(x) \in \mathbb{R}^{d_x \times d_x}$ denotes the Hessian of $y_t^*(x)$.

We apply the Cole–Hopf transformation and introduce the new function

$$v_t^*(x) := \exp(-y_t^*(x)) \tag{11}$$

in order to obtain the transformed HJB equation

$$-\partial_t v_t^* = b \cdot \nabla_x v_t^* + \frac{1}{2} \Sigma : D_x^2 v_t^* \tag{12a}$$

$$- \left(c + \frac{1}{2} \| \sigma^T \nabla_x \log v_t^* \|^2 + \min_u \left(\frac{1}{2} \| u \|_R^2 - Gu \cdot \nabla_x \log v_t^* \right) \right) v_t^* \tag{12b}$$

$$= b \cdot \nabla_x v_t^* + \frac{1}{2} \Sigma : D_x^2 v_t^* \tag{12c}$$

$$- \left(c + \frac{1}{2} \| \sigma^T \nabla_x \log v_t^* \|^2 - \frac{1}{2} \| RG^T \nabla_x \log v_t^* \|_R^2 \right) v_t^*. \tag{12d}$$

The terminal condition is $v_T^* = \exp(-f)$. Here we have used

$$\partial_t y_t^* = \frac{-1}{v_t^*} \partial_t v_t^*, \qquad \nabla_x y_t^* = \frac{-1}{v_t^*} \nabla_x v_t^*, \tag{13}$$

and

$$\Sigma : D_x^2 y_t^* = \frac{-1}{v_t^*} \Sigma : D_x^2 v_t^* + \| \sigma^T \nabla_x \log v_t^* \|^2. \tag{14}$$

The optimal control is now characterized by

$$u_t^*(x) = RG(x)^T \nabla_x \log v_t^*(x). \tag{15}$$

Remark 1 The transformed HJB equation (12) simplifies to

$$-\partial_t v_t^* = b \cdot \nabla_x v_t^* + \frac{1}{2} \Sigma : D_x^2 v_t^* - c v_t^* \tag{16}$$

for the special case $G = \sigma$ and $R = I$ and (12) becomes linear in v_t^*. This well-known fact has been exploited in the numerical work of [16]. Furthermore, the transformed HJB equation arising from the further simplification $c(x) \equiv 0$ leads to the standard backward Kolmogorov equation [19].

3 McKean–Vlasov Formulation

In this section, we extend the McKean–Vlasov forward-reverse time approach to optimal control from [16] to more general control problems defined by (1) and (2). The first step is to choose an appropriate, potentially time-dependent convex cost function $\alpha_t(x)$ and to formulate a suitable evolution equation for the density

$$\tilde{\pi}_t := Z_t^{-1} v_t^* \bar{\pi}_t, \tag{17}$$

where v_t^* satisfies (12) and $\bar{\pi}_t$ the forward evolution equation

$$\partial_t \bar{\pi}_t = -\nabla_x \cdot \left(\bar{\pi}_t \bar{b}_t \right) - \bar{\pi}_t (\alpha_t - \bar{\pi}_t[\alpha_t]) \tag{18}$$

with initial distribution $\bar{\pi}_0 = \delta_{x_0}$ and modified drift function

$$\bar{b}_t(x) := b(x) - \frac{1}{2} \left(\nabla_x \cdot \Sigma(x) + \Sigma(x) \nabla_x \log \bar{\pi}_t(x) \right). \tag{19}$$

The normalization constant Z_t is given by

$$Z_t = \bar{\pi}_t[v_t^*] \tag{20}$$

and δ_{x_0} denotes the Dirac delta function centered at x_0. We note that

$$-\nabla_x \cdot \left(\bar{\pi}_t \bar{b}_t \right) = -\nabla_x \cdot (\bar{\pi}_t b) + \frac{1}{2} \nabla_x \cdot (\nabla_x \cdot (\bar{\pi}_t \Sigma)). \tag{21}$$

We next need to find an evolution equation for the probability density $\tilde{\pi}_t$ defined by (17). Relying on

$$\nabla_x \log \tilde{\pi}_t(x) = \nabla_x \log v_t^* + \nabla_x \log \bar{\pi}_t, \tag{22}$$

we introduce the modified drift

$$\tilde{b}_t(x) := -\bar{b}_t(x) - \frac{1}{2} \Sigma(x) \nabla_x \log v_t^*(x) \tag{23a}$$

$$= -\bar{b}_t(x) - \frac{1}{2} \Sigma(x) \left(\nabla_x \log \tilde{\pi}_t(x) + \nabla_x \log \bar{\pi}_t(x) \right) \tag{23b}$$

$$= -b(x) + \nabla_x \cdot \Sigma(x) + \Sigma(x) \nabla_x \log \bar{\pi}_t(x) \tag{23c}$$

$$- \frac{1}{2} \left(\nabla_x \cdot \Sigma(x) + \Sigma(x) \nabla_x \cdot \log \tilde{\pi}_t(x) \right). \tag{23d}$$

We note that

$$- \nabla_x \cdot (\tilde{\pi}_t \tilde{b}_t) = \nabla_x \cdot (\tilde{\pi}_t \, (b - \nabla_x \cdot \Sigma - \Sigma \nabla_x \log \tilde{\pi}_t)) + \frac{1}{2} \nabla_x \cdot (\nabla_x \cdot (\tilde{\pi}_t \Sigma)), \quad (24)$$

which leads naturally to the interpretation in terms of a reverse time SDE with McKean–Vlasov-type drift function. Before investigating this aspect in more detail, we state the following lemma, which links the modified drift $\tilde{b}_t$ with the time evolution of the probability density $\tilde{\pi}_t$ defined by (17).

Lemma 1 *Given the forward evolution Eq. (18) and the HJB equation (12), the probability density defined by (17) satisfies the reverse time evolution equation*

$$- \partial_t \tilde{\pi}_t = -\nabla_x \cdot \left(\tilde{\pi}_t \tilde{b}_t \right) \tag{25a}$$

$$- \tilde{\pi}_t \left(c - \alpha_t + \frac{1}{2} \| \sigma^T \nabla_x \log v_t^* \|^2 - \frac{1}{2} \| R G^T \nabla_x \log v_t^* \|_R^2 - \zeta_t \right) \tag{25b}$$

with terminal condition $\tilde{\pi}_T = Z_T^{-1} \exp(-f) \bar{\pi}_T$, *where* ζ_t *is an appropriate normalization constant.*

Proof Since (23) and

$$\nabla_x \cdot (\Sigma \nabla_x v_t^*) = \nabla_x v_t^* \cdot (\nabla_x \cdot \Sigma) + \Sigma : D_x^2 v_t^*, \tag{26}$$

and assuming that (17) holds, it follows that

$$- \nabla_x \cdot (\tilde{\pi}_t \tilde{b}_t) = \nabla_x \cdot \left(\tilde{\pi}_t \left(\bar{b}_t + \frac{1}{2} \Sigma \nabla_x \log v_t^* \right) \right) \tag{27a}$$

$$= \frac{v_t^*}{Z_t} \nabla_x \cdot (\bar{\pi}_t \bar{b}_t) + \frac{\bar{\pi}_t}{Z_t} (\Sigma \nabla_x v_t^*) \cdot \bar{b}_t + \frac{1}{2 Z_t} \nabla_x \cdot (\bar{\pi}_t \Sigma \nabla_x v_t^*) \tag{27b}$$

$$= \frac{v_t^*}{Z_t} \nabla_x \cdot (\bar{\pi}_t \bar{b}_t) + \frac{\bar{\pi}_t}{Z_t} \left(b \cdot \nabla_x v_t^* + \frac{1}{2} \Sigma : D_x^2 v_t^* \right). \tag{27c}$$

Hence it holds indeed that

$$\partial_t \tilde{\pi}_t = \frac{v_t^*}{Z_t} \partial_t \bar{\pi}_t + \frac{\bar{\pi}_t}{Z_t} \partial_t v_t^* - \frac{\bar{\pi}}{Z_t} \frac{dZ_t}{dt} \tag{28}$$

for the partial time derivatives given by (25), (18), and (12), respectively. Furthermore, $\tilde{\pi}_T = v_T^* \bar{\pi}_T / Z_T$ at final time and, hence, (17) holds for all times $t \in (0, T]$.

Lemma 1 implies that we can solve the forward evolution Eq. (18) together with the backward evolution Eq. (25) instead of the HJB equation (12). Throughout the remainder of this chapter, we use

$$\alpha_t(x) = c(x) \tag{29}$$

in line with the previous work [16]. However, other choices could be explored. See, for example, [22], which allows one to incorporate the terminal cost $f(x)$.

Remark 2 While [17] and [16] form the basis for our work, we mention the alternative approach put forward in [14], where v_t^* is viewed directly as an unnormalized probability density. This approach leads to the interpretation of the HJB equation (12) in terms of a nonlinear Fokker–Planck equation which in turn can be approximated using interacting particles and EnKF-type approximations as in our work. Only deterministic control problems are considered in [14]. Furthermore, it is not obvious whether the value function v_t^* is always normalizable with respect to the Lebesgue measure on $\mathbb{R}^{d_x}$.

The final step is to turn (18) and (25), respectively, into forward and reverse McKean–Vlasov SDEs [17]:

$$d\bar{X}_t = \bar{f}_t^\epsilon(\bar{X}_t)dt + \sqrt{\epsilon}\sigma(\bar{X}_t)dB_t^+, \qquad \bar{X}_0 = x, \tag{30a}$$

$$-d\tilde{X}_t = \tilde{f}_t^\epsilon(\tilde{X}_t)dt + \sqrt{\epsilon}\sigma(\tilde{X}_t)dB_t^-, \qquad \tilde{X}_T \sim \tilde{\pi}_T. \tag{30b}$$

Here B_t^+ denotes Brownian motion adapted to forward time, B_t^- Brownian motion adapted to reverse time, and $\epsilon \in [0, 1]$ is a free parameter determining the noise level added to the McKean–Vlasov equations.

The drift functions are defined as follows:

$$\bar{f}_t^\epsilon(x) := b(x) - \bar{g}_t(x) - \frac{1-\epsilon}{2}\left(\nabla_x \cdot \Sigma(x) + \Sigma(x)\nabla_x \log \bar{\pi}_t(x)\right) \tag{31}$$

with $\bar{g}_t(x)$ satisfying

$$\nabla_x \cdot (\bar{\pi}_t \bar{g}_t) = -\bar{\pi}_t(c - \bar{\pi}_t[c]), \tag{32}$$

and

$$\tilde{f}_t^\epsilon(x) := -b(x) + \nabla_x \cdot \Sigma(x) + \Sigma(x)\nabla_x \log \tilde{\pi}_t(x) - \tilde{g}_t(x) \tag{33a}$$

$$- \frac{1-\epsilon}{2}\left(\nabla_x \cdot \Sigma(x) + \Sigma(x)\nabla_x \log \tilde{\pi}_t(x)\right) \tag{33b}$$

with $\tilde{g}_t(x)$ satisfying

$$\nabla_x \cdot (\tilde{\pi}_t \tilde{g}_t) = -\tilde{\pi}_t\left(\frac{1}{2}\|\sigma^T\nabla_x \log v_t^*\|^2 - \frac{1}{2}\|RG^T\nabla_x \log v_t^*\|_R^2 - \zeta_t\right). \tag{34}$$

Lemma 2 *The two diffusion processes defined by (30) satisfy $\bar{X}_t \sim \bar{\pi}_t$ and $\tilde{X}_t \sim \tilde{\pi}_t$, respectively. Given $\tilde{\pi}_t$ and $\bar{\pi}_t$, the optimal control law $u_t^*(x)$ is provided by*

$$u_t^*(x) = RG(x)^\mathsf{T} \left(\nabla_x \log \tilde{\pi}_t(x) - \nabla_x \log \bar{\pi}_t(x) \right). \tag{35}$$

Proof The lemma follows immediately from writing down the associated (nonlinear) Fokker–Planck equations for the two McKean–Vlasov SDEs (30). The stated formula for the optimal control follows from (17).

The choice $\epsilon = 0$ leads to fully deterministic evolution equations and the following intriguing representation of (30):

$$\frac{d\bar{X}_t}{dt} = b(\bar{X}_t) - \bar{g}_t(\bar{X}_t) - \frac{1}{2} \left(\nabla_x \cdot \Sigma(\bar{X}_t) + \Sigma(\bar{X}_t)\nabla_x \log \bar{\pi}_t(\bar{X}_t) \right) \tag{36a}$$

$$-\frac{d\tilde{X}_t}{dt} = -\tilde{g}_t(\tilde{X}_t) - \bar{g}_t(\tilde{X}_t) - \frac{d\bar{X}_t}{dt}(\tilde{X}_t) + \frac{1}{2}\Sigma(\tilde{X}_t)\nabla_x y_t^*(\tilde{X}_t). \tag{36b}$$

We will return to this formulation in Appendix.

4 A Brief Diversion: Diffusion-Based Generative Modeling

Before discussing numerical implementation of the proposed forward-reverse McKean–Vlasov equations (30), we demonstrate how the core idea of diffusion-based generative modeling [23, 24] arises as a special instance of (30). Recall that the goal of diffusion-based generative modeling is to use given samples from an unknown distribution π_{data} to produce more samples from that distribution. Instead of density estimation, diffusion-based generative models are built upon grad-log density estimation and SDEs.

We start from the control SDE formulation

$$dX_t = -\frac{1}{2}X_t dt + u_t dt + dB_t \tag{37}$$

with initial conditions $X_0 \sim \pi_0 = \text{N}(0, I)$. The cost function (2) is implicitly given by

$$e^{-f(x)} \propto \frac{\pi_{\text{data}}(x)}{\pi_0(x)} \tag{38}$$

with running cost $c = 0$ and $R = I$. We also note that π_0 is the invariant distribution of (37) for $u_t = 0$ and that our forward SDE (30a) simply becomes

$$d\bar{X}_t = -\left(\frac{1}{2}\bar{X}_t - \frac{1-\epsilon}{2}\bar{C}_t^{-1}(\bar{X}_t - \bar{m}_t) \right) dt + \sqrt{\epsilon}dB_t^+, \qquad \bar{X}_0 \sim \text{N}(0, I). \tag{39}$$

with $\bar{C}_t = I$ and $\bar{m}_t = 0$ for all $t \geq 0$. Furthermore, setting $\epsilon = 0$, leads to

$$\frac{d\bar{X}_t}{dt} = 0. \tag{40}$$

Similarly, the reverse SDE (30b) reduces to

$$-d\tilde{X}_t = -\left(\frac{1}{2}\tilde{X}_t + \frac{1-\epsilon}{2}\nabla_x \log \tilde{\pi}_t(\tilde{X}_t)\right)dt + \sqrt{\epsilon}dB_t^-, \qquad \tilde{X}_T \sim \pi_{\text{data}}, \tag{41}$$

which is of the standard form used in diffusion modeling for $\epsilon = 1$. The desired control term is finally provided by

$$u_t^*(x) = \nabla_x \log \tilde{\pi}_t(x) + x \tag{42}$$

and the generative SDE model (37) turns into

$$dX_t = \frac{1}{2}X_t dt + \nabla_x \log \tilde{\pi}_t(X_t)dt - \frac{1-\epsilon}{2}\nabla_x \log \pi_t(X_t)dt + \sqrt{\epsilon}dB_t, \tag{43}$$

$X_0 \sim N(0, I)$, which again reduces to the standard diffusion-based generative model for $\epsilon = 1$. A more detailed discussion on the connection between diffusion-based generative modeling and stochastic optimal control can be found in [3].

5   Numerical Implementations

In this section, we discuss the numerical implementation of the proposed forward-reverse McKean–Vlasov SDEs (30). We start with Gaussian and EnKF-type approximations [4, 11] before also utilizing diffusion maps [7, 8] in order to approximate the required grad-log density terms.

5.1   EnKF Approximation

We develop a numerical implementation of (30) based on the EnKF methodology [4, 11]. In particular, we approximate a drift g_t, which should satisfy

$$\nabla_x \cdot (\pi_t g_t) = -\pi_t(\|\psi\|_B^2 - \pi_t[\|\psi\|_B^2]) \tag{44}$$

for given function $\psi(x)$ and density π_t, in the following manner. We introduce the mean, m_t^ψ, of $\psi(x)$ and the covariance matrix, $C_t^{x\psi}$, between x and $\psi(x)$ under π_t. Then an approximative drift term g_t^{KF} is defined by

$$g_t^{KF}(x) := \frac{1}{2} C_t^{x\psi} B^{-1} \left(\psi(x) + m_t^{\psi} \right).$$

(45)

This approximation becomes exact for Gaussian density π_t and linear function $\psi(x)$. See, for example, [20, 21] for the general methodology and [14] for an application to optimal control.

We assume that the running cost $c(x)$ is of the form (3). Hence, following (32) and (45), the drift term $\bar{g}_t$ is approximated by

$$\bar{g}_t^{KF}(x) := \frac{1}{2} \bar{C}_t^{xh} S^{-1} \left(h(x) + \bar{m}_t^h \right).$$

(46)

Here $\bar{C}_t^{xh}$ denotes the covariance matrix between x and $h(x)$ with respect to $\bar{\pi}_t$ and $\bar{m}_t^h$ the mean of $h(x)$ under the same distribution.

The transformation from $\bar{X}_T$ to $\tilde{X}_T$ under the terminal cost (4) is performed by the stochastic EnKF [4, 11]; that is,

$$\tilde{X}_T = \bar{X}_T - \bar{C}_T^{x\xi} \left(\bar{C}_T^{\xi\xi} + V \right)^{-1} (\xi(\bar{X}_T) + V^{1/2}\Xi), \qquad \Xi \sim N(0, I).$$

(47)

Following the desired control *ansatz* (5), we also approximate $\nabla_x \log v_t^*$ as a linear function using the first two moments of $\tilde{\pi}_t$ and $\bar{\pi}_t$, respectively; that is,

$$\nabla_x \log v_t^{KF}(x) := \tilde{C}_t^{-1}(x - \tilde{m}_t^x) - \bar{C}_t^{-1}(x - \bar{m}_t^x)$$

(48a)

$$= (\tilde{C}_t^{-1} - \bar{C}_t^{-1})x + \left(\tilde{C}_t^{-1}\tilde{m}_t^x - \bar{C}_t^{-1}\bar{m}_t^x \right) = A_t x + c_t$$

(48b)

with

$$A_t := \tilde{C}_t^{-1} - \bar{C}_t^{-1}, \qquad c_t := \tilde{C}_t^{-1}\tilde{m}_t^x - \bar{C}_t^{-1}\bar{m}_t^x.$$

(49)

This approximation leads to the further approximation

$$\tilde{g}_t^{KF}(x) := \frac{1}{2} \tilde{C}_t A_t \left(\Sigma(\tilde{m}_t^x) - G(\tilde{m}_t^x) R G(\tilde{m}_t^x)^T \right) (A_t x + A_t \tilde{m}_t^x + 2c_t)$$

(50)

for the drift term $\tilde{g}_t$ arising from (34).

We now summarize our approximations to the drift terms in the forward-reverse McKean–Vlasov equations (30):

$$\bar{f}_t^{\epsilon}(x) := b(x) - \frac{1-\epsilon}{2} \left(\nabla_x \cdot \Sigma(x) - \Sigma(x)\bar{C}_t^{-1}(x - \bar{m}_t^x) \right)$$

(51a)

$$- \frac{1}{2} \bar{C}_t^{xh} S^{-1} \left(h(x) + \bar{m}_t^h \right)$$

(51b)

and

$$\tilde{f}_t^{\epsilon}(x) := -b(x) + \nabla_x \cdot \Sigma(x) - \Sigma(x)\bar{C}_t^{-1}(x - \bar{m}_t^x) \tag{52a}$$

$$- \frac{1-\epsilon}{2}\left(\nabla_x \cdot \Sigma(x) - \Sigma(x)\tilde{C}_t^{-1}(x - \tilde{m}_t^x)\right) \tag{52b}$$

$$- \frac{1}{2}\tilde{C}_t A_t \left(\Sigma(\tilde{m}_t^x) - G(\tilde{m}_t^x)RG(\tilde{m}_t^x)^{\mathsf{T}}\right)\left(A_t x + A_t \tilde{m}_t^x + 2c_t\right). \tag{52c}$$

The forward Eq. (30a) is solved from the initial condition $\bar{X}_0 = x_0$. The terminal $\bar{X}_T$ is transformed into the terminal condition $\tilde{X}_T$ using (47). Equation (30b) is solved from $t = T$ to $t = 0$. The desired approximation to the optimal control u_t^* is provided by (5) with A_t and c_t given by (49). We note that A_t is symmetric negative-definite whenever $\tilde{C}_t\bar{C}_t^{-1} < I$. In other words, the covariance matrix of the reverse process has to be strictly smaller than the covariance matrix of the forward process in order for the associated control (5) to act in a stabilizing manner.

We also note that the McKean–Vlasov contribution (52c) stabilizes the reverse dynamics provided

$$\Sigma(x) > G(x)RG(x)^{\mathsf{T}} \tag{53}$$

and destabilizes it otherwise. The overall reverse dynamics can still be stable due to the contributions from (52a).

5.2 Combined Diffusion Map and EnKF Approximation

In this subsection, we propose another implementation of the McKean–Vlasov formulation (30) combining the EnKF-type approximations for the drift terms $\bar{g}_t(x)$ and $\tilde{g}_t(x)$, respectively, while using diffusion maps [7, 8] for estimating grad-log density terms.

We first consider the forward McKean–Vlasov equations

$$d\bar{X}_t = b(\bar{X}_t)dt - \frac{1}{2}\bar{C}_t^{xh}\left(h(\bar{X}_t) + \bar{m}_t^h\right)dt + \sigma(\bar{X}_t)dB_t^+. \tag{54}$$

The law $\bar{\pi}_t$ of $\bar{X}_t$ induces the generator $\bar{\mathcal{L}}_t$ at time t, which is defined by

$$\bar{\mathcal{L}}_t f := \frac{1}{\bar{\pi}_t}\nabla_x \cdot (\bar{\pi}_t \Sigma \nabla_x f). \tag{55}$$

Here we assume that $\Sigma(x)$ has full rank. It is easy to verify that

$$\bar{\mathcal{L}}_t \mathrm{Id} = \nabla_x \cdot \Sigma + \Sigma \nabla_x \log \bar{\pi}_t, \tag{56}$$

where Id : $\mathbb{R}^{d_x} \to \mathbb{R}^{d_x}$ denotes the identity map; that is, $\text{Id}(x) = x$. Equation (56) suggests the approximation

$$\nabla_x \cdot \Sigma(x) + \Sigma(x) \nabla_x \log \bar{\pi}_t(x) \approx \frac{\exp(\varepsilon \bar{\mathcal{L}}_t)\text{Id} - \text{Id}}{\varepsilon}(x) \tag{57}$$

for $\varepsilon > 0$ sufficiently small, where the semi-group $\exp(\varepsilon \mathcal{L}_t)$ will be later replaced by the normalized diffusion map approximation as investigated in [25]. We introduce the conditional mean

$$\bar{m}_t^\varepsilon(x) := \exp(\varepsilon \bar{\mathcal{L}}_t)\text{Id}(x) \tag{58}$$

and obtain the compact representation

$$\frac{\exp(\varepsilon \bar{\mathcal{L}}_t)\text{Id} - \text{Id}}{\varepsilon}(x) = \varepsilon^{-1}(\bar{m}_t^\varepsilon(x) - x). \tag{59}$$

Approximation (57) is plugged into the reverse McKean–Vlasov equation to yield

$$- d\tilde{X}_t = -b(\tilde{X}_t)dt + \varepsilon^{-1}(\bar{m}_t^\varepsilon(\tilde{X}_t) - \tilde{X}_t)dt - \tilde{g}_t^{\text{KF}}(\tilde{X}_t)dt \tag{60a}$$

$$- \frac{1-\epsilon}{2}\left(\nabla_x \cdot \Sigma(\tilde{X}_t) - \Sigma(\tilde{X}_t)\tilde{C}_t^{-1}(\tilde{X}_t - \tilde{m}_t^x)\right) \tag{60b}$$

$$+ \sqrt{\epsilon}\sigma(\tilde{X}_t)dB_t^-, \tag{60c}$$

where $\tilde{g}_t^{\text{KF}}(x)$ is defined by (50) as before and $\epsilon \in [0, 1]$.

Approximation (57) can also be used in the forward McKean–Vlasov equation and (54) gets replaced by

$$d\bar{X}_t = b(\bar{X}_t)dt - \frac{1}{2}\bar{C}_t^{xh}\left(h(\bar{X}_t) + \bar{m}_t^h\right)dt \tag{61a}$$

$$- \frac{1-\epsilon}{2\varepsilon}(\bar{m}_t^\varepsilon(\bar{X}_t) - \bar{X}_t)dt + \sqrt{\epsilon}\sigma(\bar{X}_t)dB_t^+. \tag{61b}$$

Furthermore, the law $\tilde{\pi}_t$ of $\tilde{X}_t$ induces the generator $\tilde{\mathcal{L}}_t$ at time t, which is defined by

$$\tilde{\mathcal{L}}_t f := \frac{1}{\tilde{\pi}_t}\nabla_x \cdot (\tilde{\pi}_t \Sigma \nabla_x f), \tag{62}$$

and which can be used to approximate

$$\nabla_x \cdot \Sigma(x) + \Sigma(x)\nabla_x \log \tilde{\pi}_t(x) \approx \varepsilon^{-1}(\tilde{m}_t^\varepsilon(x) - x), \quad \tilde{m}_t^\varepsilon(x) := \exp(\varepsilon \tilde{\mathcal{L}}_t)\text{Id}(x). \tag{63}$$

in the reverse SDE drift function (33b). In other words, (60) gets replaced by

$$- \mathrm{d}\tilde{X}_t = -b(\tilde{X}_t)\mathrm{d}t + \varepsilon^{-1}(\tilde{m}_t^\varepsilon(\tilde{X}_t) - \tilde{X}_t)\mathrm{d}t - \tilde{g}_t^{\mathrm{KF}}(\tilde{X}_t)\mathrm{d}t \tag{64a}$$

$$- \frac{1-\epsilon}{2\varepsilon}(\tilde{m}_t^\varepsilon(\tilde{X}_t) - \tilde{X}_t) + \sqrt{\epsilon}\sigma(\tilde{X}_t)\mathrm{d}B_t^-. \tag{64b}$$

We note that the McKean–Vlasov equations (61) and (64) become deterministic under the choice $\epsilon = 0$.

5.3   Numerical Implementation Details

We numerically implement the McKean–Vlasov equations (30) with drift terms given by (51) and (52) using a Monte Carlo approach; that is, we propagate an ensemble of M particles $\bar{X}_t^{(i)}$, $i = 1, \ldots, M$, forward in time, $t \in [0, T]$, and an equally sized ensemble of particles $\tilde{X}_t^{(i)}$ backward in time. The required mean values and covariance matrices are replaced by their empirical estimators. A covariance inflation of δI, $\delta > 0$, is added to the empirical covariance matrices in order to ensure that they remain non-singular [11]. For the purpose of this chapter, we apply a simple Euler–Maruyama time-stepping method with step-size Δt both in forward and reverse time [15]. More robust time-stepping methods can be based on the formulations proposed and investigated in [1].

When running the EnKF-type formulation from Sect. 5.1, we set the initial conditions to $\bar{X}_0^{(i)} = x_0$ in the forward equation and use $\epsilon > 0$ for the first time-step in order to diffuse these identical particles. All subsequent time-steps employ then $\epsilon = 0$ (deterministic dynamics). The terminal ensemble $\tilde{X}_T^{(i)}$, $i = 1, \ldots, M$, is computed using the forward ensemble $\bar{X}_T^{(i)}$ at final time and a standard ensemble implementation of the EnKF update (47) [4, 11]. The reverse McKean–Vlasov equations are solved with $\epsilon = 0$ (deterministic dynamics).

The reverse McKean–Vlasov equation (60) also requires the approximation of the semi-group $\exp(\epsilon \bar{\mathcal{L}}_t)$. We now describe an implementation which follows ideas from [13]. Let us assume, for simplicity, that $\Sigma(x)$ has full rank. Based on the forward-in-time samples $\{\bar{X}_t^{(i)}\}$, we first define the diffusion map approximation

$$P_t^\varepsilon = D(v_t^\varepsilon)R_t^\varepsilon D(v_t^\varepsilon), \tag{65}$$

where the matrix $R_t^\varepsilon \in \mathbb{R}^{M \times M}$ has entries

$$(R_t^\varepsilon)_{ij} = \exp\left(\frac{-1}{2\varepsilon}(\bar{X}_t^{(i)} - \bar{X}_t^{(j)})^{\mathrm{T}}\left(\Sigma(\bar{X}_t^{(i)}) + \Sigma(\bar{X}_t^{(j)})\right)^{-1}(\bar{X}_t^{(i)} - \bar{X}_t^{(j)})\right), \tag{66}$$

$D(v) \in \mathbb{R}^{M \times M}$ denotes the diagonal matrix with diagonal entries given by the vector $v \in \mathbb{R}^M$, and the vector $v_t^\varepsilon \in \mathbb{R}_+^M$ is chosen such that

$$\sum_{i=1}^{M} (P_t^\varepsilon)_{ij} = \sum_{j=1}^{M} (P_t^\varepsilon)_{ij} = \frac{1}{M}. \tag{67}$$

The vector v_t^ε can be computed efficiently using the iterative algorithm from [25]. More precisely, one iterates the fixed point formulation

$$u_t^\varepsilon = 1/(M R_t^\varepsilon v_t^\varepsilon), \qquad v_t^\varepsilon = \sqrt{u_t^\varepsilon/(M R_t^\varepsilon u_t^\varepsilon)} \tag{68}$$

to self-consistency, where division and taking square root have to be performed entry-wise.

We define a probability vector $p_t^\varepsilon(x) \in \mathbb{R}^M$ for all $x \in \mathbb{R}^{d_x}$ as follows. First we introduce the vector $r_t^\varepsilon(x) \in \mathbb{R}^M$ with entries

$$(r_t^\varepsilon)_i(x) = \exp\left(\frac{-1}{2\varepsilon}(\bar{X}_t^{(i)} - x)^{\mathrm{T}} \left(\Sigma(\bar{X}_t^{(i)}) + \Sigma(x)\right)^{-1} (\bar{X}_t^{(i)} - x)\right) \tag{69}$$

for $i = 1, \ldots, M$. Next we compute the vector v_t^ε in (65), which in turn is used to define

$$p_t^\varepsilon(x) = \frac{D(v_t^\varepsilon)r_t^\varepsilon(x)}{(v_t^\varepsilon)^{\mathrm{T}}r_t^\varepsilon(x)}. \tag{70}$$

Setting $\epsilon = \Delta t$, we finally obtain the approximation

$$\bar{m}_t^{\Delta t}(x) = \exp(\Delta t \bar{\mathcal{L}}_t)\mathrm{Id}(x) \approx \bar{\mathcal{X}}_t p_t^{\Delta t}(x) \tag{71}$$

with

$$\bar{\mathcal{X}}_t = \left(\bar{X}_t^{(1)}, \bar{X}_t^{(2)}, \ldots, \bar{X}_t^{(M)}\right) \in \mathbb{R}^{d_x \times M}. \tag{72}$$

The reverse McKean–Vlasov equation (60), here with $\epsilon = 1$ for simplicity, is integrated backward in time using the following split-step scheme:

$$\tilde{X}_{t_{n-1/2}}^{(i)} = \tilde{X}_{t_n}^{(i)} - \Delta t b(\tilde{X}_{t_n}^{(i)}) - \Delta t \tilde{g}_{t_n}^{\mathrm{KF}}(\tilde{X}_{t_n}^{(i)})dt + \sqrt{\Delta t}\sigma(\tilde{X}_{t_n}^{(i)})\Xi_{t_n}^{(i)}, \tag{73a}$$

$$\tilde{X}_{t_{n-1}}^{(i)} = \bar{\mathcal{X}}_{t_{n-1}} p_{t_{n-1}}^{\Delta t}(\tilde{X}_{t_{n-1/2}}^{(i)}), \tag{73b}$$

$i = 1, \ldots, M$, where $\Xi_{t_n}^{(i)}$ denote independent standard Gaussian random variables with mean zero and identity covariance matrix, and $t_{n+1} = t_n + \Delta t$. This implementation guarantees that any reverse time solution $\tilde{X}_{t_n}$ is contained in the convex hull generated by the forward samples $\{\bar{X}_{t_n}^{(i)}\}$ [13], which is a desirable property in terms of $\tilde{\pi}_t \propto v_t^* \bar{\pi}_t \ll \bar{\pi}_t$. The approximation (71) can also be applied in the forward McKean–Vlasov equation (61) in case $\epsilon < 1$.

Please note that we propose to still approximate the optimal control $u_t^*(x)$ by (5) with A_t and c_t given by (49). However, diffusion map approximations could also be used in this context. See also [17].

It should be noted that the diffusion map approximation requires $M \gg d_x$, which is in contrast to the EnKF-type approximation from Sect. 5.1, which can be implemented with as little as $M = d_x + 1$ particles in order to render the empirical covariance matrices non-singular and, hence, to obtain well-defined evolution equations at the particle level. This desirable property is verified in the following section. However, EnKF-type approximations can fail due to stability and accuracy reasons and need to then be augmented by diffusion map approximations as we also demonstrate in the following section.

6 Numerical Examples

In this section, we discuss numerical findings for two simple control problems. The first control problem is to stabilize the unstable equilibrium position of a mathematical pendulum. This control problem is nonlinear in nature and linear feedback control laws will be suboptimal. However, we find that (5) is nevertheless able to drive the pendulum from the stable to the unstable equilibrium in finite time. The second control problem concerns the stabilization of an unstable equilibrium point of one-dimensional nonlinear Langevin dynamics. Here the computational challenge arises from the fact that the drift term $b(x)$ becomes strongly destabilizing when integrated backward in time which requires a diffusion map approximation of the stabilizing grad-log density term in the reverse McKean–Vlasov dynamics.

6.1 Inverted Pendulum

As a first example, we consider the inverted pendulum with control [18]. The state variable is $x = (\theta, \dot{\theta})^{\mathrm{T}} \in \mathbb{R}^2$ with equations of motion

$$d\theta = \dot{\theta} dt, \tag{74a}$$

$$d\dot{\theta} = \sin(\theta) dt - \cos(\theta) u dt + \rho dB_t, \tag{74b}$$

and $\rho = 1$. Consider the running cost

$$c(x) = \frac{10}{2} \|\dot{\theta}\|^2 \tag{75}$$

over a finite time window $t \in [0, 1]$ with final cost

$$f(x) = \frac{10^3}{2} \|x\|^2.$$ (76)

The control is scalar-valued and the penalty term in the cost function uses $R = 10$. Note that $G(x)$ is position dependent and that $(0, 0)$ is an unstable equilibrium point of the deterministic pendulum ($\rho = 0$). The noise acts only on the momentum equation. We seek a control law of the form (5) that leads us from the stable equilibrium $(\pi, 0)$ to the unstable one $(0, 0)$ at time $T = 1$. We initialize $\bar{X}_0 = (\pi, 0.1)$; that is, we give the stable equilibrium a small initial kick.

The numerical experiment uses $M = 3$ ensemble members in the EnKF-type formulation from Sect. 5.1. The time-step is set to $\Delta t = 10^{-4}$ and $\epsilon = 0.01$ for the first time-step of the forward dynamics. The additive covariance inflation factor is set to $\delta = 10^{-4}$. The results from the forward and reverse McKean–Vlasov equations can be found in Fig. 1. It can be seen that the forward dynamics stays close to the stable equilibrium point $(\pi, 0)$ over the whole time interval $[0, 1]$. The stiff final cost implied by (76) transforms the ensemble $\bar{X}_T^{(i)}$ to an ensemble $\tilde{X}_T^{(i)}$, $i = 1, \ldots, M$, which is tightly clustered about the unstable equilibrium $(0, 0)$ at $T = 1$. Solving the reverse McKean–Vlasov equations leads us gradually back to the unstable equilibrium, which is reached at time $t = 0$.

We then apply the computed control (5) to the inverted pendulum Eq. (74) with the noise set to zero ($\rho = 0$). The time evolution of the resulting solution is displayed in Fig. 2. The computed time-dependent affine control is able to drive the solution from the stable to the unstable equilibrium point over a unit time interval. The time evolution of the associated velocity indicates that strong acceleration terms are required and indeed provided by the computed control law.

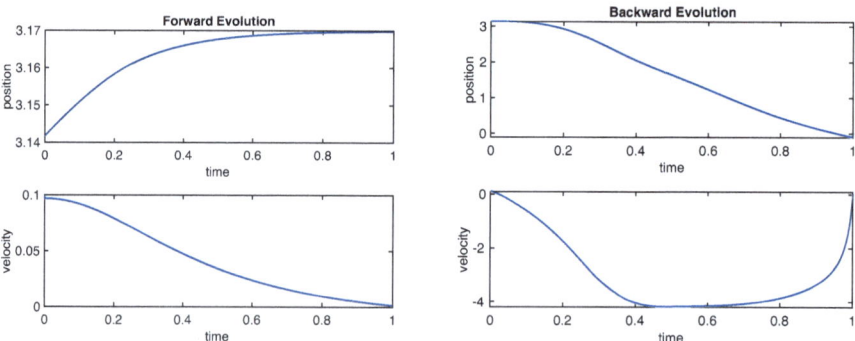

Fig. 1 Time evolution of the ensemble mean from the forward evolution (left panel) and the reverse evolution (right panel) both in terms of pendulum position and velocity. It can be seen that the reverse evolution connects the stable and unstable equilibrium points while the forward dynamics stays close to the stable equilibrium

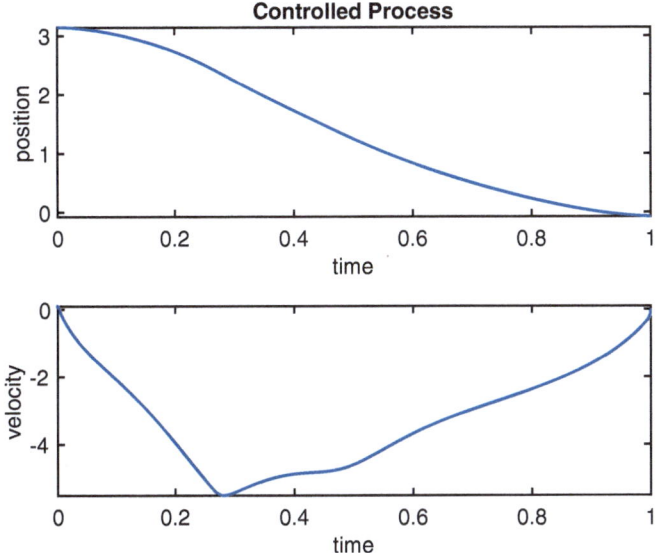

Fig. 2 Time evolution of the position and velocity of the pendulum under the computed affine control law. The pendulum leaves its initial stable solution to reach the unstable equilibrium at time $T = 1$. The initial and final velocities are essentially zero

6.2 Controlled Langevin Dynamics

As a second example, we consider the controlled Langevin dynamics

$$dX_t = -(X_t^3 - X_t)dt + u_t(X_t)dt + dB_t \tag{77}$$

with unstable equilibrium at $x = 0$ and two stable equilibria at $x = \pm 1$. The imposed running cost is $c(x) = 100x^2/2$ and the terminal cost at $T = 30$ is $f(x) = x^2/2$. We implement the combined diffusion map and EnKF scheme from Sect. 5.2 with step-size $\Delta t = 0.01$, $M = 8$ ensemble members, and $\epsilon = 0$ in the forward (61) and reverse (60) dynamics except for the first ten steps of the forward dynamics (61) where we set $\epsilon = 1$. We also employ additive ensemble inflation with $\delta = 10^{-4}$.

The diffusion map approximation of the grad-log density term in the reverse dynamics is essential for counterbalancing the strongly unstable contribution stemming from the drift term in (77) when integrated backward in time. We also find that the Gaussian approximation to the grad-log density term in the forward dynamics is insufficient and that the diffusion map approximation in (61) significantly improves the behaviour of the deterministic formulation ($\epsilon = 0$). The scale parameter in the diffusion map approximation is set to $\varepsilon = \Delta t$.

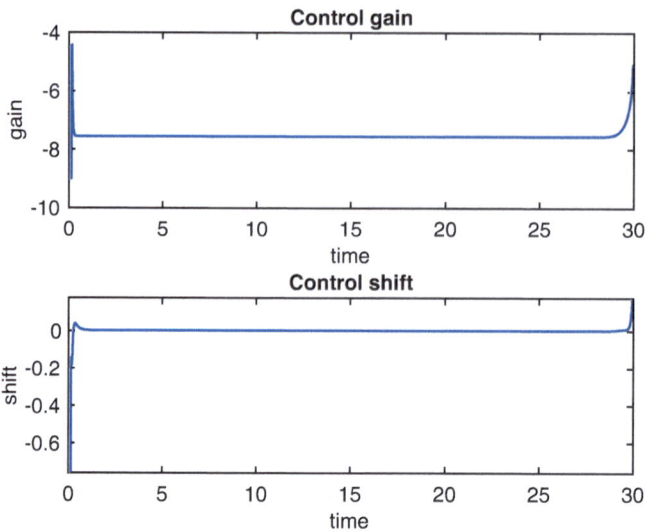

Fig. 3 Computed control gain A_t and shift c_t from the forward and reverse McKean–Vlasov evolution equations. The control is time-independent except for brief transition periods at the beginning and end of the simulation interval

Except for brief transition periods at initial and final time, the control law (5) is essentially time-independent. This behaviour corresponds to the well-known turnpike in optimal control [12]. See Fig. 3 and Appendix for a related discussion of infinite horizon optimal control problems. The effectiveness of the control is demonstrated in Fig. 4.

7   Conclusions

Solving the HJB equation numerically constitutes a challenging task. Here we have provided a new perspective by combining forward and reverse evolution McKean–Vlasov equations with the tremendously successful EnKF methodology. We have done so by building on the previous work [16] and have generalized it to a wider class of forward and reverse McKean–Vlasov equations. In order to keep those equations computationally tractable we have employed EnKF-type approximations to the McKean–Vlasov interaction terms. While not delivering optimal control laws, the resulting time-dependent affine control laws can either be sufficient in themselves or, alternatively, may provide the starting point for more accurate approximations such as the diffusion map approach outlined in Sect. 5.2. We have applied the proposed methodology to a two-dimensional nonlinear control problem using only $M = d_x + 1 = 3$ particles. Such a small ensemble size constitutes a

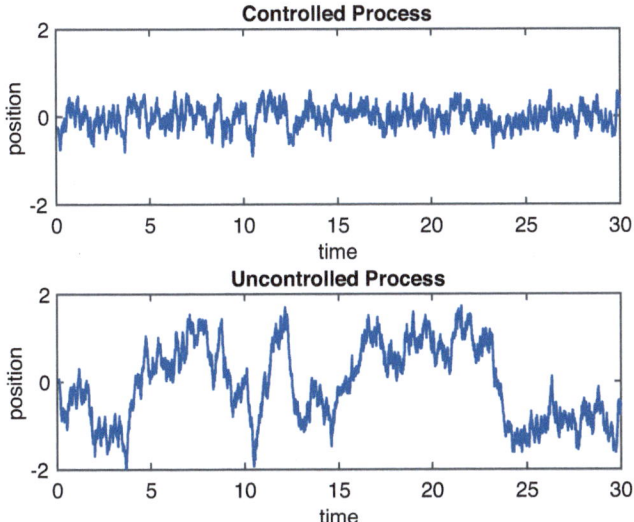

Fig. 4 Comparison of the controlled and uncontrolled Langevin dynamics. Displayed is the time evolution of a single realisation of the SDE (77) with and without control

significant improvements over the results presented in [16] and [14]. It remains to be demonstrated that the methodology can be further extended to high-dimensional control problems in the spirit of EnKF applications to data assimilation, which deliver useful approximations even with $M \ll d_x$ ensemble members [11]. At the same time, strongly nonlinear Langevin dynamics (77) requires more sophisticated approximations of the grad-log terms in terms of diffusion maps. Still the ensemble size could be kept at a moderate level ($M = 8$) in order to recover the linear control law (5) robustly.

Acknowledgments This work has been funded by Deutsche Forschungsgemeinschaft (DFG) - Project-ID 318763901 - SFB1294. The author thanks Manfred Opper for insightful discussions on the subject of this work.

Appendix

In this appendix, we discuss an extension of the proposed methodology to infinite horizon control problems with cost function

$$J_\infty(x_0, u_{0:\infty}) = \mathbb{E}\left[\int_0^\infty e^{-\gamma t}\left(c(X_t) + \frac{1}{2}\|u_t\|_R^2\right) dt\right]. \qquad (78)$$

Here $\gamma \geq 0$ denotes the discount factor. The associated transformed HJB equation becomes

$$- \partial_t v_t^* = b \cdot \nabla_x v_t^* + \frac{1}{2} \Sigma : D_x^2 v_t^* \tag{79a}$$

$$- \left(c + \gamma \log v_t^* + \frac{1}{2} \|\sigma^{\mathsf{T}} \nabla_x \log v_t^*\|^2 - \frac{1}{2} \|RG^{\mathsf{T}} \nabla_x \log v_t^*\|_R^2 \right) v_t^* \tag{79b}$$

with terminal condition $v_\infty^* = 1$.

The only modification to the finite horizon formulations from Sect. 3 concerns the drift function (33), where $\tilde{g}_t(x)$ has to now satisfy the Poisson equation

$$\nabla_x \cdot (\tilde{\pi}_t \tilde{g}_t) = -\tilde{\pi}_t \left(\gamma \log v_t^* + \frac{1}{2} \|\sigma^{\mathsf{T}} \nabla_x \log v_t^*\|^2 - \frac{1}{2} \|RG^{\mathsf{T}} \nabla_x \log v_t^*\|_R^2 - \zeta_t \right) \tag{80}$$

and the EnKF-based approximation of $\tilde{g}_t$ becomes

$$\tilde{g}_t^{\mathrm{KF}}(x) := \frac{1}{2} \tilde{C}_t \left\{ \gamma I + A_t \left(\Sigma(\tilde{m}_t^x) - G(\tilde{m}_t^x) RG(\tilde{m}_t^x)^{\mathsf{T}} \right) \right\} (A_t x + A_t \tilde{m}_t^x + 2c_t). \tag{81}$$

Please note that $\tilde{C}_t A_t = \tilde{C}_t \bar{C}_t^{-1} - I < 0$ under the assumption that $\tilde{C}_t < \bar{C}_t$. Hence, the additional drift term stabilizes the time evolution of the reverse covariance matrix $\tilde{C}_t$ towards $\bar{C}_t$.

One expects the forward process (36a) to reach an equilibrium distribution with mean $\bar{m}_{\mathrm{eq}}$ and covariance matrix $\bar{C}_{\mathrm{eq}}$ for $t > 0$ sufficiently large. Furthermore, upon setting $\epsilon = 0$ in (30) and since we are in equilibrium, the time derivative $d\bar{X}_t/dt$ will either be zero or can be assumed to be relatively small. One may then fix these quantities in the reverse process (36b)–(36c), which, in turn, is integrated backward in time till an equilibrium distribution is reached with mean $\tilde{m}_{\mathrm{eq}}$ and covariance matrix $\tilde{C}_{\mathrm{eq}}$. The optimal control is provided by

$$u_t^*(x) = RG(x)^{\mathsf{T}} \left(\bar{C}_{\mathrm{eq}}^{-1}(x - \bar{m}_{\mathrm{eq}}^x) - \tilde{C}_{\mathrm{eq}}^{-1}(x - \tilde{m}_{\mathrm{eq}}^x) \right). \tag{82}$$

Such a methodology provides an approximation to the stationary solution of the HJB equation (79).

The combined EnKF and diffusion map approximation formulation from Sect. 5.2 can be generalized to the infinite horizon optimal control problem in a similar fashion. Please note that the numerical results from Sect. 6.2 already implied an essentially time-independent control law.

Alternatively, one can follow the actor-critic methodology to stochastic optimal control [18] and introduce a family of control laws $u_\theta(x)$ parametrized by $\theta \in \mathbb{R}^{d_\theta}$ and set $\gamma = 0$ in (78). For example, using a (stationary) linear control law of the form (5), the adjustable parameters, θ, would be given by the (constant) matrix A_t and the (constant) vector c_t.

More specifically, the actor chooses parameters, θ, and considers the controlled SDE

$$dX_t = b(X_t)dt + G(X_t)u_\theta(X_t)dt + \sigma(X_t)dB_t. \tag{83}$$

The generator of this SDE is denoted by $\mathcal{L}_\theta$. It is assumed that the SDE possesses an invariant density π_θ; that is,

$$\mathcal{L}_\theta^\dagger \pi_\theta = 0, \tag{84}$$

where $\mathcal{L}_\theta^\dagger$ denotes the adjoint of $\mathcal{L}_\theta$ [19]. The optimal θ_* is determined by

$$\theta_* = \arg\min_\theta \pi_\theta[c_\theta] \tag{85}$$

with cost

$$c_\theta(x) = c(x) + \frac{1}{2}\|u_\theta(x)\|_R^2. \tag{86}$$

The critic provides the value function y_θ, which satisfies the stationary HJB equation

$$\mathcal{L}_\theta y_\theta + c_\theta - \pi_\theta[c_\theta] = 0. \tag{87}$$

Given the value function $y_\theta(x)$, the chosen parameter, θ, can now be improved using the gradient [18]

$$\nabla_\theta \pi_\theta[c_\theta] = \pi_\theta\left[(\nabla_\theta \mathcal{L}_\theta)y_\theta\right] + \pi_\theta\left[\nabla_\theta c_\theta\right] = \pi_\theta\left[\nabla_\theta u_\theta^T\left(G^T\nabla_x y_\theta + u_\theta\right)\right] \tag{88}$$

and the optimal parameter value satisfies $\nabla_\theta \pi_{\theta_*}[c_{\theta_*}] = 0$.

In order to extend our McKean–Vlasov approach to this control setting, we replace the stationary HJB equation (87) by the forward-in-time HJB equation

$$\partial_t y_t = \mathcal{L}_\theta y_t + c_\theta - \pi_\theta[c_\theta] \tag{89}$$

and apply the transformation $v_t(x) = \exp(-y_t(x))$ to obtain the modified HJB equation

$$\partial_t v_t = \mathcal{L}_\theta v_t - \left(c_\theta + \frac{1}{2}\|\sigma^T\nabla_x \log v_t\|^2 - \zeta_t\right)v_t \tag{90}$$

for $t \geq 0$ with initial condition $v_0(x) = 1$ and ζ_t an appropriate normalization constant.

Adapting our previously developed methodology, we introduce the density

$$\tilde{\pi}_t(x) := Z_t^{-1} v_t(x) \pi_\theta(x) \tag{91}$$

with $Z_t = \pi_\theta[v_t]$ and find that

$$\nabla_x y_\theta = \nabla_x \log \pi_\theta - \lim_{t \to \infty} \nabla_x \log \tilde{\pi}_t. \tag{92}$$

Furthermore, the density $\tilde{\pi}_t$ satisfies the forward evolution equation

$$\partial_t \tilde{\pi}_t = -\nabla_x \cdot (\tilde{\pi}_t \tilde{b}_\theta) + \frac{1}{2} \nabla_x \cdot (\nabla_x \cdot (\tilde{\pi}_t \Sigma)) \tag{93a}$$

$$- \tilde{\pi}_t \left(c_\theta + \frac{1}{2} \| \sigma^{\mathsf{T}} \nabla_x \log v_t \|^2 - \tilde{\zeta}_t \right) \tag{93b}$$

$$= -\mathcal{L}_\theta \tilde{\pi}_t + \nabla_x \cdot (\tilde{\pi}_t \nabla_x \log v_t) - \tilde{\pi}_t \left(c_\theta + \frac{1}{2} \| \sigma^{\mathsf{T}} \nabla_x \log v_t \|^2 - \tilde{\zeta}_t \right), \tag{93c}$$

with modified drift function

$$\tilde{b}_\theta(x) := -b(x) - G(x) u_\theta(x) + \nabla_x \cdot \Sigma(x) + \Sigma(x) \nabla_x \log \pi_\theta(x) \tag{94}$$

and normalisation constant $\tilde{\zeta}_t$. The associated forward McKean–Vlasov evolution equation is for $\epsilon = 0$ given by

$$\frac{d\tilde{X}_t}{dt} = \tilde{b}_\theta(\tilde{X}_t) - \tilde{g}_t(\tilde{X}_t) - \frac{1}{2} (\nabla_x \cdot \Sigma(x) + \Sigma(x) \nabla_x \log \tilde{\pi}_t(x)), \tag{95}$$

with initial $\tilde{X}_0 \sim \pi_\theta$ and and the McKean–Vlasov drift term $\tilde{g}_t(x)$ has to now satisfy

$$\nabla_x \cdot (\tilde{\pi}_t \tilde{g}_t) = -\tilde{\pi}_t \left(c_\theta + \frac{1}{2} \| \sigma^{\mathsf{T}} \nabla_x \log v_t \|^2 - \tilde{\zeta}_t \right). \tag{96}$$

We may assume that (93) possesses an invariant density $\tilde{\pi}_\theta$. Then (92) reduces to

$$\nabla_x y_\theta = \nabla_x \log \pi_\theta - \nabla_x \log \tilde{\pi}_\theta. \tag{97}$$

Using (92), the chosen parameters can be improved via standard gradient descent based upon the gradient (88) giving rise to time-dependent parameters, θ_t, which, under suitable assumptions, converge to the optimal θ_*. Furthermore, provided the parameters are adjusted slowly enough in time, one can make the assumption that $\mathcal{L}_{\theta_t} \pi_{\theta_t} \approx 0$. These assumptions suggest the coupled set of forward McKean–Vlasov evolution equations

$$\frac{dX_t}{dt} = b(X_t) + G(X_t)u_{\theta_t}(X_t) - \frac{1}{2}\nabla_x \cdot \Sigma(X_t) - \frac{1}{2}\Sigma(X_t)\nabla_x \log \pi_t(X_t),$$

(98a)

$$\frac{d\tilde{X}_t}{dt} = -b(\tilde{X}_t) - G(\tilde{X}_t)u_\theta(\tilde{X}_t) + \frac{1}{2}\nabla_x \cdot \Sigma(\tilde{X}_t) + \Sigma(\tilde{X}_t)\nabla_x \log \pi_t(\tilde{X}_t) \quad (98b)$$

$$- \tilde{g}_t(\tilde{X}_t) - \frac{1}{2}\Sigma(\tilde{X}_t)\nabla_x \log \tilde{\pi}_t(\tilde{X}_t),$$

(98c)

$$\frac{d\theta_t}{dt} = -\delta \left(\pi_t \left[(\nabla_\theta \mathcal{L}_{\theta_t})y_t\right] + \pi_t \left[\nabla_\theta c_{\theta_t}\right]\right)$$

(98d)

$$= -\delta \pi_t \left[\nabla_\theta u_{\theta_t}^{\mathsf{T}} \left(G^{\mathsf{T}}\nabla_x y_t + u_{\theta_t}\right)\right]$$

(98e)

where $\delta > 0$ is sufficiently small, $\tilde{g}_t(x)$ satisfies (96), and

$$\nabla_x y_t(x) := \nabla_x \log \pi_t(x) - \nabla_x \log \tilde{\pi}_t(x).$$

(99)

Here π_t denotes the law of X_t and $\tilde{\pi}_t$ the law of $\tilde{X}_t$. The numerical approximations introduced in Sect. 5 can now be applied to this system of McKean–Vlasov SDEs as well.

In line with the previously stated (36), we note that (98b) can be rewritten in the form

$$\frac{d\tilde{X}_t}{dt} = -\tilde{g}_t(\tilde{X}_t) - \frac{dX_t}{dt}(\tilde{X}_t) + \frac{1}{2}\Sigma(\tilde{X}_t)\nabla_x y_t(\tilde{X}_t).$$

(100)

In this context, it is worthwhile to consider the special case $G = R = I$, $\Sigma = 2I$, $b(x) = -\nabla_x U(x)$, and $u_\theta(x) = -\nabla_x \Psi_\theta(x)$ in more detail. Here $U(x) : \mathbb{R}^{d_x} \to \mathbb{R}$ and $\Psi_\theta(x) : \mathbb{R}^{d_x} \to \mathbb{R}$ are given functions. Under these assumptions, the density π_θ is explicitly known and

$$\nabla_x \log \pi_\theta(x) = -\nabla_x U(x) - \nabla_x \Psi_\theta(x).$$

(101)

Furthermore, we may assume that (98a) is in equilibrium and we can set

$$\frac{dX_t}{dt} \equiv 0$$

(102)

in (100). Let $\tilde{g}_t(x) = \nabla_x V_t(x)$ denote the solution of (96) for appropriate potential $V_t(x)$, then it follows from (100) and $d\tilde{X}_t/dt \approx 0$ that

$$\nabla_x y_t(x) \approx \nabla_x V_t(x).$$

(103)

Hence,

$$\pi_t \left[(\nabla_\theta \mathcal{L}_{\theta_t}) y_t \right] + \pi_t \left[\nabla_\theta c_{\theta_t} \right] \approx \pi_{\theta_t} \left[\nabla_\theta (\nabla_x \Psi_{\theta_t})^\mathsf{T} \left(\nabla_x V_t + \nabla_x \Psi_{\theta_t} \right) \right] \tag{104}$$

and the optimal parameter choice, θ_*, satisfies

$$0 = \pi_{\theta_*} \left[\nabla_\theta (\nabla_x \Psi_{\theta_*})^\mathsf{T} \left(\nabla_x V_{\theta_*} + \nabla_x \Psi_{\theta_*} \right) \right] \tag{105}$$

subject to the potential $V_{\theta_*}(x)$ satisfying the Poisson equation

$$\nabla_x \cdot (\tilde{\pi}_{\theta_*} \nabla_x V_{\theta_*}) = -\tilde{\pi}_{\theta_*} \left(c_{\theta_*} + \| \nabla_x V_{\theta_*} \|^2 - \zeta_* \right) \tag{106}$$

with $\tilde{\pi}_{\theta_*} \propto e^{-V_{\theta_*}} \pi_{\theta_*}$. The approach proposed in this chapter can now be viewed as providing a dynamic particle-based algorithm for solving the nonlinear Eqs. (105)–(106). It is also worth noting that (106) is equivalent to

$$\mathcal{L}_{\theta_*} V_{\theta_*} = -c_{\theta_*} + \pi_{\theta_*}[c_{\theta_*}], \tag{107}$$

which implies $\nabla_x y_{\theta_*} = \nabla_x V_{\theta_*}$ as desired.

References

1. J. Amezcua, E. Kalnay, K. Ide, and S. Reich. Ensemble transform Kalman-Bucy filters. *Q.J.R. Meteor. Soc.*, 140:995–1004, 2014.
2. B. D. Anderson. Reverse-time diffusion equation models. *Stochastic Processes Applications*, 12:313–326, 1982.
3. J. Berner, L. Richter, and K. Ullrich. An optimal control perspective on diffusion-based generative modeling. *preprint arXiv:2211.01364*, 2023.
4. E. Calvello, S. Reich, and A. M. Stuart. Ensemble Kalman methods: A mean field perspective. *preprint arXiv:2209.11371*, 2022.
5. R. Carmona. *Lectures on BSDEs, Stochastic Control, and Stochastic Differential Games with Financial Applications*. SIAM, Philadelphia, 2016.
6. J. Chessari, R. Kawai, Y. Shinozaki, and T. Yamada. Numerical methods for backward stochastic differential equations: A survey. *preprint arXiv:2101.08936*, 2021.
7. R. R. Coifman and S. Lafon. Diffusion maps. *Applied and Computational Harmonic Analysis*, 21(1):5–30, 2006. ISSN 1063-5203. Special Issue: Diffusion Maps and Wavelets.
8. R. R. Coifman, S. Lafon, A. B. Lee, M. Maggioni, B. Nadler, F. Warner, and S. W. Zucker. Geometric diffusions as a tool for harmonic analysis and structure definition of data: Diffusion maps. *Proceedings of the National Academy of Sciences*, 102(21):7426–7431, 2005.
9. W. E, J. Han, and A. Jentzen. Deep learning-based numerical methods for high-dimensional parabolic partial differential equations and backward stochastic differential equations. *Communications in Mathematics and Statistics*, 5:349–380, 2017.
10. W. E, J. Han, and A. Jentzen. Algorithms for solving high-dimensional PDEs: From nonlinear Monte Carlo to machine learning. *Nonlinearity*, 35:278, 2021.
11. G. Evensen, F. C. Vossepoel, and P. J. van Leeuwen. *Data Assimilation Fundamentals: A unified Formulation of the State and Parameter Estimation Problem*. Springer Nature Switzerland AG, Cham, Switzerland, 2022.

12. B. Geshkovksi and E. Zuazua. Turnpike in optimal control of PDEs, ResNets, and beyond. *Acta Numerica*, 31:135–263, 2022.
13. G. Gottwald, F. Li, Y. Marzouk, and S. Reich. Stable generative modeling using diffusion maps. *preprint arXiv:2401.04372*, 2024.
14. A. A. Joshi, A. Taghvaei, P. G. Mehta, and S. P. Meyn. Controlled interacting particle algorithms for simulation-based reinforcement learning. *Systems & Control Letters*, 170: 105392, 2022.
15. P. Kloeden and E. Platen. *Numerical methods for stochastic differential equations*. Springer, New York, 1991.
16. D. Maoutsa and M. Opper. Deterministic particle flows for constraining stochastic nonlinear systems. *Phys. Rev. Res.*, 4:043035, 2022.
17. D. Maoutsa, S. Reich, and M. Opper. Interacting particle solutions of Fokker–Planck equations through gradient-log-density estimation. *Entropy*, 22(8), 2020.
18. S. Meyn. *Control Systems and Reinforcement Learning*. Cambridge University Press, Cambridge, 2022.
19. G. A. Pavliotis. *Stochastic Processes and Applications*. Springer Verlag, New York, 2016.
20. S. Reich. A dynamical systems framework for intermittent data assimilation. *BIT Numerical Mathematics*, 51(1):235–249, 2011.
21. S. Reich. Data assimilation: The Schrödinger perspective. *Acta Numerica*, 28:635–711, 2019.
22. S. Reich. Data assimilation: A dynamic homotopy-based coupling approach. In B. Chapron, D. Crisan, D. Holm, E. Mémin, and A. Radomska, editors, *Stochastic Transport in Upper Ocean Dynamics II*, pages 261–280, Cham, 2024. Springer Nature Switzerland.
23. J. Sohl-Dickstein, E. Weiss, N. Maheswaranathan, and S. Ganguli. Deep unsupervised learning using nonequilibrium thermodynamics. In *International Conference on Machine Learning*, pages 2256–2265. PMLR, 2015.
24. Y. Song, J. Sohl-Dickstein, D. P. Kingma, A. Kumar, S. Ermon, and B. Poole. Score-based generative modeling through stochastic differential equations. In *International Conference on Learning Representations*, 2021. URL https://openreview.net/forum?id=PxTIG12RRHS.
25. C. Wormell and S. Reich. Spectral convergence of diffusion maps: Improved error bounds and an alternative normalisation. *SIAM J. Numer. Anal.*, 59:1687–1734, 2021.

Maximum Likelihood Estimation of Subgrid Flows from Tracer Image Sequences

Valentin Resseguier

1 Model and Assumptions

1.1 Objectives

Let us consider several observed oceanic tracers q_i:

$$\frac{Dq_i}{Dt} = \dot{Q}_i, \tag{1.1}$$

where $\dot{Q}_i$ is a source term smooth in time (e.g. $\dot{Q}_i = \nu_i \Delta q_i$). For instance, q_1 would be the SST and q_2 the SSS on a bounded spatial domain $\Omega \subset \mathbb{R}^2$. We can probably use Ocean colour or SSH as well. We assume that we can observe a set of snapshots of these tracers $(q_i(\boldsymbol{x}, t_k))_{1 \leqslant k \leqslant N}$. We aim at estimate the two-dimensional velocity field $\boldsymbol{v}$—which transports the tracer—from those tracer snapshots.

First, we use a classical method [e.g. optical flow 8, 11, 13, 35] to estimate a (two-dimensional) velocity field from the tracer snapshots q_i. Note that the velocity estimate is probably different for each tracer q_i, that is why we denote it $\boldsymbol{w}_i$. Say that we have S tracers, we can compute the mean drift estimate $\tilde{\boldsymbol{w}} = \frac{1}{S} \sum_i \boldsymbol{w}_i$. This estimate is hopefully accurate but limited in resolution, typically by the resolution of tracer images and by the optical flow algorithm efficiency. Therefore, we refer to this term as large-scale velocity component. This optical flow procedure is obviously of main importance but we do not address it here. A large literature already deals with it. We assume that an optical flow algorithm—says the most efficient optical flow algorithm of the literature—is applied before our method comes into play.

V. Resseguier (✉)
INRAE, OPAALE, Rennes, France
e-mail: valentin.resseguier@inrae.fr
https://www.valentinresseguier.com

© The Author(s) 2025
B. Chapron et al. (eds.), *Stochastic Transport in Upper Ocean Dynamics III*,
Mathematics of Planet Earth 13, https://doi.org/10.1007/978-3-031-70660-8_12

Then, we note that the tracer is also transported by a small-scale velocity $v' = v - \tilde{w}$. The chapter focuses on this residual velocity field. We aim at estimating it or at least estimating its statistics. For this purpose, we make use of modern machine learning and statistics informed by physics. Specifically, we consider maximum likelihood estimation, stochastic calculus, and processes statistics [27, 34, 36] guided by stochastic fluid dynamics [14, 16, 19, 25, 31, 38]. At long term we expect that our work will benefit for the recent advances in processes statistics for linear and linear SPDE with additive and multiplicative noises [1–3, 12, 15, 23].

1.2 Simplifications of the Problem

Then, we neglect the time correlations of the residual velocity field. This assumption is supported by the fact that this velocity is small scale. The time step Δt between two tracer snapshots is finite, and possibly larger that the small-scale velocity correlation time. Therefore, it is probably hopeless to estimate a time-correlated small-scale velocity field. The best we can do is probably estimating the statistics of a time-subsampled version of the small-scale velocity (subsampled at the time step Δt). And this time-subsampling version is time-uncorrelated if Δt is larger than the correlation time of the true small-scale velocity.

We will also assume that v' is Gaussian, homogeneous and isotropic in space. Therefore, we can parameterize v' as the spatial convolution of space-time white noise:

$$v' = \breve{\sigma} * \dot{B}, \qquad (1.2)$$

where $*$ denotes the 2-dimensional spatial convolution, $\breve{\sigma}$ is a 2-dimensional vector of spatial filters and $\dot{B}$ is space-time white noise. The spatial filter imposes a spatial correlation. The covariance of v' is:

$$\mathbb{E}\{v'(x, t_1)v'(y, t_2)\} = \tfrac{1}{dt_1}\delta(t_1 - t_2)(\breve{\sigma} * \tilde{\sigma})(x - y), \qquad (1.3)$$

where $\tilde{\sigma}(x) \overset{\triangle}{=} \breve{\sigma}(-x)$. To simplify the notations, we will denote:

$$a(x - y) \overset{\triangle}{=} \tfrac{1}{dt}\mathbb{E}\left\{(\breve{\sigma} * dB_t)(y + x)(\breve{\sigma} * dB_t)^T(y)\right\} = (\breve{\sigma} * \tilde{\sigma})(x - y). \quad (1.4)$$

Under some assumptions on the filter $\breve{\sigma}$, we can show that : $a(0) = a_0\mathbb{I}_d$ where $a_0 = \tfrac{1}{2}\text{tr}(a(0)) = \tfrac{1}{2}\|\breve{\sigma}\|^2_{\mathcal{L}^2(\Omega)}$ is a positive constant, sometimes called absolute diffusivity, Kubo-type formula or variance tensor [19]. It is equal to the variance of the small-scale velocity multiplied by its correlation time.

1.3   Stochastic Transport

Since $\boldsymbol{v}'$ is assumed time-uncorrelated, the transport Eq. (1.1) can be interpreted as a dynamic under Location Uncertainty (LU) [19, 29] or Stochastic Advection by Lie Transport (SALT) [14]. With Itō notations, it reads:

$$\partial_t q_i + (\tilde{\boldsymbol{w}} + \boldsymbol{v}') \cdot \nabla q_i \approx \partial_t q_i + (\boldsymbol{w}_i + \boldsymbol{v}') \cdot \nabla q_i = \tfrac{a_0}{2} \Delta q_i + \dot{Q}_i. \qquad (1.5)$$

The term $\boldsymbol{v}' \cdot \nabla q_i$ acts as a time-uncorrelated random forcing. Eulerian stochastic transport equations always involve the Lagrangian displacement Stratonovich drift as advecting velocity. This is true for both Itō and Stratonovich notations of the Eulerian SPDEs and for both SALT and LU [32]. That Stratonovich drift corresponds to the Itō drift plus a possible correction. Since the optical flow will estimate the advecting velocity, we identify $\boldsymbol{w}_i$ as the Lagrangian displacement Stratonovich drift. Note that all stochastic differential equations of this chapter are expressed with Itō notations. We refer to [32] for a comparison between SALT and LU, and to [31] for a review of SALT/LU models and calibration methods. Note that [16, 25] and references therein have also studied in details stochastic transport of passive tracers by delta-correlated velocities.

2   Quadratic Co-variation for Turbulence Amplitude Estimation

We aim at estimating the statistics of $\boldsymbol{v}'$. In this section, we treat the estimation of the variance tensor a_0. The random forcing $\boldsymbol{v}' \cdot \nabla q_i$ being delta-correlated, we can estimate the variance tensor from the following algorithm.

We compute for every point $\boldsymbol{x}$ of the grid:

1. $dq_i(\boldsymbol{x}, t) = q_i(\boldsymbol{x}, t + \Delta t) - q_i(\boldsymbol{x}, t)$, the tracer time increments,
2. $\boldsymbol{w}_i(\boldsymbol{x}, t)$ from an optical flow algorithm,
3. $d\tilde{q}_i(\boldsymbol{x}, t) = dq_i(\boldsymbol{x}, t) - \boldsymbol{w}_i \cdot \nabla q_i \Delta t$, ,
   At this step, we can also subtract some known source terms $\dot{Q}_i$ if any.
4. $\overline{d\tilde{q}_i}(\boldsymbol{x}, t)$ a local time average of $d\tilde{q}_i(\boldsymbol{x}, t)$,
5. $dq_i'(\boldsymbol{x}, t) = d\tilde{q}_i(\boldsymbol{x}, t) - \overline{d\tilde{q}_i}(\boldsymbol{x}, t)$.

This step should subtract the effect of the unknown smooth forcing $\dot{Q}_i$. It acts as a high-pass filter to keep only the highly oscillating components of $d\tilde{q}_i$. According to the stochastic transport Eq. (1.5), we should have:

$$dq_i'(\boldsymbol{x}, t) \approx -(\boldsymbol{\sigma} * \Delta B) \cdot \nabla q_i + (\text{something small}) \times \Delta t \qquad (2.1)$$

with $\Delta B = B_{t+\Delta t} - B_t \propto \sqrt{\Delta t}$ a Brownian increment.

6. At this point, we may check—by usual statistical tests—that the increments $dq_i'(\boldsymbol{x}, t)$ is approximately time uncorrelated.

In the LU-SALT theoretical framework, the delta correlation of these increments comes from the delta correlation of the subgrid velocity $\boldsymbol{v}'$. This model assumption is consistent with the fact that the subgrid velocity has short correlation time. However, in practice, the subgrid velocity correlation time is finite. It is a recurrent issue for the data driven modeling of systems combining fast and slowly evolving components [4, 5, 9, 22, 24, 28]. If spurious correlations remain among the increments, we can down-sample the data to force the noise terms to be as decorrelated as possible. The literature proposes several time subsampling rate, generally related to the correlation time of those increments. Resseguier et al. [33] applied this method in a LU context. They estimate a minimal time subsampling rate from the empirical time correlation function of the subgrid velocity. Here, the empirical time correlation function of the increments dq_i' could be used instead.

7. Now, we can compute what is called the quadratic co-variation of q_i and q_j, denoted $< q_i, q_j >$, in stochastic calculus:

$$< q_i, q_j > (\boldsymbol{x}) = \sum_k dq_i'(\boldsymbol{x}, t_k) dq_j'(\boldsymbol{x}, t_k). \tag{2.2}$$

8. We also compute the tracer gradients cross-correlations:

$$c_{ij} = \int_0^{N\Delta t} \nabla q_i \cdot \nabla q_j dt. \tag{2.3}$$

9. Finally, we obtain the variance tensor by a simple (overdetermined) linear system:

$$a_0 \approx \frac{\sum_{ijp} c_{ij}(\boldsymbol{x}_p) < q_i, q_j > (\boldsymbol{x}_p)}{\sum_{ijq} c_{ij}^2(\boldsymbol{x}_p)}. \tag{2.4}$$

Indeed, the decorrelation between two time increments yields:

$$< q_i, q_j > \approx \int_0^{N\Delta t} \nabla q_i^T (\boldsymbol{\sigma} * \Delta B)(\boldsymbol{\sigma} * \Delta B)^T \nabla q_j$$

$$\approx \int_0^{N\Delta t} \nabla q_i^T \boldsymbol{a}(0) \nabla q_j dt = a_0 c_{ij}. \tag{2.5}$$

Theoretically, only one tracer is needed here, even though we expect a higher accuracy with several observed fields.

3   Parametric Model for the Small-scale Velocity Statistics

The variance tensor, a_0, gives the "amplitude" of the small-scale velocity. But, we may want more information (e.g. correlation length, covariance or spectrum). For this purpose, we propose in this section a parametric model for the spatial "covariance", $a(x - y)$, of that velocity introduced in Eq. (1.4). This model will depend on the variance tensor a_0 (the "variance") and on some other parameters θ. Sect. 4 will propose a method to estimate these parameters by maximum likelihood. Several choices of parametric covariance are possible (e.g., Gaussian or Matérn covariance). Here, we propose a self-similar model for turbulence statistics inline with previous work related to stochastic transport [16, 19, 30, 32].

A representation in Fourier space will be convenient for our ultimate estimation procedure. Hence we limit the present study to fields with periodic boundary conditions and we introduce the unitary Fourier transform $\widehat{\zeta}(k) = \int_\Omega dx\, \zeta(x) e^{-2i\pi k \cdot x}$ for any function ζ.

3.1   Spectrum Matrix for Divergence-Free Velocity

In order to simplify the homogeneous model (1.2) while enforcing the divergence-free and spatial stationarity constraints, we can define the small-scale velocity σdB_t with its streamfunction:

$$\sigma dB_t = \nabla^\perp \psi_\sigma dB_t = \nabla^\perp \check{\psi}_\sigma * dB_t, \tag{3.1}$$

where $*$ denote a spatial convolution and $\nabla^\perp$ the two-dimensional curl. Furthermore, we consider a Matérn for the streamfunction covariance:

$$\gamma_{\psi_\sigma}(x) = \frac{1}{dt}\mathbb{E}\left\{(\psi_\sigma(y+x)dB_t)(\psi_\sigma(y)dB_t)\right\}$$

$$= \left(\check{\psi}_\sigma * \check{\psi}_\sigma\right)(x) = D g_{\frac{\beta+1}{2}}(2\pi\kappa_m \|x\|), \tag{3.2}$$

where $g_\nu(r) = r^\nu \mathcal{K}_\nu(r)$, $\mathcal{K}_\nu$ is the modified Bessel function of second kind, $1/2\pi\kappa_m$ is the correlation length, and D is a constant defined in Appendix. We will show further below that $-\beta$ is the velocity spectrum slope. This covariance choice is physically relevant since it highlights an important symmetry of turbulence: the self-similar distribution of energy. Indeed, the corresponding streamfunction spectrum is [17, 37]:

$$S_{\psi_\sigma}(k) = \left|\widehat{\check{\psi}}_\sigma(k)\right|^2 = \widehat{\gamma}_{\psi_\sigma}(k) = S_{\psi_\sigma}(0)\left(1 + \left(\frac{k}{\kappa_m}\right)^2\right)^{-\frac{\beta+3}{2}} \tag{3.3}$$

with a constant $S_{\psi_\sigma}(0)$ defined in Appendix and the wavenumber $k = \|\mathbf{k}\|$. The small-scale velocity spectrum matrix is

$$\widehat{a}(\mathbf{k}) = \frac{1}{dt}\mathbb{E}\left\{\widehat{(\sigma\, dB_t)}(\mathbf{k})\widehat{(\sigma\, dB_t)}^H(\mathbf{k})\right\}$$

$$= (2\pi i\mathbf{k}^\perp)(2\pi i\mathbf{k}^\perp)^H S_{\psi_\sigma} = \mathbf{J}(2\pi\mathbf{k})(2\pi\mathbf{k})^T \mathbf{J}^T S_{\psi_\sigma}, \qquad (3.4)$$

where $\mathbf{J}$ is the matrix which performs a $\frac{\pi}{2}$-rotation. Equations (3.3) and (3.4) confirm that the velocity spectrum slope is $-\beta$. Note that $\widehat{a}(\mathbf{k})$ is of rank 1. Therefore, it is not invertible and even its pseudo inverse, $\widehat{a}(\mathbf{k})^\dagger$, is not defined. This singularity will induce major difficulties in the following estimation procedure (see Sect. 4 below). In other words, our methodology cannot be applied with solely a solenoidal small-scale velocity. A workaround is the consideration of a divergent component.

3.2 Spectrum Matrix from Helmholtz Decomposition of the Small-scale Velocity

Now, we do not impose divergence free anymore. From Helmholtz-Hodge theorem, we can write the hidden small-scale velocity as a sum of a solenoidal and a potential components:

$$\sigma\dot{\mathbf{B}} = \nabla^\perp\breve{\psi}_\sigma * \dot{B}^{\nabla^\perp} + \nabla\breve{\phi}_\sigma * \dot{B}^\nabla, \qquad (3.5)$$

where $\dot{B}^{\nabla^\perp}$ and $\dot{B}^\nabla$ are two independent white noises. Here, we have implicitly assumed that the two components are independent from one another. We stick to Matérn covariances, using it from both components:

$$S_{\psi_\sigma}(\mathbf{k}) = S_{\psi_\sigma}(0)\left(1 + \left(\frac{k}{\kappa_m^{\nabla^\perp}}\right)^2\right)^{-\frac{\beta^{\nabla^\perp}+3}{2}}, \qquad (3.6)$$

$$S_{\phi_\sigma}(\mathbf{k}) = S_{\phi_\sigma}(0)\left(1 + \left(\frac{k}{\kappa_m^\nabla}\right)^2\right)^{-\frac{\beta^\nabla+3}{2}}. \qquad (3.7)$$

To simplify, we may choose $\kappa_m^{\nabla^\perp} = \kappa_m^\nabla = \kappa_m$ and set $2\pi\kappa_m$ to be the smallest well resolved scale contained in the large-scale velocities $\tilde{w}$. Then, the diagonalization of the spectrum matrix is straightforward :

$$\widehat{a}(\mathbf{k}) = (2\pi\mathbf{k}^\perp)(2\pi\mathbf{k}^\perp)^T S_{\psi_\sigma} + (2\pi\mathbf{k})(2\pi\mathbf{k})^T S_{\phi_\sigma} = \mathbf{K}\mathbf{S}\mathbf{K}^T, \qquad (3.8)$$

denoting $K = \begin{bmatrix} \tilde{k} & \tilde{k}^{\perp} \end{bmatrix}$, $S = \begin{bmatrix} S^{\nabla} & 0 \\ 0 & S^{\nabla^{\perp}} \end{bmatrix} = \begin{bmatrix} (2\pi k)^2 S_{\phi_{\sigma}} & 0 \\ 0 & (2\pi k)^2 S_{\psi_{\sigma}} \end{bmatrix}$, with

$\tilde{k} = k/k$, $S^{\nabla} = (2\pi k)^2 S_{\phi_{\sigma}}$, and $S^{\nabla^{\perp}} = (2\pi k)^2 S_{\psi_{\sigma}}$, the normalized wave-vector, the divergent and the solenoidal subgrid velocity spectra respectively. Note that the spectrum matrix is now full rank.

3.3 Velocity Covariance

Once the parameters optimized, we can use of the known subgrid spectra (3.8) and covariance:

$$a(x) = J H_{\gamma_{\psi_{\sigma}}} J^T + H_{\gamma_{\phi_{\sigma}}}, \tag{3.9}$$

$$= (2\pi\kappa_m)^2 D^{\nabla^{\perp}} \left(g_{\frac{\beta^{\nabla^{\perp}}-1}{2}} (2\pi\kappa_m \|x\|) \, \mathbb{I}_d \right.$$

$$\left. - g_{\frac{\beta^{\nabla^{\perp}}-3}{2}} (2\pi\kappa_m \|x\|) \left((x^{\perp})(x^{\perp})^T \right) \right)$$

$$+ (2\pi\kappa_m)^2 D^{\nabla} \left(g_{\frac{\beta^{\nabla}-1}{2}} (2\pi\kappa_m \|x\|) \, \mathbb{I}_d - g_{\frac{\beta^{\nabla}-3}{2}} (2\pi\kappa_m \|x\|) \left(xx^T \right) \right),$$

$$\tag{3.10}$$

with constants $D^{\nabla^{\perp}}$, D^{∇} defined in Appendix as functions of the respective variance tensors $a_0^{\nabla^{\perp}}$ and a_0^{∇}, turbulence kinetic energy spectrum slope $\beta^{\nabla^{\perp}}$ and β^{∇}, and the correlation length $1/\kappa_m$. For synthetic notations, we also introduce $e^{\alpha} = a_0^{\nabla^{\perp}}/a_0^{\nabla}$ the variance ratio between the solenoidal and potential components.

3.4 Parametric Model Summary

We have introduced a parametric model for the subgrid velocity depending on the variance tensor a_0 and 3 other parameters $\theta = (\beta^{\nabla^{\perp}}, \beta^{\nabla^{\perp}}, \alpha)$. Finally, our parametric model can be summarized as follow:

$$\hat{a}(k) = K S K^T = \begin{bmatrix} \tilde{k} & \tilde{k}^{\perp} \end{bmatrix} \begin{bmatrix} S^{\nabla}(k) & 0 \\ 0 & S^{\nabla^{\perp}}(k) \end{bmatrix} \begin{bmatrix} \tilde{k} & \tilde{k}^{\perp} \end{bmatrix}^T, \tag{3.11}$$

$$S^{\nabla}(k) = a_0 \left(\frac{1}{1+e^{\alpha}} \right) \left(\frac{2\pi((\beta^{\nabla})^2-1)}{(2\pi\kappa_m)^2} \right) \left(\frac{k}{\kappa_m} \right)^2 \left(1 + \left(\frac{k}{\kappa_m} \right)^2 \right)^{-\frac{\beta^{\nabla}+3}{2}}, \tag{3.12}$$

$$S^{\nabla^\perp}(k) = a_0 \left(\frac{e^\alpha}{1 + e^\alpha} \right) \left(\frac{2\pi ((\beta^{\nabla^\perp})^2 - 1)}{(2\pi \kappa_m)^2} \right) \left(\frac{k}{\kappa_m} \right)^2 \left(1 + \left(\frac{k}{\kappa_m} \right)^2 \right)^{-\frac{\beta^{\nabla^\perp} + 3}{2}}.$$

$$(3.13)$$

Such parametrization naturally ensures that the total variance is always a_0 and that the solenoidal and potential variances are $a_0^{\nabla^\perp} = a_0 \left(\frac{e^\alpha}{1 + e^\alpha} \right)$ and $a_0^{\nabla} = a_0 \left(\frac{1}{1 + e^\alpha} \right)$ respectively, both restricted to $[0, a_0]$.

4   Maximum Likelihood Estimation of Turbulence Correlations

The variance tensor a_0 (the "variance") can be estimated by quadratic variation as explained in Sect. 2. The other parameters $\boldsymbol{\theta}$ will be estimated by maximum likelihood. However, due to the multiplicative structure of the random forcing in (1.5), the tracers q_i are not Gaussian. So, the tracers likelihood is not trivial. Fortunately, the Girsanov theorem gives the expression of the mutual likelihood of the whole set $(q_i(\boldsymbol{x}_q, t_k))_{\substack{1 \leqslant q \leqslant M \\ 1 \leqslant k \leqslant N}}$. Note that most of the results of this section are relatively general. They can be applied to most homogeneous subgrid spatial covariance parametrisations in stochastic transport contexts.

Piterbarg and Rozovskii [25, 26] have already proposed surface current estimations from satellite image by maximum likelihood estimation. However, their algorithm estimates a constant velocity field only. Then, they repeat the operation on a multitude of patches to eventually obtain a gridded velocity field. We believe that such coarse-grid velocity field is probably better estimated by state-of-the-art optical flow methods, and we do not address this issue here. We rather try to extract additional statistical information from the residue of coarse-scale current estimations.

4.1   Girsanov Theorem

In the literature of processes statistics, a lot of parametric estimation methods rely on likelihood. Indeed, denoting $\boldsymbol{\theta}$ the parameters, even for a non-Gaussian process $\boldsymbol{Q}(t)$, such as

$$\frac{\mathrm{d}}{\mathrm{d}t} \boldsymbol{Q}(t) = \boldsymbol{F}(\boldsymbol{Q}(t)|\boldsymbol{\theta}) + \boldsymbol{G}(\boldsymbol{Q}(t)|\boldsymbol{\theta}) \dot{\boldsymbol{W}}, \qquad (4.1)$$

where $\dot{W}$ is a vector of independent white noise, there is a simple expression of the joint likelihood, $p\left(\{Q_t|0 \leqslant t \leqslant N\Delta t\}|\theta\right)$. Here above $\dot{W}$ is to be interpreted in a Itō sense. We assume that GG^T is invertible where $G = G(Q(t)|\theta)$. We will discuss the validity of this assumption later in the chapter. The Girsanov theorem [21] leads as explained in [27] (Eq. (3.3.2) page 147) to the following log-likelihood:

$$l\left(\{Q(t)|0 \leqslant t \leqslant N\Delta t\}|\theta\right) = \int_0^{N\Delta t} \log|\Sigma(Q(t)|\theta)|^{-\frac{1}{2}} dt$$

$$+ \int_0^{N\Delta t} \left(F^T\Sigma^{-1}\right)(Q(t)|\theta) dQ(t)$$

$$- \frac{1}{2} \int_0^{N\Delta t} \left(F^T\Sigma^{-1}F\right)(Q(t)|\theta) dt. \quad (4.2)$$

where

$$\Sigma^{-1} = G[[G^TG]^\dagger]^2 G^T = G[[G^TG]^\dagger]^2 G^T[GG^T][GG^T][GG^T]^{-2}$$

$$= GG^T[GG^T]^{-2} = [GG^T]^{-1}, \quad (4.3)$$

i.e. $\Sigma = GG^T = [GG^T](Q(t)|\theta)$ is the noise conditional covariance given the current state $Q(t)$ and the parameter θ. Here, we have added the normalizing constant logarithm, $\int_0^{N\Delta t} \log|\Sigma|^{-\frac{1}{2}} dt$, since Σ depends on the parameters θ to be estimated.

Note that the Girsanov theorem does not give us the conditional probability density function of $Q(t)$ at time t but only the joint probability density function of $\{Q(t)|0 \leqslant t \leqslant t\}$. The above formula is widely used to perform maximum likelihood estimations and Bayesian estimations in finance [27, 34, 36] and more recently to linear and nonlinear SPDE in biology and reaction-diffusion systems with additive noise [1–3, 12, 23] and multiplicative noise [15]. Note that the Girsanov theorem is also valid in infinite dimension [10, 18]. Janák and Reiß [15] treat the variance estimator aside, before performing MLE. This prevents theoretical estimation issues. That is why we operate similarly here: first estimating the variance tensor and then estimating the other covariance parameters.

As already discussed in Sect. 2, another important issue to deal with is the finite time correlation time of the observed increments. It is a common problem that biases MLE [4, 5, 9, 22, 24, 28]. As explained Sect. 2, we address this issue by subsampling the data.

An alternative estimation method can also be derived from the discretized-in-time version of the stochastic differential Eq. (4.1). Q is not Gaussian but $dQ(t)$ and thus $Q(t + dt)$ are conditionally Gaussian given $Q(t)$ and θ. By factorizing the conditional Gaussian distribution from $t = 0$ to $t = N\Delta t$, we obtain a similar likelihood expression.

4.2   Application of the Girsanov Theorem

After spatial discretization, the stochastic Eq. (1.5) reduces to the form (4.1). A sequence of satellite images of a tracer could hence be used to estimate a parametrization of the stochastic model. In this case,

$$\boldsymbol{Q}_{pi}(t) = q_i(\boldsymbol{x}_p, t), \tag{4.4}$$

$$\boldsymbol{F}_{pi}(t) = \left(-\boldsymbol{w}_i \cdot \nabla q_i + \tfrac{a_0}{2}\Delta q_i + \dot{Q}_i\right)(\boldsymbol{x}_p, t), \tag{4.5}$$

$$(\boldsymbol{G}\dot{\boldsymbol{W}})_{pi}(t) = \left(-\boldsymbol{v}' \cdot \nabla q_i\right)(\boldsymbol{x}_p, t), \tag{4.6}$$

$$\Sigma_{pi,rj}(t) = (\boldsymbol{G}\boldsymbol{G}^T)_{pi,rj}(t) = \nabla q_i^T(\boldsymbol{x}_p, t)\boldsymbol{a}(\boldsymbol{x}_p - \boldsymbol{x}_r)\nabla q_j(\boldsymbol{x}_r, t), \tag{4.7}$$

$$\forall \boldsymbol{Q}', (\boldsymbol{\Sigma}\boldsymbol{Q}')_{pi}(t) = \nabla q_i^T(\boldsymbol{x}_p, t)\left(\boldsymbol{a} * \left(\sum_j \nabla q_j q_j'\right)\right)(\boldsymbol{x}_p, t)$$

$$\text{where } q_j'(\boldsymbol{x}_r) \overset{\Delta}{=} \boldsymbol{Q}'_{jr}$$

$$\tag{4.8}$$

In practice, we shall use finite-dimensional approximations for every calculations steps. For any d-dimensional vector fields $\boldsymbol{\zeta}$, we must define spatially-discretized version. We represent $\boldsymbol{\zeta}$ by $M \times d$-dimensional vectors, $\boldsymbol{Z}$. More precisely, we denote $Z_{pi} = \zeta_i(\boldsymbol{x}_p)$, We may also give a matrix notation $\boldsymbol{A} = \boldsymbol{A}(\boldsymbol{\theta})$ to the convolution and – with a slight abuse of notations – we identify:

$$\forall \boldsymbol{Z}, \ \boldsymbol{A}(\boldsymbol{\theta})\boldsymbol{Z} = \boldsymbol{a}(\boldsymbol{\theta}) * \boldsymbol{\zeta} \tag{4.9}$$

To simplify, we will consider periodic boundary conditions and a discretisation over a uniform spatial grid. It prevents technical problems with convolutions and Fourier transform. Then, we introduce the local matrix $\boldsymbol{Y}_{is}(\boldsymbol{x}, t) = \partial_{x_s} q_i(\boldsymbol{x}, t)$ and the associated block-diagonal matrix $\mathbb{Y}$ for the point-wise application of the small matrix $\boldsymbol{Y}$:

$$\mathbb{Y}_{rj,pi} = \mathbb{Y}_{rj,pi}(\boldsymbol{Q}(t)) = Y_{ji}(\boldsymbol{x}_r, t)\delta_{rp} = \partial_i q_j(\boldsymbol{x}_r, t)\delta_{rp}. \tag{4.10}$$

We can eventually rewrite the operator $\boldsymbol{\Sigma} = \boldsymbol{\Sigma}(\boldsymbol{Q}(t)|\boldsymbol{\theta})$ as :

$$\forall \boldsymbol{Q}', \ \boldsymbol{\Sigma}\boldsymbol{Q}' = \mathbb{Y}\boldsymbol{A}(\boldsymbol{\theta})\mathbb{Y}^T \boldsymbol{Q}', \tag{4.11}$$

i.e. $\boldsymbol{\Sigma} = \mathbb{Y}\boldsymbol{A}(\boldsymbol{\theta})\mathbb{Y}^T$ where $\mathbb{Y} = \mathbb{Y}(\boldsymbol{Q}(t))$.

4.3 Inversion of the Operator Σ

If the matrix Y is everywhere sufficiently well conditioned (in particular if we observe enough tracers and if the fronts of different tracers are not aligned), we can locally consider the pseudo-inverse of Y, that we will denote $Y^\dagger$. Accordingly, we can obtain a approximate inverse of the operator Σ:

$$\Sigma^{-1} = (\mathbb{Y}^\dagger)^T A^{-1}(\theta) \mathbb{Y}^\dagger. \tag{4.12}$$

where $A^{-1}(\theta)$ is a deconvolution operation. It can be computed in Fourier space or using other usual deconvolution methods. Needless to say that $\mathbb{Y}^\dagger_{rj,pi} = Y^\dagger_{ji}(x_r, t)\delta_{rp}$ is also block-diagonal, enabling such large matrix computation is reasonable time. Using a given parametric form for a (see Sect. 3), we can compute A^{-1} in Fourier space. For continuous Fourier transform, it would read:

$$\forall \zeta, \ \widehat{[a*]^{-1}\zeta} = \widehat{[a*]^{-1}}\widehat{\zeta} = \widehat{a}^{-1}\widehat{\zeta}, \tag{4.13}$$

assuming that the small-scale spectrum matrix $\widehat{a}$ has full rank. From (3.8), we have an explicit expression of the inverse:

$$\widehat{a}^{-1} = K S^{-1} K^T. \tag{4.14}$$

In practice, we shall use the Fast Fourier Transform (FFT). For an uniform spatial grid of M points, we represent $\widehat{\zeta}$ by $\widehat{Z} = P^H Z$. More precisely, we denote $\widehat{Z}_{pi} = \widehat{\zeta}_i(k_p)$, $P_{pi,rj} = \delta_{ij} e^{2i\pi x_p \cdot k_r}$ and $P^H = \overline{P}^T$ its conjugate transpose, i.e. $P^H_{rj,pi} = \delta_{ij} e^{-2i\pi x_p \cdot k_r}$. The inverse discrete Fourier transform is defined by matrix $\frac{1}{M} P$ since $(P^H)^{-1} = \frac{1}{M} P$. We can now express the deconvolution (4.13) with the block-diagonal matrix $\widehat{A}^{-1}(\theta) = P^{-1} A^{-1}(\theta) P$

$$\forall Z, \ P^H (A^{-1}(\theta) Z) = (P^{-1} A^{-1}(\theta) P) \widehat{Z} \tag{4.15}$$

where $\left(\widehat{A}^{-1}(\theta)\right)_{rj,pi} = (P^{-1} A^{-1}(\theta) P)_{rj,pi} = \frac{1}{M}(\widehat{a}^{-1})_{ji}(k_r, \theta))\delta_{rp}$. Finally, from (4.12) we obtain a simple matrix form for the inverse covariance:

$$\Sigma^{-1} = \frac{1}{M}(\mathbb{Y}^\dagger)^T P \widehat{A}^{-1}(\theta) P^H \mathbb{Y}^\dagger. \tag{4.16}$$

The efficiency of the FFT algorithm together with the block-diagonal structures of the other matrices ensure a low computational cost. With the expression of F and Σ^{-1}, we can now compute the expression of the log-likelihood (4.2).

4.4 Gradient of the Likelihood

In order to estimate the parameters θ_r, we will need to maximize the log-likelihood, by e.g. gradient descent. Such algorithm necessitates the log-likelihood derivative along each parameters. Since F does not depend on θ, we have $\partial_{\theta_r} l = \int_0^{N\Delta t} d\partial_{\theta_r} l$ with :

$$d\partial_{\theta_r} l = -\frac{1}{2}\partial_{\theta_r} \log|\Sigma| dt + F^T \partial_{\theta_r}(\Sigma^{-1}) dQ(t) - \frac{1}{2}F^T \partial_{\theta_r}(\Sigma^{-1})F dt, \quad (4.17)$$

where

$$\partial_{\theta_r}(\Sigma^{-1}) = \frac{1}{M}(\mathbb{Y}^\dagger)^T P \partial_{\theta_r}(\widehat{A}^{-1}) P^H \mathbb{Y}^\dagger, \quad (4.18)$$

$$\partial_{\theta_r}\left(\widehat{A}^{-1}\right)_{rj,pi} = \frac{1}{M}\partial_{\theta_r}(\widehat{a}^{-1})_{ji}(k_r)\delta_{rp}, \quad (4.19)$$

$$\partial_{\theta_r}\left(\widehat{a}^{-1}\right) = K\partial_{\theta_r}(S^{-1})K^T = K\begin{bmatrix} \partial_{\theta_r}(1/S^\nabla) & 0 \\ 0 & \partial_{\theta_r}(1/S^{\nabla^\perp}) \end{bmatrix}K^T \quad (4.20)$$

The normalizing constant can be differentiate with Jacobi formula:

$$\partial_{\theta_r}\log|\Sigma| = \text{tr}\left(\Sigma^{-1}\partial_{\theta_r}\Sigma\right), \quad (4.21)$$

$$= \text{tr}\left(\left(\frac{1}{M}(\mathbb{Y}^\dagger)^T P\widehat{A}^{-1}P^H\mathbb{Y}^\dagger\right)\left(\frac{1}{M}\mathbb{Y}P\partial_{\theta_r}(\widehat{A})P^H\mathbb{Y}^T\right)\right), \quad (4.22)$$

$$= \text{tr}\left(\widehat{A}^{-1}\partial_{\theta_r}(\widehat{A})\right), \quad (4.23)$$

$$= \frac{1}{M}\sum_{ir}(\widehat{a}^{-1}\partial_{\theta_r}\widehat{a})_{ii}(k_r), \quad (4.24)$$

$$= \frac{1}{M}\sum_{r}\text{tr}\left((KS^{-1}K^T)(K\partial_{\theta_r}(S)K^T)\right), \quad (4.25)$$

$$= \frac{1}{M}\sum_{r}\left(\partial_{\theta_r}\log S^\nabla(k_r) + \partial_{\theta_r}\log S^{\nabla^\perp}(k_r)\right), \quad (4.26)$$

which can evaluate analytically from (3.12)–(3.13). We skip these straightforward calculations for concision and readability. We now gather the different terms to obtain the full the log likelihood gradient to be used in the gradient descent algorithm, to eventually find the optimal covariance parameters θ:

$$d\partial_{\theta_r} l = -\frac{1}{2M}\sum_{r}\left(\partial_{\theta_r}\log S^\nabla(k_r) + \partial_{\theta_r}\log S^{\nabla^\perp}(k_r)\right) dt$$

$$+ \left(F(\boldsymbol{Q}(t)|\boldsymbol{\theta})^T \frac{1}{M} (\mathbb{Y}^{\dagger}(t))^T \boldsymbol{P} \mathbb{K} \partial_{\theta_r} (\mathbb{S}^{-1}) \mathbb{K}^T \boldsymbol{P}^H \mathbb{Y}^{\dagger}(t) \right)$$

$$\times \left(d\boldsymbol{Q}(t) - \frac{1}{2} F(\boldsymbol{Q}(t)|\boldsymbol{\theta}) dt \right), \tag{4.27}$$

where $\mathbb{K}$ and $\mathbb{S}^{-1}$ are the block-diagonal versions of $\boldsymbol{K}$ and $\boldsymbol{S}^{-1}$:

$$\mathbb{S}^{-1}_{rj,pi} = (1/S_{jj}(\boldsymbol{k}_r)) \, \delta_{ji} \delta_{rp} \quad \text{and} \quad \mathbb{K}_{rj,pi} = K_{ji}(\boldsymbol{k}_r) \delta_{rp}. \tag{4.28}$$

Since all matrices except $\boldsymbol{P}$ are block diagonal, their evaluations have a complexity $O(M)$. Only $\boldsymbol{P}$ necessities a complexity $O(M \log^2(M))$ (Fast Fourier Transform algorithm along two dimensions). Evaluating one gradient step requires the time integration of the above formula over N time steps. The computational cost of one gradient is hence $O(NM \log^2(M))$ only.

5 Conclusion

We have proposed a new approach to estimate statistics of a hidden subgrid velocity field from a sequence of tracer images. The first contribution of the chapter is to convert LU-SALT into (4.1) to then apply standard MLE techniques. The second important contribution is an efficient method to solve the MLE optimization problem through a fast log-likelihood gradient evaluation algorithm (4.27). We rely on a parametric model and Fourier-based representations of that velocity to tackle the curse of dimensionality of the problem. As statistics tools, we choose quadratic co-variation and maximum likelihood estimation for their reliable theoretical grounds. Notably, if the subgrid velocity component is solenoidal, its spectrum matrix has rank 1 locally and cannot be inverted. As such, we cannot apply our method. We have to consider both divergent and rotational terms to obtain full rank spectrum matrix, evaluate the joint tracers series distribution, and perform MLE.

Measurement errors are neglected here. True measurement errors are generally weak for satellite images of the oceanic tracers (e.g., SST) even though the image resolution is always limited. Fake measurement errors are sometimes considered to mimic the effect of observed aliased geophysical signals (e.g., nugget in altimetry optimal interpolation (i.e. kriging)). This aliasing is often filtered out by regularization. We do not consider it explicitly here, but this preprocessing could be pursued before applying our method. If we work on L4 satellite products, this would be the case. Nevertheless, by forgetting resolution issues we may increase model errors. These errors may be large and may lead to statistics robustness issues (e.g., when estimating quadratic variations, correlation times, and other parameters through MLE). Indeed, even though erroneous advection—and thus structure mislocation— is well modeled by SALT-LU, we may debate about the dynamics assumed for the partially-resolved geophysical observables. Errors in the dynamics itself may not be negligible. To alleviate this issue together with regularizing the statistics, we could probably add an additive noise in the modeled dynamics on top of the multiplicative noise. We could try to jointly learn this additional stochastic forcing terms following [26]. Alternatively, the additive white noise forcing variance could be inferred from the tracer quadratic variation (2.2). This statistic would contain the information of both the forcing variance and the subgrid velocity variance. The number of available statistics being large (one by grid point), the least square problem can be solved easily. Considering additive noise complexifies the evaluation of the inverse covariance matrix Σ but the Woodbury identity can probably yields a tractable solution.

In future work, we could combine our approach with stochastic optical flow algorithms [6, 7]. Furthermore, we will adapt our algorithm to deal with non-periodic boundary conditions. Then, we shall apply this methodology to synthetic and real satellite images of tracers like SSS or SST. Our method may also be generalized to treat not only 0-form like SSS or SST, but also more complex differential forms through stochastic geometric fluid dynamics [14, 38]. As such, we will be able to directly treat surface currents observations from high-frequency radars or wind observations from Doppler radars. More generally, we hope that our work will enable new calibration methods for SALT and LU, but also new observation capabilities from current and future satellite observations, and better physical understanding of tracer budgets from them [e.g. 20].

Acknowledgments We thank Bernard Delyon for pointing out the issues arising in estimating the variance tensor directly from MLE. We also thank Bertrand Chapron and the reviewers for helpful discussions, comments and inspiring references.

Appendix: Variance Parameters for Streamfunction, Potential and Velocity Covariances

From properties of the modified Bessel function of second kind, we get the following formula:

$$\partial_r g_\nu(r) = -r g_{\nu-1}(r), \tag{A.1}$$

$$\partial_r^2 g_\nu(r) = -g_{\nu-1}(r) + r^2 g_{\nu-2}(r), \tag{A.2}$$

$$\nabla\left(g_\nu(\|\boldsymbol{x}\|)\right) = -\boldsymbol{x} g_{\nu-1}(\|\boldsymbol{x}\|), \tag{A.3}$$

$$H\left(g_\nu(\|\boldsymbol{x}\|)\right) = -g_{\nu-1}(\|\boldsymbol{x}\|)\mathbb{I}_d + g_{\nu-2}(\|\boldsymbol{x}\|)\boldsymbol{x}\boldsymbol{x}^T, \tag{A.4}$$

$$g_\nu(r) \underset{r\to 0}{\to} \frac{\Gamma(\nu)}{2^{1-\nu}}, \tag{A.5}$$

with Γ is the Gamma function. From the last equation, we obtain the normalization factor of the streamfunction and potential covariances:

$$D = \frac{2^{\frac{1-\beta}{2}}}{\Gamma(\frac{\beta+1}{2})} \gamma_{\psi_\sigma}(0). \tag{A.6}$$

From the other formula above, we can evaluate the divergence-free velocity covariance:

$$a^{\nabla^\perp}(x) = \frac{1}{dt} \mathbb{E}\left\{ \left(\sigma^{\nabla^\perp} dB_t\right)(y+x)\left(\sigma^{\nabla^\perp} dB_t\right)^T(y) \right\}, \tag{A.7}$$

$$= J H_{\gamma_{\psi_\sigma}} J^T, \tag{A.8}$$

$$= D^{\nabla^\perp}(2\pi\kappa_m)^2 \left(g_{\frac{\beta^{\nabla^\perp}-1}{2}} \mathbb{I}_d - g_{\frac{\beta^{\nabla^\perp}-3}{2}}(Jx)(Jx)^T \right). \tag{A.9}$$

Finally, the variance tensor is $a^{\nabla^\perp}(0) = a_0^{\nabla^\perp} \mathbb{I}_d$ with

$$a_0^{\nabla^\perp} = \frac{(2\pi\kappa_m)^2}{\beta^{\nabla^\perp}-1} \gamma_{\psi_\sigma}(0) \text{ and } D^{\nabla^\perp} = \frac{2^{\frac{1-\beta^{\nabla^\perp}}{2}}}{\Gamma\left(\frac{\beta^{\nabla^\perp}+1}{2}\right)} \frac{\beta^{\nabla^\perp}-1}{(2\pi\kappa_m)^2} a_0^{\nabla^\perp}, \tag{A.10}$$

About the amplitude spectrum $S_{\psi_\sigma}(0)$, we know from [17, 37] that:

$$S_{\psi_\sigma}(0) = \frac{4\pi \Gamma\left(\frac{\beta^{\nabla^\perp}+3}{2}\right)}{2\pi\kappa_m \Gamma\left(\frac{\beta^{\nabla^\perp}+1}{2}\right)} \gamma_{\psi_\sigma}(0) = \frac{2\pi(\beta^{\nabla^\perp}+1)}{(2\pi\kappa_m)^2} \gamma_{\psi_\sigma}(0), \tag{A.11}$$

$$= \frac{2\pi((\beta^{\nabla^\perp})^2-1)}{(2\pi\kappa_m)^4} a_0^{\nabla^\perp}.$$

The formula for the potential component are similar, replacing $\nabla^\perp$ by ∇, J by $\mathbb{I}_d$, and ψ_σ by ϕ_σ.

References

1. Altmeyer R, Reiß M (2021) Nonparametric estimation for linear spdes from local measurements. The Annals of Applied Probability 31(1):1–38
2. Altmeyer R, Bretschneider T, Janák J, Reiß M (2022) Parameter estimation in an spde model for cell repolarization. SIAM/ASA Journal on Uncertainty Quantification 10(1):179–199

3. Altmeyer R, Cialenco I, Pasemann G (2023) Parameter estimation for semilinear spdes from local measurements. Bernoulli 29(3):2035–2061
4. Azencott R, Beri A, Timofeyev I (2010) Adaptive sub-sampling for parametric estimation of gaussian diffusions. Journal of Statistical Physics 139(6):1066–1089
5. Azencott R, Beri A, Jain A, Timofeyev I (2013) Sub-sampling and parametric estimation for multiscale dynamics. Communications in Mathematical Sciences 11(4):939–970
6. Cai S, Mémin E, Dérian P, Xu C (2018) Motion estimation under location uncertainty for turbulent fluid flows. Experiments in Fluids 59(1):8
7. Corpetti T, Mémin E (2012) Stochastic models for local optical flow estimation. In: Scale Space and Variational Methods in Computer Vision: Third International Conference, SSVM 2011, Ein-Gedi, Israel, May 29–June 2, 2011, Revised Selected Papers 3, Springer, pp 701–712
8. Corpetti T, Heitz D, Arroyo G, Mémin E, Santa-Cruz A (2006) Fluid experimental flow estimation based on an optical-flow scheme. Experiments in fluids 40(1):80–97
9. Cotter C, Pavliotis G (2009) Estimating eddy diffusivities from noisy lagrangian observations. Communications in Mathematical Sciences 7(4):805–838
10. Da Prato G, Zabczyk J (1992) Stochastic Equations in Infinite Dimensions. Encyclopedia of Mathematics and its Applications, Cambridge University Press
11. Dérian P, Héas P, Herzet C, Mémin E (2013) Wavelets and optical flow motion estimation. Numerical Mathematics: Theory, Methods and Applications 6(1):116–137
12. Gaudlitz S, Reiß M (2023) Estimation for the reaction term in semi-linear spdes under small diffusivity. Bernoulli 29(4):3033–3058
13. Ghalenoei E, Sharifi MA, Hasanlou M (2015) Estimating and fusing optical flow, geostrophic currents and sea surface wind in the waters around kish island. The International Archives of the Photogrammetry, Remote Sensing and Spatial Information Sciences 40:221–226
14. Holm D (2015) Variational principles for stochastic fluid dynamics. Proceedings of the Royal Society of London A: Mathematical, Physical and Engineering Sciences 471(2176)
15. Janák J, Reiß M (2023) Parameter estimation for the stochastic heat equation with multiplicative noise from local measurements. arXiv preprint arXiv:230300074
16. Kraichnan R (1994) Anomalous scaling of a randomly advected passive scalar. Physical review letters 72(7):1016
17. Lim S, Teo L (2009) Generalized whittle–matérn random field as a model of correlated fluctuations. Journal of Physics A: Mathematical and Theoretical 42(10):105,202
18. Liu W, Röckner M (2015) Stochastic partial differential equations: an introduction. Springer
19. Mémin E (2014) Fluid flow dynamics under location uncertainty. Geophysical & Astrophysical Fluid Dynamics 108(2):119–146, DOI 10.1080/03091929.2013.836190
20. Michel S, Chapron B, Tournadre J, Reul N (2007) Sea surface salinity variability from a simplified mixed layer model of the global ocean. Ocean Science Discussions 4(1):41–106
21. Oksendal B (1998) Stochastic differential equations. Spinger-Verlag
22. Papavasiliou A, Pavliotis G, Stuart A (2009) Maximum likelihood drift estimation for multiscale diffusions. Stochastic Processes and their Applications 119(10):3173–3210
23. Pasemann G, Stannat W (2020) Drift estimation for stochastic reaction-diffusion systems. Electronic Journal of Statistics 14:547–579
24. Pavliotis GA, Stuart A (2007) Parameter estimation for multiscale diffusions. Journal of Statistical Physics 127:741–781
25. Piterbarg L, Ostrovskii A (1997) Advection and diffusion in random media: implications for sea surface temperature anomalies. Kluwer Academic
26. Piterbarg L, Rozovskii B (1996) Maximum likelihood estimators in the equations of physical oceanography. In: Stochastic modelling in physical oceanography, Springer, pp 397–421
27. Rao P (1999) Statistical inference for diffusion type processes. Arnold
28. Reich S (2023) Frequentist perspective on robust parameter estimation using the ensemble kalman filter. Stochastic Transport in Upper Ocean Dynamics p 237

29. Resseguier V, Mémin E, Chapron B (2017) Geophysical flows under location uncertainty, part I random transport and general models. Geophysical & Astrophysical Fluid Dynamics 111(3):149–176
30. Resseguier V, Mémin E, Chapron B (2017) Geophysical flows under location uncertainty, part II quasi-geostrophy and efficient ensemble spreading. Geophysical & Astrophysical Fluid Dynamics 111(3):177–208
31. Resseguier V, Li L, Jouan G, Dérian P, Mémin E, Bertrand C (2020) New trends in ensemble forecast strategy: uncertainty quantification for coarse-grid computational fluid dynamics. Archives of Computational Methods in Engineering pp 1–82
32. Resseguier V, Pan W, Fox-Kemper B (2020) Data-driven versus self-similar parameterizations for stochastic advection by lie transport and location uncertainty. Nonlinear Processes in Geophysics 27(2):209–234
33. Resseguier V, Picard AM, Memin E, Chapron B (2021) Quantifying truncation-related uncertainties in unsteady fluid dynamics reduced order models. SIAM/ASA Journal on Uncertainty Quantification 9(3):1152–1183
34. Sørensen H (2004) Parametric inference for diffusion processes observed at discrete points in time: a survey. International Statistical Review 72(3):337–354
35. Sun H, Song Q, Shao R, Schlicke T (2016) Estimation of sea surface currents based on ocean colour remote-sensing image analysis. International journal of remote sensing 37(21):5105–5121
36. van Waaij J, van Zanten H (2016) Gaussian process methods for one-dimensional diffusions: optimal rates and adaptation. Electronic Journal of Statistics 10(1):628–645
37. Williams CK, Rasmussen CE (2006) Gaussian processes for machine learning. the MIT Press 2(3):4
38. Zhen Y, Resseguier V, Chapron B (2023) Physically constrained covariance inflation from location uncertainty. EGUsphere 2023:1

Transport Noise Defined from Wavelet Transform for Model-based Stochastic Ocean Models

Francesco L. Tucciarone, Long Li, Etienne Mémin, and Louis Thiry

1 Introduction

The global climate system is strongly depending on the Ocean's state, as the interaction with the Atmosphere in the forms of mutual exchange of energy fluxes of different natures and global heat redistribution plays a crucial role in the climate regulation [1]. While observations are crucial for understanding the current state of the global ocean, numerical simulation remains the only way to forecast the system and assess future states. This is fundamental for predicting meteorological and climatological events and related hazards. Large-scale simulations of the Ocean (as well as of the Atmosphere) remain the primary tool of investigation while high resolution simulations can be obtained only for small geographical domains or short integration periods. The complex interdependence of mesoscale and sub-mesoscale dynamics, however, is lost in state-of-the-art simulations when performed at scales that are too large to capture these phenomena. Most of the modeling challenges arise from the representation of these effects in a parametrized manner [2]. A novel research trend involves incorporating perturbations and noise terms into the dynamics. The goal is to enhance the variability and parameterize sub-grid processes, turbulence, boundary value uncertainty, and account for numerical and discretization errors. Along this path, two companion methodologies have been introduced by Mémin [3] and Holm [4], providing rigorously justified methodologies to define stochastic large scales representations of the Navier-Stokes equations [5] conserving energy and circulation, respectively. These two models

F. L. Tucciarone (✉) · L. Li · E. Mémin
Centre Inria de l'Université de Rennes – UMR CNRS 6625, Rennes, France
e-mail: francesco.tucciarone@inria.fr

L. Thiry
Centre Inria de Paris – CS 42112, Paris, France

© The Author(s) 2025
B. Chapron et al. (eds.), *Stochastic Transport in Upper Ocean Dynamics III*,
Mathematics of Planet Earth 13, https://doi.org/10.1007/978-3-031-70660-8_13

rely on a stochastic decomposition of the Lagrangian trajectory into a smooth-in-time component induced by the large-scale velocity and a random fast-evolving uncorrelated displacement noise, following ideas proposed by [6–8]. The solid theoretical background allows the definition of a large-scale representation with a stochastic component representing the subgrid contribution, introducing additional degrees of freedom to be exploited in the modelling of specific phenomena (such as large scale components [10, 11], small-scale turbulence [12, 13], boundary layer effects [14] or convection processes [15]) or to devise intermediate models [13, 14, 17–19]. The Location Uncertainty (LU) model [3] has been applied to the barotropic quasi-geostrophic model [17, 20], the baroclinic quasi-geostrophic model [9], the single-layered shallow water model [22], the surface quasi-geostrophic [21], hydrostatic primitive equations [10, 11] and recently non hydrostatic Boussinesq equations [16], proving its efficacy in structuring the large-scale flow [17], reproducing long-term statistics [20] and providing a good trade-off between model error representation and ensemble spread [21, 22]. In this work, the efficacy of a wavelet representation [23, 24] for the small scale turbulence is assessed in the context of stochastic hydrostatic primitive equations following [10, 11] and in a novel stochastic multi-layered shallow water model, based on the derivation of [22] and a modified implementation of [25].

2 Location Uncertainty (LU)

Location Uncertainty is based on a stochastic decomposition of the Lagrangian trajectory $\mathbf{X}_t$ of the fluid particle, so that the displacement is represented by means of the stochastic differential equation (SDE)

$$\mathrm{d}\mathbf{X}_t = \mathbf{v}_t\,\mathrm{d}t + \boldsymbol{\sigma}_t\mathrm{d}\mathbf{B}_t, \tag{1}$$

where $\mathbf{X}\colon \mathcal{S} \times \mathbb{R}^+ \to \Omega$ is the fluid flow map, i.e. the trajectory followed by fluid particles starting at initial map $\mathbf{X}|_{t=0} = \mathbf{x}_0$ of the bounded domain $\mathcal{S} \subset \mathbb{R}^d$ ($d = 2, 3$). The trajectory is thus split into a smooth-in-time (Lagrangian) velocity, $\mathbf{v}_t$, and a stochastic contribution $\boldsymbol{\sigma}_t\mathrm{d}\mathbf{B}_t$, referred to as noise, that is non-smooth in time. The first component in Eq. (1) is associated to the resolved velocity in the integration of the equations of motions, while the second component accounts for processes that are either neglected or not representable at a given resolution. In order to specify the characteristic of this last (martingale) term, let H be the Hilbert space, $H = \left(L^2\left(\mathcal{S}\right), \mathbb{R}^d\right)$, the space of square integrable functions over $\mathcal{S}$ with value in $\mathbb{R}^d$, with inner product $\langle \boldsymbol{f}, \boldsymbol{g}\rangle_H = \int_{\mathcal{S}}(\boldsymbol{f}^{\mathsf{T}}\boldsymbol{g})\,\mathrm{d}\mathbf{x}$ and induced norm $\|\boldsymbol{f}\|_H = \langle \boldsymbol{f}, \boldsymbol{f}\rangle_H^{1/2}$ and let T be a finite time, $T < +\infty$. In this context, $\{\mathbf{B}_t\}_{0\leq t\leq T}$ is a cylindrical Wiener process defined on H [26]:

$$\mathbf{B}_t = \sum_{i\in\mathbb{N}} \hat{\beta}^i \mathbf{e}_i, \tag{2}$$

where $(\mathbf{e}_i)_{i \in \mathbb{N}}$ is a Hilbertian orthonormal basis of the space H and $(\hat{\beta}_i)_{i \in \mathbb{N}}$ is a sequence of independent standard Brownian motions on a stochastic basis $(\Omega, \mathcal{F}, (\mathcal{F}_t)_{t \in [0,T]}, \mathbb{P})$. The application of a Hilbert-Schmidt symmetric integral kernel $\boldsymbol{\sigma}_t f (\mathbf{x}) = \int_S \breve{\sigma} (\mathbf{x}, \mathbf{y}, t) f (\mathbf{y}) \, d\mathbf{y}$ to the Wiener process $\mathbf{B}$ on H constitutes the theoretical definition of the noise term:

$$(\boldsymbol{\sigma}_t d\mathbf{B}_t)^i (\mathbf{x}) = \int_S \breve{\sigma}_{ik} (\mathbf{x}, \mathbf{y}, t) \, d B_t^k (\mathbf{y}) \, d\mathbf{y}, \tag{3}$$

where the Einstein convention for summation over repeated indices is adopted. The kernel $\breve{\sigma}$ is a Hilbert-Schmidt integration kernel, assumed to be bounded in space and time. It follows that the convolution of $\breve{\sigma}$ with $\mathbf{B}_t$ is Hilbert-Schmidt, compact, self-adjoint, positive definite and thus, by Mercer's theorem, it admits eigenfunctions and eigenvalues decreasing toward zero. This defines a centred Gaussian process

$$\int_0^t \boldsymbol{\sigma}_s d\mathbf{B}_s (\mathbf{X}_s) \sim \mathcal{N} \left(0, \int_0^t \mathbf{Q} (\mathbf{X}_s, \mathbf{X}_s, s, s) \, ds \right), \tag{4}$$

where the covariance tensor $\mathbf{Q}$ is defined as

$$Q_{ij} (\mathbf{x}, \mathbf{y}, t, s) = \mathbb{E} \left[(\boldsymbol{\sigma}_t d\mathbf{B}_t (\mathbf{x}))^i (\boldsymbol{\sigma}_t d\mathbf{B}_s (\mathbf{y}))^j \right]$$

$$= \delta (t - s) \, dt \int_S \breve{\sigma}_{ik} (\mathbf{x}, \mathbf{z}, t) \, \breve{\sigma}_{kj} (\mathbf{z}, \mathbf{y}, s) \, d\mathbf{z},$$

with the integral kernel $\breve{\sigma}$ modelled in such a way that a spatial correlation to the fast/small scale components is imposed. The strength of the noise is measured by the diagonal components of the covariance tensor per unit of time, $\mathbf{a}(\mathbf{x}, t)\delta(t - t')dt = \mathbf{Q}(\mathbf{x}, \mathbf{x}, t, t')$, also referred to as the variance tensor. Notably, the variance tensor has the dimension of a viscosity in $m^2 s^{-1}$ and is symmetric and positive definite. Furthermore, the covariance operator $\mathbf{Q}$ is a compact self-adjoint positive definite operator on H, that thus admits a set of orthonormal eigenfunctions $\{\boldsymbol{\xi}_n (\cdot, t), n \in \mathbb{N}\}$ with (strictly) positive eigenvalues $\lambda_n (t)$ decreasing toward zero and satisfying $\sum_{n \in \mathbb{N}} \lambda_n (t) < +\infty$. Consequently, the noise term and the variance tensor can be expressed with respect to the basis provided by the eigenfunctions randomized by a series of scalar Brownian variables, $\beta_{t,n}$, as

$$\boldsymbol{\sigma}_t d\mathbf{B}_t (\mathbf{x}) = \sum_{n \in \mathbb{N}} \lambda^{1/2} (t) \boldsymbol{\xi}_n (\mathbf{x}, t) \, d\beta_{t,n}, \tag{5}$$

$$\mathbf{a} (\mathbf{x}, t) = \sum_{n \in \mathbb{N}} \lambda (t) \boldsymbol{\xi}_n (\mathbf{x}, t) \boldsymbol{\xi}_n^{\mathrm{T}} (\mathbf{x}, t). \tag{6}$$

The noise term defined above is centred, but as introduced in [9, 27] and applied in [11, 28], a modification can be applied through Girsanov transformation in order to consider a Lagrangian displacement of the form

$$d\mathbf{X}_t = [\mathbf{v}_t - \boldsymbol{\sigma}_t Y_t]\,dt + \boldsymbol{\sigma}_t d\mathbf{B}_t, \tag{7}$$

where a correlated component $\boldsymbol{\sigma}_t Y_t$ can be introduced to model phenomena displaying a non-zero time average like in the case of ocean eddies and gyres.

The transition from the Lagrangian point of view to the Eulerian point of view is provided by the stochastic Reynolds transport theorem (SRTT), introduced in [3]. It describes the rate of change of a random scalar q transported by the stochastic flow (1) within a flow volume V_t:

$$d \int_{V_t} q\,(\mathbf{x}, t)\,d\mathbf{x} = \int_{V_t} \left\{ D_t q + q \nabla \cdot \left[\mathbf{v}^\star\,dt + \boldsymbol{\sigma}_t d\mathbf{B}_t \right] \right\} (\mathbf{x}, t)\,d\mathbf{x}, \tag{8}$$

with the operator

$$D_t q = d_t q + \left[\mathbf{v}^\star\,dt + \boldsymbol{\sigma}_t\,d\mathbf{B}_t \right] \cdot \nabla q - \frac{1}{2} \nabla \cdot (\mathbf{a} \nabla q)\,dt, \tag{9}$$

defining the stochastic transport operator. Each term of this operator has a physical interpretation. Proceeding in order, the first term of the right-hand side of (9) is the *increment in time* at a fixed location of the random process q, that is $d_t q = q\,(\mathbf{x}, t + dt) - q\,(\mathbf{x}, t)$. This contribution plays the role of the partial time derivative for a process that is not time differentiable. In the square brackets it is enclosed the *stochastic advection displacement*. It involves a time correlated modified advection velocity,

$$\mathbf{v}^\star = \mathbf{v} - \frac{1}{2} \nabla \cdot \mathbf{a} + \boldsymbol{\sigma}_t^{\mathrm{T}} (\nabla \cdot \boldsymbol{\sigma}_t), \tag{10}$$

and a fast evolving, uncorrelated noise $\boldsymbol{\sigma}_t\,d\mathbf{B}_t$. The advection of the process q by this term leads to a *multiplicative noise* which is non Gaussian. This noise is referred to as *transport noise* in the literature. The second term in Eq. (10) represents the effective transport velocity induced by statistical inhomogeneities of the noise term, and it is referred to as *Itô-Stokes drift* in [17]. In the following it is denoted as $\mathbf{v}^s = \frac{1}{2} \nabla \cdot \mathbf{a}$. The last term of the transport operator is a dissipation term that depicts the mixing mechanism due to the unresolved scales. In the following, the Location Uncertainty principle will be applied to a set of two-dimensional equations, the Shallow Water system, and to a set of three-dimensional equations, the Primitive Equations model. The stochastic transport operator D_t has thus to be intended as built with two-dimensional differential operators in the former case, and with three dimensional differential operators in the latter.

3   Noise Modelling with Wavelets

The modelling of the noise is chosen to enhance the accuracy and the variability of
a (large-scale) simulation in representing the effect of the truncated scales through
random variables. Many data-driven approaches referenced previously have been
proposed to that end (see for instance [9–11, 21]). Here, our goal is to propose a
model-based approach for the noise definition relying only on the current state of
the simulation. Opposite to data-driven technique, the noise hence depends only
on the solution. It is important to outline that this does not violate any principle
of the LU derivation. Let us note however, that the noise needs to remain smooth
enough in space to guarantee the existence of martingale solution [5]. A *wavelet* is a
compactly supported wave-like oscillation that is localized in time [23, 24]. Wavelet
processing has the characteristic of combining data processing in the time domain
and in the frequency domain, with a reasonable trade-off. The forward wavelet
transform decomposes the signal $\mathbf{u}$ from the time domain to its representation in
the wavelet basis, an oscillatory waveform that reveal many signal properties and
provide a sparse representation. Conversely, the inverse transform reconstructs the
signal from its wavelet representation back to the time domain. The result of this
operation is a set of *details* $\langle \mathbf{u}, \boldsymbol{\psi}_{j,k} \rangle_{L^2}$ and a large scale component $\langle \mathbf{u}, \boldsymbol{\phi}_{C,k} \rangle_{L^2}$.
These fields are then randomised with a Brownian field $\mathbf{B}_t$ defined on each point of
the computational wavelet coefficients grid, so that the noise wavelet ansatz can be
defined as

$$
\sigma_t d\mathbf{B}_t(\mathbf{x}) = \sum_{k=0}^{2^C-1} \langle \mathbf{u}^{(n)}, \boldsymbol{\phi}_{C,k} \rangle_{L^2} d\mathbf{B}_{t,C,k} \, \boldsymbol{\phi}_{C,k}(\mathbf{x})
$$

$$
+ \sum_{j=C}^{F} \sum_{k=0}^{2^j-1} \langle \mathbf{u}^{(n)}, \boldsymbol{\psi}_{j,k} \rangle_{L^2} d\mathbf{B}_{t,j,k} \, \boldsymbol{\psi}_{j,k}(\mathbf{x}). \tag{11}
$$

In the previous equation, F and C are indexes that divide the details and the large
scale component. The superscript (n) emphasizes that the wavelet processing is
applied to the current-state n of the simulation. The first component of the noise
represents the randomised large scale dynamics, and is set to zero to represent
the small scale features only and perform a spatial Reynolds-like decomposition.
The definition of the variance tensor can then be based on the definition of the
details. Such type of noise terms can easily be shown to be well defined. They are
spatially regular and their regularity is given by the choice of the wavelet basis. The
wavelet transform conveys a natural multi-resolution structure to the noise as well
as a natural notion of spatial scale at each level of the multi-resolution hierarchy.

4 Stochastic Shallow Water Model

A sketch of the stacked shallow water system is depicted in Fig. 1. In the framework of location uncertainty, the governing equations for the k-th layer ($k = 1, \ldots, N$) are formulated as follows:

Horizontal momentum:

$$D_t u_k + f u_k^\perp \, dt = (-\nabla P_k + F_k) \, dt, \tag{12}$$

Mass conservation :

$$D_t h_k + h_k \nabla \cdot u_k \, dt = 0, \tag{13}$$

where $u_k = (u_k, v_k)$ denotes the horizontal velocity with $u_k^\perp = (-v_k, u_k)$, h_k stands for the variable layer thickness, f is the Coriolis frequency, $P_k = \sum_{\ell=0}^{k-1} g_\ell' \eta_\ell$ is the Montgomery potential, $\eta_\ell = \eta_b + \sum_{j=\ell+1}^{N} h_j$ represents the vertical position of interface ℓ with η_b indicating the position of the bottom topography, $g_\ell' = g(\rho_{\ell+1} - \rho_\ell)/\rho_1$ is the reduced gravity with layer density ρ_ℓ and gravity value g, and $F_k = \partial_z \tau|_k \approx (\tau_{k-1} - \tau_k)/h_k$ is the vertical stress divergence. In particular, we consider only a steady surface wind stress τ_0 and a linear bottom drag stress τ_N. Moreover, when discussing the shallow water model the stochastic transport operator has to be understood as a two-dimensional operator. Derivation of this model can be found in [22] while a discussion of its analytical properties has been done in [29].

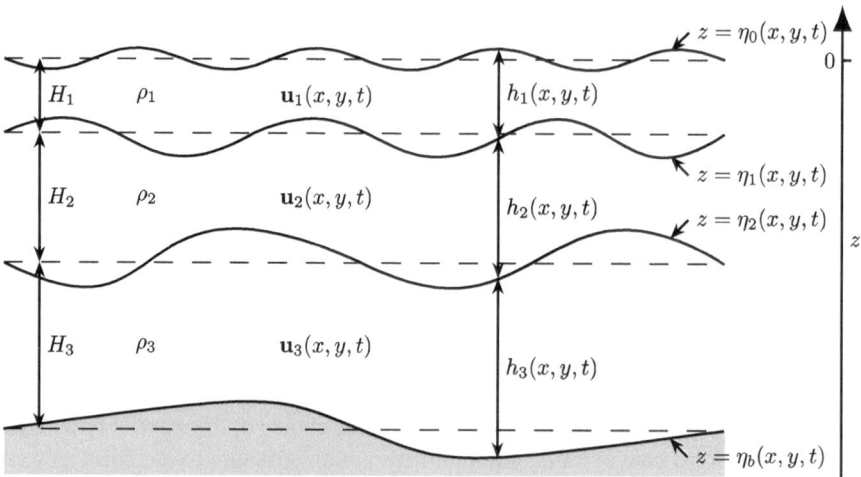

Fig. 1 Illustration of a three-layer ocean shallow water model. Each layer k has a uniform density ρ_k and background height H_k, a horizontal momentum u_k and a variable thickness $h_k = \eta_{k-1} - \eta_k$, where η_k represents the position of the interface between layers k and $k+1$

Table 1 Common parameters for all the models

Parameters	Value	Description
$X \times Y$	$(5120, 5120)$ km	Domain size
H_k	$(400, 1100, 2600)$ m	Mean layer thickness
g'_k	$(0.025, 0.0125)$ m s^{-2}	Reduced gravity
f_0	9.375×10^{-5} s^{-1}	Mean Coriolis
β	1.754×10^{-11} m^{-1} s^{-1}	Coriolis gradient
τ_0	0.08 Pa	Wind stress magnitude
δ_{ek}	2 m	Bottom Ekman layer thickness

For numerical studies, we consider a three-layer shallow water system with a steady symmetric zonal wind stress $\tau_0^x = (\tau_0/\rho_1)\cos(2\pi(y - Y/2)/Y)$, a flat bottom $\eta_b = -\sum_{k=1}^{N} H_k$, and a linear bottom drag $\boldsymbol{\tau}_N = (\delta_{ek} f_0/2)\boldsymbol{u}_N$. The common parameters for all the simulations are listed in Table 1. Time integration is performed with a third order Strong Stability Preserving Runge-Kutta (SSPRK3) method [30] for the deterministic part and a Milstein scheme (without Levy area) for the stochastic part [31, 32]. The time step is set to $0.6\Delta x/\sqrt{-g\eta_b}$ for a given grid spacing Δx. The unresolved external gravity waves are filtered using the method proposed by [33]. Advection of deterministic fluxes is performed with a fifth order Weighted Essentially Non-Oscillatory (WENO) scheme, while a second order centred scheme is applied to transport noise. The numerical implementation of this configuration follows tightly that of MQgeometry-1.0 [25].

Three simulation are performed following a spin-up run as described below: a deterministic high resolution simulation at 5 km, filtered and subsampled at 10 km resolution, is taken as a reference (and thus named REF); a deterministic coarse simulation at 10 km (named DET) is taken as a reference for the low resolution model. Finally, a stochastic simulation (named STO) is performed at 10 km. Considering Fig. 2 it is noticeable that the proposed localised basis enhances the presence of filaments and small eddies along the meandering eastward jet.

This result can be further highlighted by the temporal standard deviation of the surface relative vorticity ($\omega_1 = \nabla \times \boldsymbol{u}_1$), as shown in the top row of Fig. 3. We observe that the STO model produces greater low-frequency variability in the most energetic zonal jet region than the DET model at the same resolution. However, the latter allows the jet to extend further east than the former. To maintain the jet further east for the STO model, a time-correlated unresolved flow component can be added onto the uncorrelated noise through Girsanov transformation, as successfully demonstrated in our previous works [9–11, 27]. This could be performed in future work. Additionally, as illustrated in the bottom row of Fig. 3, we also observe a homogenization effect of the ocean middle layer potential (PV) in the central area for both models, which corresponds well to oceanographic theory [37, 38]. Note that the PV in the middle layer is defined as $q_2 = H_2(\omega_2 + f)/h_2$, and the magnitude of the gradient of its temporal mean ($|\nabla \overline{q_2}|$) is evaluated to measure the homogeneity.

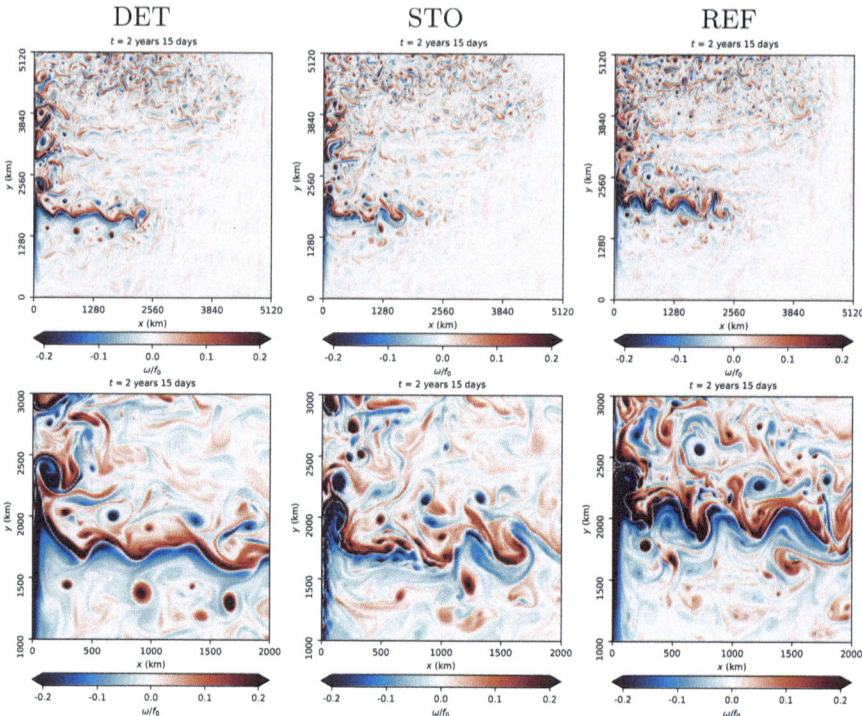

Fig. 2 Comparison of instantaneous surface vorticity (top) with the zoomed version in the jet region (bottom) provided by different models at 10 km

We then investigate the ensemble statistical properties of the proposed stochastic model by performing 20 random realizations. Figure 4 shows the establishment of an enstrophy transfer mechanism from the large scale mean flow towards the small scale turbulent eddies. This can be seen from the top row where the progressive decrease of the ensemble average of surface vorticity is associated to an increase of its ensemble variance.

We next focus on the ensemble decomposition of kinetic energy (KE) and available potential energy (APE) for the random shallow water system. Recall that the KE and APE densities for the k-th layer ($k = 1, \ldots, N$) and ℓ-th interface ($\ell = 0, \ldots, N - 1$) are defined as follows:

$$\mathrm{KE}_k = \frac{1}{2} h_k |\boldsymbol{u}_k|^2, \quad \mathrm{APE}_\ell = \frac{1}{2} g'_\ell \zeta_\ell^2, \quad \zeta_\ell = \sum_{j=N}^{\ell+1} (h_j - H_j), \tag{14}$$

where ζ_ℓ represents the deviation of the interface. We decompose the random thickness into $h_k = \overline{h_k} + h'_k$, where $\overline{h} = \mathbb{E}[h]$ denotes the ensemble mean thickness. Consequently, $\zeta_\ell = \overline{\zeta_\ell} + \zeta'_\ell$, allowing us to define the mean potential energy (MPE)

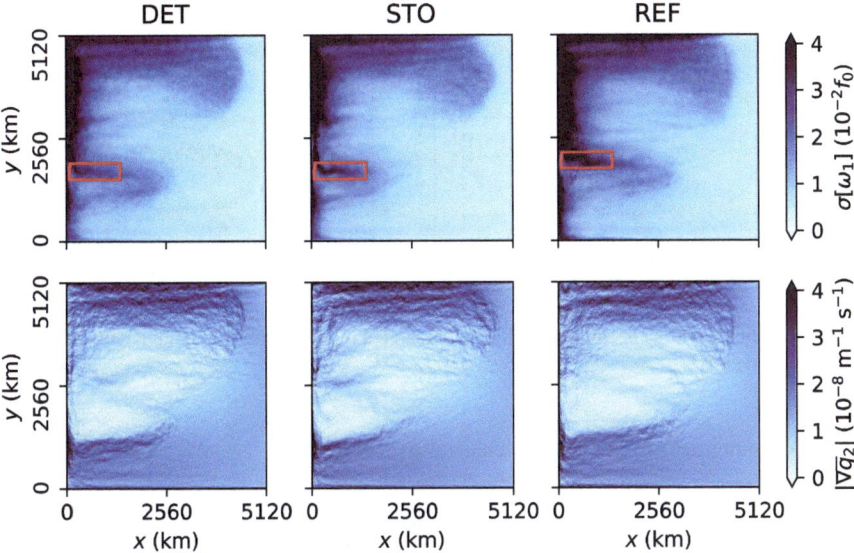

Fig. 3 Comparison of the (top row) temporal standard deviation of surface layer relative vorticity and (bottom row) homogenization of time-averaged potential vorticity in the middle layer, using 10-year data provided by different models (grouped by columns). The area-integrated values of $\sigma(\omega_1)/f_0$ in the most energetic zonal jet regions (highlighted by red boxes) for the DET, STO and REF models are 0.024, 0.025 and 0.032, respectively

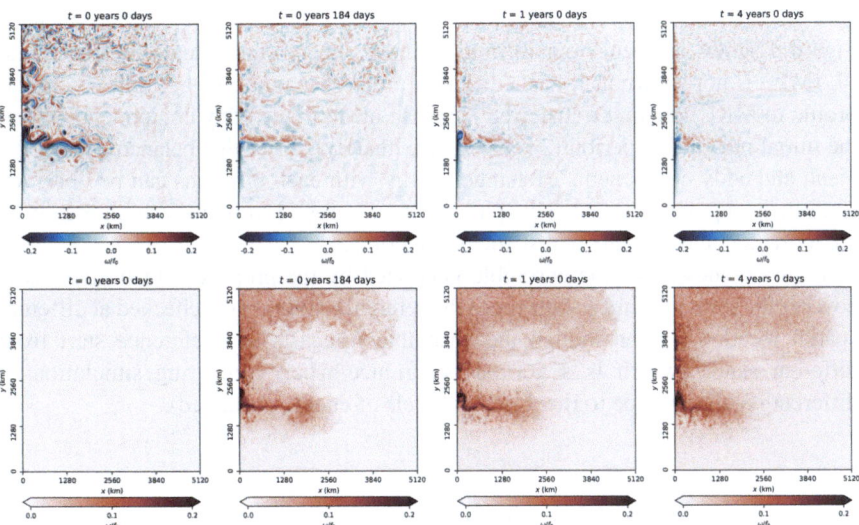

Fig. 4 Time evolution (from left to right) of ensemble mean (top) and standard deviation (bottom) of surface vorticity provided by the stochastic model at 10 km with 20 realizations

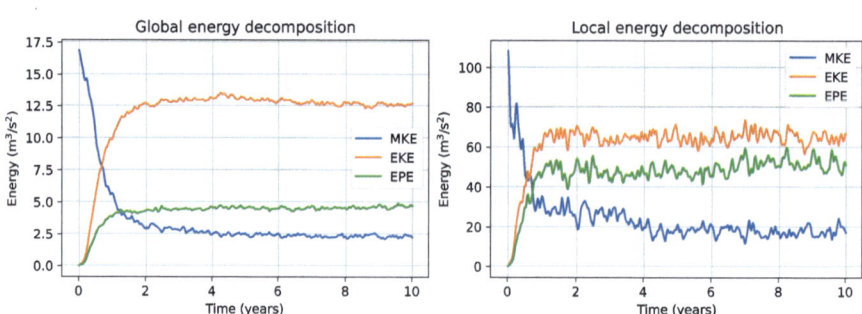

Fig. 5 Ensemble energy decomposition. Left: integrated over the whole domain; Right: integrated over the jet region

and eddy potential energy (EPE) densities as follows:

$$\text{MPE}_\ell = \frac{1}{2} g'_\ell \overline{\zeta_\ell}^2, \quad \text{EPE}_\ell = \frac{1}{2} g'_\ell (\zeta'_\ell)^2. \tag{15}$$

Decomposing next the momentum by $\boldsymbol{u}_k = \widehat{\boldsymbol{u}}_k + \boldsymbol{u}''_k$ with $\widehat{\boldsymbol{u}}_k = \overline{h_k \boldsymbol{u}_k}/\overline{h_k}$ the thickness-weighted momentum, we define the mean kinetic energy (MKE) and eddy kinetic energy (EKE) densities by

$$\text{MKE}_k = \frac{1}{2} \overline{h_k} |\widehat{\boldsymbol{u}}_k|^2, \quad \text{EKE}_k = \frac{1}{2} \overline{h_k |\boldsymbol{u}''_k|^2}. \tag{16}$$

Figure 5 shows the behaviour in time of these energy components. Both KE and PE (MPE is not shown as it has a different order of magnitude, but follows similar profile to MKE) are first transferred from the mean to the eddy components within the initial integration period (2 years approximately). After this balancing time, the mean and eddy components exchange energy with each other (as can be observed from their opposite phases). This phenomenon is found to be valid both locally in the jet region and globally across the entire domain.

Figure 6 shows that the ensemble generated by the proposed stochastic model covers efficiently (within a short time) the reference solution (as checked at different spatial locations), even though the ensemble forecasts and reference start from different states (which is a normal occurrence when comparing simulation of different resolution, due to the different levels of energy sustained).

5 Stochastic Primitive Model

Within the stochastic framework of location uncertainty the Boussinesq equations can be written as

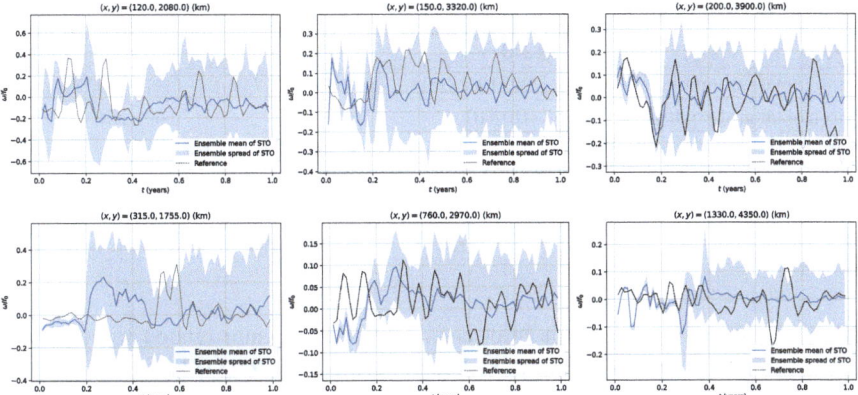

Fig. 6 Spread of moderate-term forecast for ensemble runs (20 members, 10 km) compared to reference run (5 km)

Horizontal momentum:

$$D_t \mathbf{u} + f \mathbf{e}_3 \times \left(\mathbf{u}\, dt + \sigma_t d\mathbf{B}_t^{\mathrm{H}} \right) = \nabla_{\mathrm{H}} \left(-p' + \frac{\nu}{3} \nabla \cdot \mathbf{v} \right) dt - \nabla_{\mathrm{H}} dp_t^{\sigma}, \quad (17)$$

Vertical momentum:

$$D_t w = \frac{\partial}{\partial z} \left(-p' + \frac{\nu}{3} \nabla \cdot \mathbf{v} \right) dt - \frac{\partial}{\partial z} dp_t^{\sigma} + b\, dt, \quad (18)$$

Temperature and salinity:

$$D_t T = \kappa_T \Delta T\, dt, \quad (19)$$

$$D_t S = \kappa_S \Delta S\, dt, \quad (20)$$

Incompressibility:

$$\nabla \cdot \left[\mathbf{v} - \mathbf{v}^s \right] = 0, \qquad \nabla \cdot \sigma_t d\mathbf{B}_t = 0, \quad (21)$$

Equation of state:

$$b = b\left(T, S, z \right), \quad (22)$$

with the convention $\mathbf{v} = (\mathbf{u}, w)$ and with the buoyancy defined as $b = -g\frac{\rho - \rho_0}{\rho_0}$. As opposed to the shallow water model, within the discussion of the primitive equations the stochastic transport operator has to be intended as three-dimensional. These equations where derived in [10] using asymptotic analysis starting from the stochastic Navier-Stokes of [3]. A more recent derivation starting from compressible Navier-Stokes is provided in [16]. Temperature T and Salinity S are considered active tracers transported by the stochastic flow, impacting the momentum equation through the (deterministic) equation of state. Consistency between the left hand side and the forcing is provided by the term dp_t^{σ} in Eqs. (17) and (18), a martingale

Table 2 Parameters of the model experiments

	R27d	R9d	R9LU
Horizontal resolution	1/27° (3.9 km)	~ 1/9° (11.8 km)	~ 1/9° (11.8 km)
Horizontal grid points	540×810	160×256	160×256
Vertical levels	30	30	30
Time step	5 min	15 min	15 min
Eddy viscosity	$5\times10^{-9}\,\mathrm{m^4\,s^{-1}}$	$5\times10^{-9}\,\mathrm{m^4\,s^{-1}}$	$5\times10^{-9}\,\mathrm{m^4\,s^{-1}}$
Eddy diffusivity	$5\times10^{-10}\,\mathrm{m^4\,s^{-1}}$	$5\times10^{-9}\,\mathrm{m^4\,s^{-1}}$	$5\times10^{-9}\,\mathrm{m^4\,s^{-1}}$

correction corresponding to a zero-mean turbulent pressure related to the noise, termed *stochastic pressure*. Primitive equations are then obtained from Boussinesq equations introducing the hydrostatic hypothesis on the vertical acceleration, that provides

$$\left[\boldsymbol{\sigma}_t \mathrm{d}\mathbf{B}_t - \mathbf{u}^s \, \mathrm{d}t\right] \cdot \nabla w - \frac{1}{2}\nabla \cdot (\mathbf{a}\nabla w) \, \mathrm{d}t = -\frac{\partial p}{\partial z}\, \mathrm{d}t - \frac{\partial \mathrm{d}p_t^\sigma}{\partial z} + b \tag{23}$$

so that the pressure and stochastic pressure can be defined in relation to the vertical component of the diagnosed large scale velocity as

$$p'(\mathbf{x}) = \int_{\eta_b}^{z} b + \mathbf{u}^s \cdot \nabla w + \frac{1}{2}\nabla \cdot (\mathbf{a}\nabla w) \, \mathrm{d}\zeta, \tag{24}$$

$$\mathrm{d}p_t^\sigma(\mathbf{x}) = \int_{\eta_b}^{z} \boldsymbol{\sigma}_t \mathrm{d}\mathbf{B}_t \cdot \nabla w \, \mathrm{d}\zeta. \tag{25}$$

The implementation of the stochastic Primitive Equations has been done in the level-coordinate free-surface primitive equations model NEMO [34] in a wind-forced double-gyre configuration. This setting, that has already been used in previous works on stochastic parameterization [10, 11], consists of a 45° degrees rotated beta plane centred at ~ 30°N, 3180 km long, 2120 km wide and 4 km deep, bounded by vertical walls and with a flat bottom and is fully described in [35, 36]. Table 2 summarizes the physical parameters used in the simulation, in agreement to the parameters of original chapters. It has to be noticed that the resolution of the R9 simulation is slightly different from that of the original chapter [35] as the wavelet noise requires the domain to be a multiple of a power of 2 when an MPI z-pencil domain decomposition strategy is employed. To assess the benefits of the stochastic parametrization two purely deterministic simulations were run at two different resolutions: 1/27° (R27d), a high resolution reference, and 1/9° (R9d), the deterministic reference. These two deterministic simulations are compared to a stochastic 1/9° simulation (R9LU). The R27d simulation has been spun-up for 100 years before collecting data for the LU framework. An initial condition for R9 has been generated starting from this simulation by filtering, downsampling and

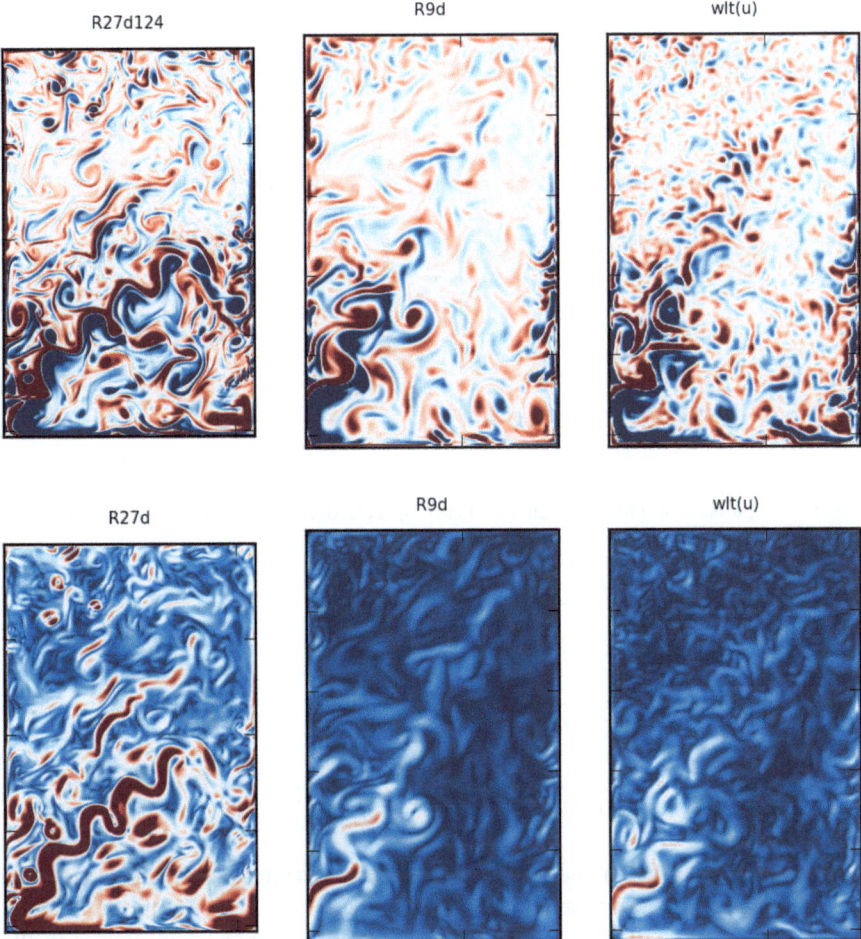

Fig. 7 Snapshot of vorticity (top) and strain rate (bottom)

running a 10 years adjustment period. Each simulation consists of 5 years of data, collected every 5 days and averaged over the 5 days.

The effect of the stochastic parametrization is assessed on the gradient of horizontal velocity. Considering Fig. 7 it can be noticed that there is an increase of the small scale structures along the jet structure and in the southern gyre, where the turbulence generated by the boundary is more intense. This effect, which is a consequence of the stochastic parametrization and the associated enhancement of the small scales variability, can be assessed with the symmetric part of the velocity gradient, the strain tensor. Recall the classical decomposition of the velocity gradient as $\nabla \mathbf{u} = \frac{1}{2}(\nabla \mathbf{u} + (\nabla \mathbf{u})^{\mathrm{T}}) + \frac{1}{2}(\nabla \mathbf{u} - (\nabla \mathbf{u})^{\mathrm{T}})$. From the symmetric part, the normalised strain rate can be defined for the mean flow and the fluctuations

Table 3 Experimental values for total strain rate and turbulent strain rate

	R27d	R9d	R9LU	I
$\overline{\|S\|}^{t}$	6.7295×10^{-5}	2.0773×10^{-5}	2.5314×10^{-5}	9.7%
$\overline{\|s\|}^{t}$	6.7295×10^{-5}	1.7570×10^{-5}	2.2328×10^{-5}	9.5%

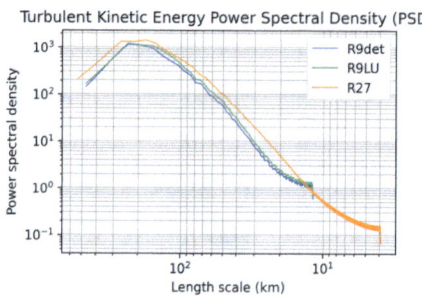 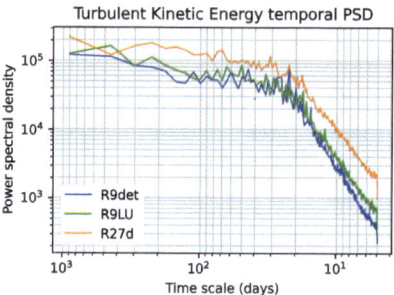

Fig. 8 Comparison of turbulent kinetic energy spectra over (left) spatial scales and (right) temporal scales, provided by different models

respectively, as

$$\|S\| = \frac{1}{\|\mathbf{u}\|_{\infty}} \left(\left|\frac{\partial u}{\partial x}\right| + \left|\frac{\partial v}{\partial x}\right| + \left|\frac{\partial u}{\partial y}\right| + \left|\frac{\partial v}{\partial y}\right| \right), \tag{26}$$

$$\|s\| = \frac{1}{\|\mathbf{u}'\|_{\infty}} \left(\left|\frac{\partial u'}{\partial x}\right| + \left|\frac{\partial v'}{\partial x}\right| + \left|\frac{\partial u'}{\partial y}\right| + \left|\frac{\partial v'}{\partial y}\right| \right). \tag{27}$$

The integrated total strain rate $\overline{\|S\|}^{t}$ and turbulent strain rate $\overline{\|s\|}^{t}$ provide a metric to assess the effects of the parametrization along the total duration of the simulation. Table 3 summarizes the numerical estimation of this improvement, that is of the order of 10% when bounded by the two deterministic formulations:

$$I(f) = \frac{f_{\text{R9LU}} - f_{\text{R9d}}}{f_{\text{R27d}} - f_{\text{R9d}}}. \tag{28}$$

The left panel of Fig. 8 shows that this increase in variability is well captured by the model across a wide range of spatial scales, leading to an increase of the turbulent energy content of the flow. Additionally, the stochastic model with the proposed noise parameterization enhances the intrinsic variability of the flow at different temporal scales, as demonstrated in the right panel of Fig. 8. The deterministic large-scale simulation exhibits prominent peaks at certain frequencies (around 20 and 25 days), indicative of an over-representation of certain eddies. Additionally, the inertial slope appears to be steeper, suggesting a poor representation of the eddies' distribution within the inertial range. These aspects are clearly rectified in the stochastic simulation: the slope is significantly weaker, almost reaching the slope of the reference. Furthermore, the anomalous peaks are attenuated, indicating a more balanced distribution of eddies within the inertial range. Eddies of larger frequencies are also better represented, suggesting a more pronounced inverse cascade since no energy is injected at this scale by the noise.

6 Conclusions

A novel, wavelet based, stochastic parametrization has been implemented in two different models to test its strengths and weaknesses. The general outcome of this study is that the addition of this model-based noise term, (that depends on the current state of the simulation and not on external data), can be beneficial in facilitating the energy transfer from large scale to small scale. Both the hydrostatic primitive equations model and the shallow water model appear to support turbulent dynamics at scales smaller than those sustained by the deterministic model. This enhancement of variability is shown to be successfully exploited in ensemble-run simulations to create a larger envelope for the spread of the shallow water model. Similar ensemble experiments with the primitive equations model will be considered in the future. For future research, we plan to investigate and incorporate a time-correlated unresolved barotropic flow component by applying Girsanov transformation to the uncorrelated noise. This addition, coupled with available observational data, aims to further enhance the accuracy and reliability of the current random model.

Acknowledgments The authors acknowledge the support of the ERC EU project 856408-STUOD.

References

1. Minobe, S., Kuwano-Yoshida, A., Komori, N. et al.: Influence of the Gulf Stream on the troposphere. Nature 452, 206–209 (2008). https://doi.org/10.1038/nature06690
2. Hewitt, H.T., Roberts, M., Mathiot, P. et al.: Resolving and Parameterising the Ocean Mesoscale in Earth System Models. Curr Clim Change Rep 6, 137–152 (2020).
3. Mémin, E: Fluid flow dynamics under location uncertainty. Geophysical and Astrophysical Fluid Dynamics 108, 119–197 (2014).
4. Holm, D. D.: Variational principles for stochastic fluid dynamics. Proceedings of the Royal Society A: Mathematical, Physical and Engineering Sciences, 471(20140963), 2015.
5. Debussche, A., Hug, B., Mémin, E.: A consistent stochastic large-scale representation of the Navier-Stokes equations, accepted for publication in Journal of Mathematical Fluid Mechanic, 2023.
6. Brzeźniak, Z., Capiński, M., Flandoli, F.: Stochastic partial differential equations and turbulence. Mathematical Models and Methods in Applied Sciences 1, 41–59 (1991).
7. Brzeźniak, Z., Capiński, M., Flandoli, F.: Stochastic partial differential equations and turbulence. Stochastic Analysis and Applications 10, 523–532 (1992).
8. Mikulevicius, R., Rozovskii, B.L.: Stochastic Navier-Stokes equations for turbulent flows. SIAM J. Math. Anal. 4, 1250–1310 (2004).

9. Li, L., Deremble, B., Lahaye, N., Mémin, E.:. Stochastic data-driven parameterization of unresolved eddy effects in baroclinic quasi-geostrophic model. Journal of Advances in Modeling Earth Systems, 2022

10. Tucciarone F.L., Mémin, E., Li, L.: Primitive Equations Under Location Uncertainty: Analytical Description and Model Development Stochastic Transport in Upper Ocean Dynamics, Springer, 2022, pp.287–300.

11. Tucciarone F.L., Mémin, E., Li, L.: Data driven stochastic primitive equations with dynamic modes decomposition. In Stochastic Transport in Upper Ocean Dynamics, Springer, pp: 321–336, 2023.

12. Chandramouli, P., Mémin, E., Heitz, D.: 4D large scale variational data assimilation of a turbulent flow with a dynamics error model. Journal of Computational Physics, Volume 412,2020

13. Kadri Harouna, S., Mémin, E.: Stochastic representation of the Reynolds transport theorem: Revisiting large-scale modelling. Computers and Fluids 156, 456–469 (2017).

14. Pinier, B., Mémin, E., Laizet, S., Lewandowski R.: Stochastic flow approach to model the mean velocity profile of wall-bounded flows. Phys. Rev. E, 99(6):063101, 2019.

15. Quentin, J., Mémin, E., Dumas, F., Li, L., Garreau, P., Toward a Stochastic Parameterization for Oceanic Deep Convection. In Stochastic Transport in Upper Ocean Dynamics II. 2023.

16. Tissot, G., Mémin, E., Jamet, Q.: Stochastic Compressible Navier–Stokes Equations Under Location Uncertainty. In Stochastic Transport in Upper Ocean Dynamics, Springer, pp: 293–319, 2023.

17. Bauer, W., Chandramouli, P., Chapron, B., Li, L., Mémin, E.: Deciphering the role of small-scale inhomogeneity on geophysical flow structuration: a stochastic approach. Journal of Physical Oceanography, (2020).

18. Chapron, B., Dérian, P., Mémin, E., Resseguier, V.: Large-scale flows under location uncertainty: a consistent stochastic framework. QJRMS, 144(710):251–260, 2018.

19. Cintolesi, C., Mémin, E.: Stochastic Modelling of Turbulent Flows for Numerical Simulations. Fluids 5, (2020).

20. Bauer, W., Chandramouli, P., Li, L., Mémin, E.: Stochastic representation of mesoscale eddy effects in coarse-resolution barotropic models Ocean Modelling, 2020, 151, pp.1–50.

21. Resseguier, V., Li, L., Jouan, G., Dérian, P., Mémin, E., Chapron, B.: New trends in ensemble forecast strategy: uncertainty quantification for coarse-grid computational fluid dynamics Archives of Computational Methods in Engineering, 2021, 28 (1), pp.215–261. <10.1007/s11831-020-09437-x>

22. Brecht, R. , Li, L., Bauer, W., Mémin, E.: Rotating shallow water flow under location uncertainty with a structure-preserving discretization. Journal of Advances in Modeling Earth Systems, American Geophysical Union, 2021, 13 (12)

23. Daubechies I.: Ten Lectures on Wavelets. Society for Industrial and Applied Mathematics (SIAM), 1992.

24. Mallat S.: A Wavelet Tour of Signal Processing (Third Edition). Academic Press. 2009.

25. Thiry, L., Li, L., Roullet, G., and Mémin, E.: MQGeometry-1.0: a multi-layer quasi-geostrophic solver on non-rectangular geometries, Revised manuscript accepted for GMD, 2024, preprint available on EGUsphere, https://doi.org/10.5194/egusphere-2023-1715, 2024.

26. G. D. Prato and J. Zabczyk. *Stochastic equations in infinite dimensions*. Cambridge University Press, 1992.

27. Li, L., Deremble, B., Lahaye, N., Mémin, E. (2023). Stochastic data-driven parameterization of unresolved eddy effects in a baroclinic quasi-geostrophic model. Journal of Advances in Modeling Earth Systems, 15, e2022MS003297.

28. Dufée, B., Mémin, and Crisan, D., Observation-Based Noise Calibration: An Efficient Dynamics for the Ensemble Kalman Filter. Stochastic Transport in Upper Ocean Dynamics 10, Springer, 43–56, 2023.

29. Lang, O., Crisan, D., Mémin, E.: Analytical Properties for a Stochastic Rotating Shallow Water Model Under Location Uncertainty. Journal of Mathematical Fluid Mechanics, (2023).

30. Gottlieb, S., Shu, C.-W., Tadmor, E., Strong stability-preserving high-order time discretization methods. SIAM Rev. 43, 89–112 (2001)
31. Fiorini, C., Boulvard, P. M., Li, L., Mémin, E. A two-step numerical scheme in time for surface quasi geostrophic equations under location uncertainty. In Stochastic Transport in Upper Ocean Dynamics, Springer, 2022.
32. Boulvard, P. M., Mémin, E.: Diagnostic of the Lévy area for geophysical flow models in view of defining high order stochastic discrete-time schemes. Foundations of Data Science. 2023.
33. Roullet, G. and Madec, G.: Salt conservation, free surface, and varying levels: A new formulation for ocean general circulation models. Journal of Geophysical Research, 2000.
34. Madec, G., Bourdallé-Badie, R., Chanut, J., Clementi, E., Coward, A.,Ethé, C., Iovino, D., Lea, D., Lévy, C., Lovato, T., Martin, N., Masson, S., Mocavero, S., Rousset, C., Storkey, D., Vancoppenolle, M., Müller, S., Nurser, G., Bell, M., Samson, G.: Nemo ocean engine, Oct. 2019.
35. Lévy, M. , Klein, P., Tréguier, A.-M., Iovino, D. , Madec, G., Masson, S., Takahashi, K.: Modifications of gyre circulation by sub-mesoscale physics. Ocean Modelling, 34(1–2):1–15, 2010.
36. Lévy, M. Resplandy, L., Klein, P., Capet, X., Iovino, D., Ethé, C.: Grid degradation of submesoscale resolving ocean models: Benefits for offline passive tracer transport. Ocean Modelling, 48:1–9, 2012.
37. Rhines, P., Young, W.: Homogenization of potential vorticity in planetary gyres. Journal of Fluid Mechanics, 122:347–367, 1982.
38. Holland, W., Keffer, T., Rhines, P.: Dynamics of the oceanic general circulation: the potential vorticity field. Nature, 308:698–705, 1984.

Stochastic Fluids with Transport Noise: Approximating Diffusion from Data Using SVD and Ensemble Forecast Back-Propagation

James Woodfield

1 Introduction

1.1 History and Motivation

Motivated by the need to model the effect of viscosity not present in the inviscid vortex method, Chorin [10] proposed a constant (Itô) noise in the particle trajectory map. Chorin's stochastic parameterisation represented the diffusion effect present in the corresponding Fokker-Plank equation (the deterministic Navier-Stokes equation). Numerical methodology based on the idea of a stochastic particle trajectory map was later proven convergent in [33], and are sometimes called computational vortex methods [34].

Computational vortex methods are numerical methods based on tracking the particle trajectories of a finite number of discrete points of (potential) vorticity, making use of the Biot-Savart Law to both close the system and define the velocity elsewhere. Typically, one requires the approximation of an integral, the regularisation of a kernel, and the closure as a collocation method [34]. Possible advantages to computational vortex methods include not needing to allocate computational resources to regions with little or no vorticity, the absence of a pressure solve, little to no numerical viscosity, a less stringent timestep requirement, and access to the velocity globally.

More recently, in 2015, Holm proposed a different type of stochastic parameterisation of the particle trajectory map [25], rather than modelling diffusion, the introduced stochastic parameterisation aims at representing uncertainty associated with additional transport. In this setting a family of spatially dependent vector fields

J. Woodfield (✉)
Department of Mathematics, Imperial College, London, UK
e-mail: james14641@gmail.com

© The Author(s) 2025
B. Chapron et al. (eds.), *Stochastic Transport in Upper Ocean Dynamics III*,
Mathematics of Planet Earth 13, https://doi.org/10.1007/978-3-031-70660-8_14

$\{\boldsymbol{\xi}_p(\boldsymbol{x})\}_{p=1}^P$ act (stochastically) on the particle trajectory map. The basis of vector fields are integrated (in the Stratonovich sense) against a P dimensional Brownian motion, as to remain consistent with the variational principle, preserve Kelvin's theorem, and preserve infinite integral quantities known as Casimir's, such ideas are presented in [4, 25, 26]. In practice one still requires methods for estimating the vector-fields $\{\boldsymbol{\xi}_p(\boldsymbol{x})\}_{p=1}^P$, as to take into account the uncertainty associated with unresolved or unrepresented transport, doing so is the problem we tackle in this chapter. This task will colloquially be described as calibration or as an offline batch data assimilation technique, the aim is to present a calibrated stochastic forward model capable of producing an ensemble which represents statistics of the data, or more generally the statistics of the hidden distribution from which the data was sampled.

In Cotter et al. [11] and Crisan et al. [13], vector fields are calibrated from weather station positional data using an SVD/PCA/EOF decomposition of a Data-Anomaly-Matrix (DAM) formed in the context of stochastic coarsegraining of a high-resolution deterministic model, see also [37, 38] for application and variants of this methodology to other stochastic fluid models. In these works [11, 13, 37, 38] stochastic parameterisation of the coarse-graining operator have been proposed in the context of stochastic model reduction. We instead consider parameterisation between a reference dataset and the proposed stochastic forward model, including when the data arises as a realisation of another stochastic model, not necessarily the proposed stochastic forward model. In this work we use a similar truncated SVD approach to calibrating a basis from weather-station data, however amongst other minor details we differ in the construction of the data anomaly matrix, the resulting calibrated proposed stochastic equation, and the numerical methods used. We test the new calibration technique using a twin experiment framework where the reference data is a known parameterised SDE/SPDE realisation (whose parameters are known). We also test in an extended twin experiment in which the distribution from which the data is sampled is known, and more SPDE/SDE realisations are generated for hidden testing datasets.

We also introduce preliminary results regarding, a loss-based approach to the calibration problem. The motivation for another calibration method stems from the need for the calibration/assimilation of other types of data such as drifter data or simply state-valued data in the forward model, and is motivated by trying to alleviate the expensive interpolation from weather-station cost in the forward ensemble model.

1.2 Outline

1. Section 1 contained the motivation and history.
2. Section 2 contains a review of the stochastic fluid modelling assumptions (Sect. 2.1) and the 2D computational vortex methods of interest (Sect. 2.2).

3. Section 3 introduces several approaches to calibration.

   (a) Section 3.1 contains the two proposed calibration methodologies. The first calibration approach focuses on the SVD decomposition of a space-time data anomaly matrix. In the second calibration approach, we treat an entire forecast ensemble method (over a space-time window) as a function, and a Continuous Ranked Probability Score estimator is used as a loss. The loss is minimised with respect to the parameterised basis of vector fields.
   (b) Section 3.3 contains a list of the different ensemble forwards models that we will test, including benchmark ensembles.

4. Section 4 contains the details and results of several numerical experiments.

   (a) We describe a twin experiment in which the reference data comes from a parameterised SDE approximation of an SPDE, with a fixed known single basis and a single known Brownian motion.
   (b) We describe a twin experiment in which the data comes from a parameterised SDE approximation of an SPDE, with 5 fixed known single basis functions and 5 i.i.d. Brownian motion realisations.
   (c) We describe a twin experiment in which the synthetic data is generated from a realisation of a parameterised SDE approximation of an SPDE with an additional Ito-Stratonovich drift.
   (d) We describe a twin experiment in which the synthetic data is generated as a realisation of an SDE system with a larger physical drift term representing the effect of unrepresented dynamics.
   (e) We finally present results comparing Continuous Rank Probability Score (CRPS) and relative skill scores (CRPSS) for various calibration techniques proposed in this chapter against some proposed benchmarks.

5. Section 5 concludes and summarises key results and preliminarily discusses the application of this calibration methodology to less idealised data.

2 Governing Equations and Numerical Method

2.1 Governing Equations

Various stochastic parameterisations of the particle-trajectory mapping in fluid mechanics have been proposed [10, 12, 25]. In this work, we are interested in the two-dimensional case where the initial label $X = (X, Y)^T \in \mathbb{R}^2$ is evolved to current configuration $x = (x, y)^T$ by the parameterised Stratonovich stochastic ordinary differential equation

$$x(X, t) = x(X, 0) + \int_0^t u(x(X, s), s)ds$$

$$+ \sum_{p=1}^{P} \int_{0}^{t} \theta_p \boldsymbol{\xi}_p(\boldsymbol{x}(\boldsymbol{X}, s)) \circ dW^P(s); \quad \boldsymbol{x}(\boldsymbol{X}, 0) = \boldsymbol{X}. \tag{1}$$

Where $\boldsymbol{u}(\boldsymbol{x}, t)$ is the drift velocity, and $\{\theta_p \boldsymbol{\xi}_p(\boldsymbol{x})\}_{p=1}^{P}$, is a set of velocity fields associated with the stochastic component of the flow. $\circ dW^p$ denotes Stratonovich integration against the p-th component of a P-dimensional Brownian motion, (see [30]). Each vector field basis is multiplied by a parameter value $\theta_p \in \mathbb{R}$. We assume that the drift stream function ψ is related to vorticity ω by a yet specified differential relationship, solvable by convolving the Green's functions against the vorticity as follows

$$\psi(\boldsymbol{x}, t) = \int_{\mathbb{R}^2} G(\boldsymbol{x} - \boldsymbol{x}')\omega(\boldsymbol{x}')d\boldsymbol{x}'. \tag{2}$$

and that the negative skew gradient $\boldsymbol{u} = -\nabla^{\perp}\psi$ relates the velocity $\boldsymbol{u}$ to the vorticity ω by another convolution against the kernel K

$$\boldsymbol{u}(\boldsymbol{x}, t) = \int_{\mathbb{R}^2} K(\boldsymbol{x} - \boldsymbol{x}')\omega(\boldsymbol{x}')d\boldsymbol{x}', \tag{3}$$

this relationship is known as the Biot-Savart law, and when substituted into Eq. (1) describes the particle trajectory map integrodifferentially. Many fluid models have such a formulation, some important examples are discussed in the appendix Examples B.3–B.1, and the ones used explicitly in this chapter are discussed below (Examples 2.1 and 2.2). To completely define the above infinite dimensional SDE system, (equivalent to the solution of an SPDE fluid model) one typically defines an initial vorticity field ω_0, and notes that this quantity is invariant along solution trajectories. See [15] for additional information about deriving such a model from a variational principle.

Example 2.1 (2D Euler) Euler on $\mathbb{R}^2$ has the following differential relationship between the drift stream function and vorticity, $\psi = (-\Delta)^{-1}\omega$, with Green's function given by $G(\boldsymbol{x}) = -(2\pi)^{-1}\log(\|\boldsymbol{x}\|_2)$, and kernel by $K(\boldsymbol{x}) = (2\pi)^{-1}\boldsymbol{x}^{\perp}\|\boldsymbol{x}\|^{-2}$. The corresponding SPDE is a stochastic version of Euler's equation given below in vorticity form

$$d\omega_t + (\boldsymbol{u} \cdot \nabla)\omega_t dt + \sum_{p=1}^{P}(\theta_p \boldsymbol{\xi}_p \cdot \nabla)\omega_t \circ dW^p, \quad \boldsymbol{u} = -\nabla^{\perp}\psi, \quad \psi = (-\Delta)^{-1}\omega,$$

$$\tag{4}$$

where $\omega = \text{curl}(\boldsymbol{u})$, is the vorticity.

Example 2.2 (Regularised Euler) It is often the case that the kernel K possesses a singularity (at $x = 0$), making numerical methods approximating the velocity field from a delta function initial condition ansatz and the Biot-Savart kernel Eq. (3) inaccurate.[1] The kernel K is instead typically regularised by component-wise convolution with a parameterised mollifier function ϕ_δ, $\delta \in \mathbb{R}^{>0}$, such that the resulting kernel

$$K_\delta = K * \phi_\delta = \int_{\mathbb{R}^2} K(x - y)\phi_\delta(y)dy, \tag{5}$$

is desingularised, and the regularised Biot-Savart law is given by

$$u^\delta(x, t) = \int_{\mathbb{R}^2} K_\delta(x - x')\omega(x')dx' = \int_{\mathbb{R}^2} K(x - x')\omega_\delta(x')dx'. \tag{6}$$

Where in the last line, the associative property of convolution has been used to give interpretation as a regularised Euler vorticity $\omega_\delta = \omega * \phi_\delta$.

In this work, we are interested in a manner of determining sensible proposals for $\{\theta_p \boldsymbol{\xi}_p\}_{p=1}^P$ from data, such that a forward model numerical method can produce a probabilistic forecast with skill. More specifically in this work, we are interested in parameterising the difference between reference data and the forward model, a Stochastic Advection by Lie Transport (SALT) inviscid vortex dynamics solver. We will test on idealised reference data arising from realisations of similar known stochastic forward models. This is an idealised setting in which we can test the calibration methodologies, without modelling error concerns. However, the application of the methodology can be speculated to be applicable in other modelling scenarios such as stochastic model reduction as in [11, 13, 37, 38]. We will also do testing in the setting in which there is a known modelling discrepancy. Namely, we suppose that there exists an additional constant drift velocity between the forward model and the data. One example of such a model discrepancy would be interpreting the stochastic integration in a different setting, i.e. Itô-Stratonovich, or Wong-Zakai anomaly type drift [15]. Another likely motivation is the assumption that in a time-averaged scenario, models simply differ by a drift from real-world data.

---

[1] In particular, Beale and Majda 1985 [7], show that the point vortex method for the Euler equation has poorer convergence properties as the number of points increases, as compared with methods employing a regularised kernel. Furthermore, they show point vortex methods have a larger error when evaluating velocities not on point vortex trajectories, as compared with their vortex blob counterparts.

2.2   Numerical Method

Point vortex methods model the initial vorticity by a field whose vorticity is concentrated at a finite sum of delta functions whose strength is denoted $\Gamma_i \in \mathbb{R}$ as follows

$$\omega(x) = \sum_i \Gamma_i \delta(x - x_i). \tag{7}$$

If the vorticity is assumed a finite sum of delta functions, using a regularised kernel K_δ, is equivalent to approximating the vorticity with "vortex-blobs" with finite width $\omega(x) = \sum_i \Gamma_i \phi_\delta(x - x_i)$, $\phi_\delta = \phi_\delta * \delta$ and using the unregularised Euler kernel K. The mollifier ϕ_δ used in vortex blob regularisation's are typically constructed with specific smoothness and moment boundedness properties (pg227)[34] (pg190)[7], required for convergence and stability estimates (sec 6.4 and sec 6.6 [34]). In this work, we consider inviscid vortex methods, which approximate the regularised stochastic integrodifferential equation for 2d Euler on $\mathbb{R}^2$, and essentially use a stochastic version of the deterministic discretisation strategy proposed in [7], outlined below.

Let the multi-index $i = (i_1, i_2)$, belong to a finite index set $\wedge_0^i$ spanning (labelling) the dynamically evolving points $x_i = (x_{(i_1, i_2)}, y_{(i_1, i_2)})$ with non zero initial vorticity, these points will initially be defined on a Cartesian mesh in $[x_{\min}, x_{\max}] \times [y_{\min}, y_{\max}]$ with uniform width and height h_x, h_y. One assumes that the deterministic part of the regularised vorticity field $\omega_\delta(x, t)$, velocity field $u_\delta(x, t)$ and stream function $\psi_\delta(x, t)$ can be reconstructed globally on $\mathbb{R}^2$, $\forall t \in [0, T]$ in the following way

$$\omega^\delta(x, t) = \sum_{i \in \wedge_0^i} \phi_\delta(x - x_i(t))\omega_0(X_i)h_x h_y, \tag{8}$$

$$u^\delta(x, t) = \sum_{i \in \wedge_0^i} K_\delta(x - x_i(t))\omega_0(X_i)h_x h_y, \tag{9}$$

$$\psi^\delta(x, t) = \sum_{i \in \wedge_0^i} G_\delta(x - x_i(t))\omega_0(X_i)h_x h_y, \tag{10}$$

from the finite set of evolving points $x_i(t) \in \wedge_0(t)$. Where $G_\delta = G * \phi_\delta$, denotes the convolution of the Green's function with the mollifier, $K_\delta = K * \phi_\delta$ denotes the convolution of the Euler kernel with the mollifier, and in Eq. (8) the mollifier convolves a delta function to define the vortex blob function $\phi_\delta = \phi_\delta * \delta$ at the positions x_i. Noting that (potential)vorticity is preserved along solution trajectories $\omega_0(X_i) = \omega(x_i(t))$, it is possible to interpret Eqs. (9) and (10), as discretisations of the convolutions in Eqs. (2) and (3), with either vortex blob initial conditions, or with regularised convolutions.

Numerically, in practice Eqs. (8)–(10) are not evaluated globally, but will be evaluated at fixed weather-station positions for data denoted $x_d \in \wedge_d$, and moving vortex positions $x_i(t) \in \wedge_0(t)$.

Upon appropriate vectorisation of the initial mesh $\wedge_0$, and identification of the "point" vortex strength $\Gamma_i = \omega_0(X_i)h_x h_y$, $\forall i \in \wedge_0^i$, one can use Eq. (9) to close the system as a finite-dimensional system,

$$x_i(t) = x_i(0) + \int_0^t u^\delta(x_i(s), s)ds + \sum_{p=1}^P \int_0^t \theta_p \xi_p(x_i(s)) \circ dW_s^p, \quad \forall i \in \wedge_0,$$

(11)

$$u^\delta(x_i(s), s) = \sum_{j \in \wedge_0^i, j \neq i} \Gamma_j K_\delta(x_i(s) - x_j(s)),$$

(12)

where each vortex does not self-induce a velocity.

Various mollifiers can be used in inviscid vortex methods. For the simulation of the Euler equation example (2.1), Rosenhead [39], Krasney [32] and Chorin [10], all introduced mathematical equivalents to mollification of the Euler kernel, preventing division by zero and cutting off the singularity. In 1979 Hald [22] proved second order convergence for 2D deterministic vortex methods when using a specific mollifier over an arbitrary time interval. However, the specific form of mollifier (compact locally three times differentiable) required the regularisation parameter δ to be larger than the mesh spacing h. Beale and Majda in [6] introduce smoother mollifiers allowing smaller regularisation parameter $\delta = O(h)$ and proved arbitrary order convergence. In [7] Beale and Majda introduce convenient additional explicit higher order kernels, we adopt one such family of mollifiers in this work, and the effect on the Biot-Savart kernel can be described in the following manner,

$$(u^\delta(x, y), v^\delta(x, y))^T = \sum_{i \in \wedge_0} \frac{\Gamma_i(-(y - y_i), (x - x_i))^T}{2\pi \|x - x_i\|_2^2}$$

$$\times (1 - L_p(\|x - x_i\|_2^2/\delta^2)\exp(-\|x - x_i\|_2^2/\delta^2)),$$

(13)

where L_p is the p-th order Laguerre polynomial, and can be found in [7]. This scheme (in the deterministic setting) has been shown to have the property that if $\delta = h^q$ for $q \in (0, 1)$, the order of convergence to the solution of the Euler Equation is given by $O(h^{(2p+2)q})$ see [7] (or sec 6[34]).

To deal with the stochastic Stratonovich term, we discretise in time with the stochastic generalisation of the SSP33 scheme of Shu and Osher (can be found in [43]), where the forward Euler scheme is replaced with Euler Maruyama scheme in the Shu Osher representation. This time-stepping is applied to Eq. (11), the scheme is weak order 1, strong order 0.5, as can be found by Taylor expanding (see [40] for the strict generalisation of this result) and ([31] for definitions of convergence).

3 Calibration Methodology

3.1 Procedures and Methodology: In the Estimation of Basis Functions and Parameters

This section details two methods used in the estimation of $\{\theta_p\}_{p=1}^P$, the recovery of basis functions $\{\sigma_p\tilde{\xi}\}_{p=1}^P$, the recovery of the time mean difference $\bar{v}^r$ and the recovery of paths $\{\Delta W_p^r\}_{p=1}^P$.

The first method takes inspiration from the coarse-graining parameterisation approaches taken in [11, 13] in the use of the SVD. However, the aim of the model here is to parameterise the difference from the reference data and the forward model. This is done by using the Biot-Savart kernel to "access" the drift component of the velocity directly in the creation of a data anomaly matrix. In practice the details of the algorithm are given below.

Method 3.1 (TSVD∘WSD) Truncated Singular Value Decomposition of weather-station data.

1. Data Collection; We assume that over the discrete time interval $\mathcal{T} := \{t_n\}_{n=0}^{n_t}$, we have recorded a velocity field $\boldsymbol{u}_d^m = \boldsymbol{u}^m(\boldsymbol{x}_d)$, measured at the fixed weather station positions $\boldsymbol{x}_d = (x_d, y_d)^T \in \wedge_d$, and have a record of the positions of the dynamically evolving point vortices $\boldsymbol{x}_i \in \{\wedge_0(t_n)\}_{n=0}^{n_t}$, with know vorticity Γ_i. This is the reference solution and data.

2. The drift components of velocity $\boldsymbol{u}^\delta(\boldsymbol{x}_d, t_n)$ are estimated at weather stations by the Biot-Savart kernel Eq. (9) using known observed positions $\boldsymbol{x}_i \in \wedge_0(t_n)$ of the point vortices at times t_n.

$$\boldsymbol{u}^\delta(\boldsymbol{x}_d, t_n) = \sum_{i \in \wedge_0} K_\delta(\boldsymbol{x}_d - \boldsymbol{x}_i(t_n))\omega_0(X_i)h_x h_y, \quad \forall t_n \in \mathcal{T}, \quad \forall \boldsymbol{x}_d \in \wedge_d.$$

$$(14)$$

3. The difference $(\boldsymbol{u}_d^m - \boldsymbol{u}^\delta(\boldsymbol{x}_d, t_n))\Delta t$ between the measured velocity, and the drift reconstructed velocity at weather stations is taken $\forall t \in \mathcal{T}_n$. Upon appropriate (invertible) vectorisation this is turned into a $n_t \times n_d$ matrix $M \in \mathbb{R}^{n_t \times n_d}$ where there are n_d weather stations, and n_t observation instances in time. Our specific vectorisation in space is the following operation,

$$M_{n,:} = [\text{vec}(\text{vec}(u_d^m(\boldsymbol{x}_d, t_n)), \text{vec}(v_d^m(\boldsymbol{x}_d, t_n)))$$

$$- \text{vec}(\text{vec}(u^\delta(\boldsymbol{x}_d, t_n)), \text{vec}(v^\delta(\boldsymbol{x}_d, t_n)))]\Delta t.$$

This M is the data anomaly matrix, representing the effect of the stochastic velocity and driving signal on the weather stations.

4. A common post-processing step in an SVD/PCA/EOF procedure is the removal of the row or column mean, such that the matrix has zero row sum or column

sum. Since we are working with a $n_t \times n_d$ matrix, we remove the column(time mean) $TM(M) := e_{n_t} M/n_t$, from the data matrix in the following manner $M' = M - e_{n_t} \otimes (e_{n_t} M/n_t)$. The time mean difference observed from data will be denoted by $\bar{v}^r$.

5. One performs the truncated SVD([23]) of the time mean removed data anomaly matrix

$$M' = U_t \Sigma_t V_t^T = (c^{-1} U_t^c)(c \Sigma_t (V_t^c)^T).$$

We have re-scaled the construction by a constant $c \in \mathbb{R}$, such that the k-th column of $U = c^{-1} U_t^c$ has variance aligning with the timestep between observations, and the removal of time mean has normalised the data. For incremental data $\text{var}(U) = \Delta t$, and $\mathbb{E}(U) = 0$.

Here the p-th row of $V_t^T \in \mathbb{R}^{n_t \times n_d}$, forms the p-th vectorised spatial eigenvectors $\tilde{\xi}_p(x_d)$ effect on weather station data $\wedge_d$. The (p, p)-th element of Σ_t denotes the corresponding singular value σ_p. The matrix product $U_t = M' V_t^T \Sigma_t^{-1}$, gives U whose p-th column is the pth recovered path ΔW_p^r over the time window $\mathcal{T}$. For information about the SVD see Remark B.3.

6. Output: $\{\sigma_p \tilde{\xi}_p\}_{p \in [P]}$, $\bar{v}^r$, $\{W_p^r\}_{p \in [P]}$, are recovered from ΣV^T, $e_{n_t} M/n_t$, U, respectively.

In practice, reconstruction is performed as to transform the discrete set of points $x_d \in \wedge^d$ and their reconstructed evaluation $\sigma_p \tilde{\xi}_p(x_d)$, $\bar{v}^r(x_d)$, at $\forall x_d \in \wedge^d$ into continuous fields $\sigma_p \tilde{\xi}_p(x, y)$, $\bar{v}^r(x, y)$. This interpolation step is required for the evolution of unstructured points in the calibrated inviscid vortex method. We use Fourier interpolation in the understanding, that specific to this work we expect periodic smooth basis functions and assume the data remains within the weather station grid, see Remark B.4 for specifics in such an interpolation procedure. We propose two ensemble forward methods based on Method 3.1, one including the time mean drift (Method 3.5), and one without (Method 3.4).

One does not always have access to Eulerian weather-station data. It may be desirable to estimate parameters only from "tracer" positional values such as evolving buoys or inherent state values in the evolving forward model. Furthermore, the values at weather stations need to be constructed into continuous fields, which require evaluation by interpolation in the forwards ensemble model during runtime, the cost of such a procedure scales badly (squared) with the number of weather-stations used see Remark B.4.

It may be advantageous to avoid this computational runtime problem (particularly for large number of stochastic basis functions) or lack of weather-station data by setting the problem as a minimisation problem with a predefined basis. This will turn interpolation into evaluation in the forward model, resulting in a much faster ensemble method. However, the calibration problem is phrased as a significantly more costly nonlinear optimisation problem, described below (in the specific context of an inviscid vortex method).

Method 3.2 (B(SPDEE)wrtFM) Backpropagation through SPDE Ensemble with respect to diffusion parameters as to minimise a forecasting metric.

1. Data Collection; Record the positions of the dynamically evolving point vortices as data $\{x_i^* \in \wedge_0(t_n)\}_{n=0}^{n_t}$. Here the astrix superscript denotes data. This can be stored as a $n_t \times 2n_v$ matrix (positions in 2d have two components).
2. Generate an entire ensemble run, over the time window of interest, recording all state variables aligning with observation instances in time, using E_o number of proposed Ensemble members over an observation window $\mathcal{T}_o$. This is defined as a vectorised ensemble function denoted F_E, defined by going forwards in time with a forward discrete model of the following type,

$$d_t x_i(t) = u^\delta(x_i(t), t)dt$$

$$+ \sum_{p=1}^{P} \theta_p \xi_p(x_i(t)) \circ dW_e^p(t), \quad \forall i \in \wedge_0, \forall t_n \in \mathcal{T}_o, \forall e \in [E_o].$$

(15)

This is done for all time steps in an observation window $\forall t_n \in \mathcal{T}_o$, and for E_o realisations of P-dimensional Brownian motion from the initial condition. The output of the vectorised ensemble function F_E is a $2n_v \times n_{\mathcal{T}_o} \times E_o$ matrix, generated by the input of a $n_{\mathcal{T}_o} \times E_o \times P$ sized Gaussian random variable "fed" as a component into a stochastic ensemble forecast model. This can be described heuristically as follows

$$F_E : \mathbb{R}^P \times \mathbb{R}^{n_{\mathcal{T}_o} \times E_o \times P} \times \mathbb{R}^{2n_v} \times \mathbb{R} \times \mathbb{R} \times \mathfrak{X}(\mathbb{R}^2)^P \times \ldots \mapsto \mathbb{R}^{2n_v \times n_{\mathcal{T}_o} \times E_o}$$

(16)

$$F_E(\{\theta_p\}_{p=1}^P; \{\Delta W_{e,n}^p\}_{e,p,n \in [E_o],[P],[n_t]}; \{x_i\}_{i \in \wedge_0(t_0)}; \delta, h, \{\xi_p\}_{p \in [P]}, \ldots)$$

$$= \text{ensemble forecast},$$

(17)

Where we have suppressed additional inputs in this function, as all but the first component (and perhaps the second) does not improve clarity.

3. We then define the following observation averaged continuous rank probability score loss function, taking in the space-time observations and the ensemble forecast forward model

$$L : \mathbb{R}^{2n_v \times n_t} \times \mathbb{R}^{2n_v \times n_t \times E_o} \mapsto \mathbb{R},$$

$$L := \frac{1}{n_t 2n_v} \sum_{\forall n \in \mathcal{T}} \sum_{\forall x_i(t_n) \in \wedge_0(t_n)} \hat{\text{crps}}(x_{i,t_n}^*; (F_E)_{i,t_n})$$

(18)

Where $\hat{\text{crps}} : \mathbb{R}^{2n_v \times n_t} \times \mathbb{R}^{2n_v \times n_t \times E_o} \mapsto \mathbb{R}^{2n_v \times n_t}$ denotes a vectorised continuous rank probability score estimator approximating the regular CRPS value over the space and time observations. The notation $(F_E)_{i,t_n}$ is used to indicate we compare the E_o sized ensemble forecast at the position x_i, at time t_n, to the data point x^*_{i,t_n} at the same location and temporal instance. For a single observation in space-time $x \in \mathbb{R}$, $z \in \mathbb{R}^{E_o}$ this is done using the following formula

$$\hat{\text{crps}}(y, z) = \frac{1}{E_o} \sum_{e \in [E_o]} |y - z_e| - \frac{1}{E_o(E_o - 1)} \sum_{(i,j) \in [E_o] \times [E_o]} |z_i - z_j| \quad (19)$$

see [21, 46], and [17], for more insights into this estimator and its relationships to other estimators of the CRPS. See Remark B.2 for further insights and more detailed references as to the importance of the CRPS score in forecast verification.

4. It is assumed that the discrete ensemble forecast model F_E and the loss function is differentiable with respect to the parameters $\{\theta_p\}$, so one can compute the gradient, and perform (nonlinear) optimisation (e.g. gradient descent) to minimise the Loss(CRPS estimate using $\mathcal{T}_o$, E_o), through back-propagation. Should this converge, this is a methodology to minimise the CRPS average of an ensemble forecast, it is open to whether this can recover parameters such as $\{\theta\}_{p \in [P]}$ due to non-uniqueness.

Remark 3.1 When $E_o = 1$ Remark 3.2 is equivalent to minimising the mean absolute error, between a proposed solution path and the data. It is possible to interpret the above method as a stochastic version of an ensemble 4DVAR with a CRPS estimator loss.

3.2 Methodology Justification

Mathematical motivation for the generation of the DAM (in Method 3.1), and adding the time mean back in (ensemble Method 3.5), can be justified by considering Method 3.1 applied in the context of a twin experiment described below.

1. In the context of a twin experiment where the data is generated by observing a stochastic model with known parameters $\{\theta^*_p\}_{p \in [P]}$, using a normally distributed driving signal W^* assumed free of measurement error. The recorded total velocity field u_d that would be seen by an observer at a weather station x_d at time t_n is assumed to be measurable in the following form

$$u_d(x_d, t_n) = u^\delta(x_d; \{x_i\}_{i \in \wedge_0^i(t_n)})$$

$$+ \sum_{p=1}^{P} \theta^*_p \xi_p(x_d) \Delta W^*_n / \Delta t + D(x_d), \quad \forall x_d, \quad \forall t_n \in \mathcal{T}_n. \quad (20)$$

Where the Biot-Savart kernel Eq. (9) is used for the computation drift component of velocity $u^\delta(x_d)$ induced by the point vortices at $x_i \in \wedge_0(t_n)$, and direct evaluation is assumed on $\boldsymbol{\xi}_p$. Here we have divided the stochastic contribution to the velocity by Δt as to represent how such a wind field would be measured in practice. We have also included the presence of an additional (time-independent) drift term $\boldsymbol{D}$.

2. Method 3.1, forms rows of the DAM from $\Delta t(\boldsymbol{u}_d(\boldsymbol{x}_d, t_n) - \boldsymbol{u}^\delta(\boldsymbol{x}_d; \{\boldsymbol{x}_i\}_{i \in \wedge_0^i(t_n)}))$, which in the context of a twin experiment represents the discrete effect of $\boldsymbol{D}\Delta t + \sum_{p=1}^P \theta_p^* \boldsymbol{\xi}_p(\boldsymbol{x}_d) \Delta W_n^*$ for all measurement times (see Eq. (20)).

3. Let $\Delta W_n^* \in \mathbb{R}^{P \times n_t}$, be a (P, n_t)-matrix made up of the P dimensional sampled Brownian motion over $\{t_n\}_{n=1}^{n_t}$ used to generate the data. Let $\Xi \in \mathbb{R}^{d \times P}$, be a matrix whose p-th column is defined by the vertically stacked components of (vectorised) stochastic velocity contribution evaluated at the $d = n_c \times n_c$ weather stations of interest. Let $\hat{\boldsymbol{D}} \in \mathbb{R}^d$, denote the vectorised drift effect on particle positions. Let $\boldsymbol{e}_{n_t}$ denote a vector of ones length n_t, and $\otimes$ the outerproduct. Then $M = \Delta t \boldsymbol{e}_{n_t} \otimes \hat{\boldsymbol{D}} + (\Delta W^*)^T \Xi^T \in \mathbb{R}^{n_t \times d}$ is the DAM observed in the twin experiment whose nth column is the stochastic contribution of velocity at the weather stations.

4. The SVD procedure in Method 3.1 finds an alternative representation of the effect of

$$\Delta t \boldsymbol{e}_{n_t} \otimes \hat{\boldsymbol{D}} + (\Delta W^*)^T \Xi^T = M = \boldsymbol{e}_{n_t} \otimes (\boldsymbol{e}_{n_t} M / n_t) + U \Sigma V^T, \qquad (21)$$

interpretable through PCA as the reconstruction of an efficient basis to explain the covariance structure over the time window of interest.

5. A Stratonovich-Taylor expansion of the stochastic particle trajectory map, when evaluated at the weather stations reveals

$$\boldsymbol{x}_d^{n+1} = \boldsymbol{x}_d^{n+1} + \boldsymbol{u}^\delta(\boldsymbol{x}_d^n, t^n)\Delta t + \boldsymbol{D}\Delta t + (\Delta W^T(t^n)\Xi^T)^T + H.O.T, \qquad (22)$$

where $\Delta W(t^n)$ is the n-th collumn. The substitution of the alternative representation in the other experiment Eq. (21) gives

$$\boldsymbol{x}_d^{n+1} = \boldsymbol{x}_d^{n+1} + \boldsymbol{u}^\delta(\boldsymbol{x}_d^n, t^n)\Delta t + (\boldsymbol{e}_{n_t} \otimes (\boldsymbol{e}_{n_t} M / n_t) + U_n \Sigma V^T)^T + H.O.T. \qquad (23)$$

Where U_n denotes the n-th row of U. Upon appropriate time rescaling we observe justification for the addition of a time mean, seen in Eq. (25) and Method 3.5.

The time mean drift term from data $(\boldsymbol{e}_{n_t} M / n_t)$ can be well motivated to represent drifts $\hat{\boldsymbol{D}}$ not present in the underlying model. One could foresee application in compensating systematic measurement error in sensing devices, or correcting for an unrepresented Itô-Stratonovich correction. The time mean drift term from data $(\boldsymbol{e}_{n_t} M / n_t)$ could also be seen as a parameterisation technique for representing unresolved drift terms, arising from fast dynamics [15] or unresolved physics. However, if the model has no $\boldsymbol{D} = 0$, one does not necessarily get $(\boldsymbol{e}_{n_t} M / n_t) = 0$.

An additional nonphysical drift can be observed, associated with a statistical error from sampling from the data distribution. The inclusion of the observed time mean drift term may bias towards a specific realisation of the data distribution, and not necessarily improve forecast skill.

One of the objectives of this chapter will be to numerically test the potential benefit for using an observed time mean drift $(e_{n_t} M / n_t) = 0$ attained from Method 3.1. In the context of data arising from no time mean drift $D = 0$. In the context where D represents an Itô-Stratonovich sized drift term. In the context of data with a drift D representing the effect of physical processes. These ideas will be tested using datasets(1,2), datasets 3 and datasets 4 respectively, introduced later. We use a twin experiment in which the aim is to re-simulate the training data, as well as computing forecast verification metrics on hidden testing data to account for statistical error associated with sampling the data distribution.

Mathematical motivation for Methods 3.2 and 3.6 is fairly transparent. The Continuous Ranked Probability Score (CRPS) is an example of a probabilistic forecast metric commonly used to assess ensemble forecast skill, a lower CRPS score indicates better forecast skill. The CRPS is probabilistic and compares the cumulative distribution function of the forecast with the observation values. In the context of an ensemble forward model, a lower CRPS score serves as an indicator of enhanced forecasting skill. In practice, this requires estimation over many observations, as taken into account with the loss in Method 3.2. For additional detail motivation and references regarding the CRPS see Remark B.2.

3.3 Ensemble Methods

This section contains a list of ensemble methods that will be tested, these are forward models, some requiring estimation of $\{\theta_p\}_{\forall p}$, some require generation of a basis $\{\tilde{\xi}_p\}_{\forall p}$. Methodology for such estimation has been described in the previous Sect. 3.1.

Ensemble Method 3.1 (Persistence) We predict an ensemble whose particles remain at their initial conditions $\wedge_0(t) = \wedge_0$ for the entire time interval of interest.

Ensemble Method 3.2 (RIC (Random Initial Condition Perturbation)) We initially perturb each particle position by scaled samples from the normal distribution, then we run forwards with the deterministic model to generate an ensemble. (A sensible magnitude (giving small CRPS) perturbation was searched for through trial and error and found to be of the type $0.001 N(0, \mathbb{I})$.)

Ensemble Method 3.3 (Perfect Model) We propose the SPDE used to generate the synthetic data, as a forecast model. This involves knowing true parameters $\{\theta^*\}_{p \in [P]}$ and running an ensemble forecast with new samples from the normal distribution.

Ensemble Method 3.4 (TSVDWD Without Time Mean) We perform Method 3.1 to obtain a basis for noise $\{\sigma_p \tilde{\xi}_p(x)\}_{\forall p}$. We then use the SSP33 stochastic integrator to run the regularised integrodifferential model for particle trajectories

$$d_t \boldsymbol{x}_i(t) = \boldsymbol{u}^\delta(\boldsymbol{x}_i(t), t)dt + \sum_{p=1}^{P} \sigma_p \tilde{\boldsymbol{\xi}}_p(\boldsymbol{x}_i(t)) \circ dW^p(t), \quad \forall i \in \wedge_0, \qquad (24)$$

Where $\boldsymbol{u}^\delta(\boldsymbol{x}_i(t), t)$, is computed as before using Eq. (13), and Fourier interpolation (Remark B.4) is used to evaluate $\tilde{\boldsymbol{\xi}}$, at points.

Ensemble Method 3.5 (TSVDWD with Time Mean) We perform Method 3.1 to obtain a basis for noise $\{\sigma_p \tilde{\boldsymbol{\xi}}_p(\boldsymbol{x})\}_{\forall p}$, and a time mean effect $\bar{\boldsymbol{v}}(\boldsymbol{x})$. We then use the same SSP33 stochastic integrator to run the regularised integrodifferential model for particle trajectories

$$d_t \boldsymbol{x}_i(t) = \boldsymbol{u}^\delta(\boldsymbol{x}_i(t), t)dt + \bar{\boldsymbol{v}}(\boldsymbol{x}_i(t))dt + \sum_{p=1}^{P} \sigma_p \tilde{\boldsymbol{\xi}}_p(\boldsymbol{x}_i(t)) \circ dW^p(t), \quad \forall i \in \wedge_0,$$
$$(25)$$

where $\boldsymbol{u}^\delta(\boldsymbol{x}_i(t), t)$, is computed as before using Eq. (13), and the Fourier interpolation described in Remark B.4 is used is used in the evolution of the points by the additional deterministic drift term $\bar{\boldsymbol{v}}$ and stochastic terms $\{\tilde{\boldsymbol{\xi}}\}_{p \in [P]}$.

Ensemble Method 3.6 (Backpropagation Ensemble Approach) We take the parameters $\{\theta_p\}_{p \in [P]}$ that minimise the CRPS mean loss after Method 3.2 is performed over $\mathcal{T}_o$ with E_o ensemble members, and run the trained ensemble method F_E forward

$$d_t \boldsymbol{x}_i(t) = \boldsymbol{u}^\delta(\boldsymbol{x}_i(t), t)dt$$

$$+ \sum_{p=1}^{P} \theta_p \boldsymbol{\xi}_p(\boldsymbol{x}_i(t)) \circ dW_e^p(t), \quad \forall i \in \wedge_0, \forall t_n \in \mathcal{T}, \forall e \in [E]. \qquad (26)$$

Where $\boldsymbol{u}^\delta(\boldsymbol{x}_i(t), t)$, is computed as before using Eq. (13), and the vectorfields $\{\boldsymbol{\xi}_p\}_{p \in [P]}$ are directly evaluated rather than Fourier interpolated. This could potentially be interpreted as a stochastic ensemble version of 4DVAR with a CRPS loss.

4 Numerical Experiments

4.1 Twin Experiment Frameworks

In operational practice (in, say, weather prediction), the state values such as temperature come from an unknown distribution and are recorded using measurement devices with the addition of measurement noise. However, to assess the proposed data assimilation methodology a known reference dataset should be predefined beforehand. This naturally leads to the concept of a twin experiment framework, a common practice in both weather forecasting and inverse modelling communities. We will now (in the next paragraph) describe how by fixing the driving path, we can perform a twin experiment for the SVD calibration of a stochastic fluid system. We then, in the proceeding paragraph describe another method of validation, in which the underlying distribution of the observation signal is assumed known. This type of testing can alleviate errors associated with sampling data from the unknown distribution.

A reference trajectory is assumed known, and computed by fixing all parameters $\{\theta_p^*\}_{p=0}^{p=P}$, $\{\{\Delta W_p(t_n)\}_{p=0}^{p=P}\}_{n=0}^{N}$ and running the stochastic forward model over a time window $t \in [0, T]$. Synthetic measurements (at weather-stations) are then collected by sampling values from this reference trajectory. Finally, the data assimilation technique of interest Method 3.1 is implemented as to attain $\{\sigma_p^r\}_{p=0}^{p=P}$, $\{\{\Delta \tilde{W}_p(t_n)\}_{p=0}^{p=P}\}_{n=0}^{N}$, and $\{\bar{v}^r\}$. Using these "recovered" parameters, we generate a new output trajectory, for the evolution of points. We then can compare the output trajectory to the reference trajectory. This allows the SVD calibration accuracy to be assessed. These tests will be performed and assessed using datasets 1,3,4 and 2, using $P = \{1, 5\}$ respectively. In this setting the twin experiment is not viewed in the context of verifying the calibration of a stochastic model, but as an assessment of the method viewed as a data assimilation procedure in which the reference "training" trajectory is aimed at being captured as accurately as possible.

Going further than this, suppose for testing/validation purposes that we know more than just a single reference trajectory, but the entire reference distribution. Namely, we know the stochastic forward model used to generate data and its parameters (not necessarily the one proposed for modelling). In this setting, we can account for the additional sampling error associated with drawing data from the underlying distribution. To do so in practice we generate 1000 realisations of the stochastic forward model used for data. These 1000 realisations of the stochastic forward model will be treated as hidden testing datasets for which ensemble forecast verification metrics can be employed. In this scenario, CRPS scores arise for data for which the model has not been trained, this can help distinguish stochastic sample path model error associated with sampling the data distribution. This can be thought of as the verification/validation of unseen/hidden test data.

4.2 Setup: Generation of Synthetic Data

This subsection contains specific details about the generation of four reference datasets we wish to calibrate from. Dataset 1 is made with a realisation of a parametrised stochastic model with one basis function. Dataset 2 is made with a realisation of a parametrised stochastic model with 5 basis functions. Dataset 3 is made with a realisation of a stochastic model, but the model has a predefined additional drift, replicating model-data mismatch. Dataset 4 contains the same set-up as Dataset 3 but with a different predefined drift, larger in magnitude replicating more realistic model data mismatch. Per dataset, we also compute an additional 1000 corresponding realisations for testing purposes.

The initial condition $\omega_0(x, y)$, is constructed from two compactly supported circular regions with radius $R = 1/8$ of non zero vorticity in the following way

$$
\omega_0(x, y) := \begin{cases} 1/2 + 1/2 \left(1 - (\frac{r_1}{R})^2\right)^3, & r_1 = ((x - \frac{1}{2} - R)^2 + (y - \frac{1}{2})^2)^{1/2} \leq R, \\ 1/2 + 1/2 \left(1 - (\frac{r_2}{R})^2\right)^3, & r_2 = ((x - \frac{1}{2} + R)^2 + (y - \frac{1}{2})^2)^{1/2} \leq R, \\ 0 & \text{else.} \end{cases}
$$

(27)

We use a initial $n \times m = 128 \times 128$ mesh over the domain $[0, 1] \times [0, 1]$. Resulting in a mesh spacing of $h = h_x = h_y = 0.0078125$ and regularisation parameter $\delta = 1.5h^{3/4} = 0.03941702$. We remove the point vortices with non-zero vorticity from the flattened (x, y)-meshgrid of points. Resulting in $N_v = 1216$ points remaining from the 16384 initially specified on the mesh. We use $n_t = 256$ timesteps on the time interval $t \in [0, 32]$, with $\Delta t = 0.125$.

In dataset 1, we consider idealised data generated from a parametrised run of the SPDE, with a single $P = 1$ basis function $\theta_1 = 0.003$, given by

$$
\theta_1(\xi_1^x, \xi_1^y)^T = 0.003(2\pi \cos(2\pi y), -2\pi \cos(2\pi x))^T.
$$

(28)

In dataset 2, we consider data generated from a parametrised run of the SPDE with 5 basis functions given by

$$
\theta_p(\xi_p^x, \xi_p^y)^T = 0.0001(2\pi \cos(2\pi py), -2\pi \cos(2\pi px))^T, \quad p \in [5].
$$

(29)

In dataset 3, we create the data as a single realisation of the time mean included system described by Eq. (25), where we prescribe the same basis function as dataset 1 Eq. (28), however we choose the following Stratonovich-Itô correction drift

$$
\bar{v} = -0.003^2 4\pi^3 (\sin(2\pi y) \cos(2\pi x), \sin(2\pi x) \cos(2\pi y))^T,
$$

(30)

in the underlying stochastic model that generates the data. To test the importance or non-importance of the Itô-Stratonovich correction in the generation of the data anomaly matrix.

In dataset 4, we create the data as a single realisation of the time mean included stochastic system with the same basis function Eq. (28) to that of datasets 1 and 3 but use the following drift

$$
\bar{v} = 0.0003(8\pi \cos(8\pi y), -8\pi \cos(8\pi x))^T,
$$

(31)

replicating some small-scale unresolved drift velocities not proposed in the stochastic forward model, but observed by the data.

We either use $\{\Delta W_1(t_n)\}_{n \in [n_t]}$ a n_t sized sample from the normal distribution or $\{\Delta W_p(t_n)\}_{n \in [n_t], p \in [5]}$ a $(n_t \times 5)$ sized sample from the normal distribution for datasets (1,3,4) and 2 respectively. The resulting set of evolving points $\wedge_0(t)$, do not remain a Cartesian mesh after initial time. We also consider a 64×64 Cartesian meshgrid $\wedge^d$, of fixed weather centers in a closed subdomain $[0, 1] \times [0, 1]$ of $\mathbb{R}^2$ for all time. Where the additional subscript d denotes "data", and indicates that this is a weather-station in which velocity data $u^m(x_d, t_n)$ is measured and recorded (see e.g. Eq. (20)).

Snapshots at $t = (0, 16, 32)$ of the stochastic solution dataset 1 is plotted in the first row of Fig. 1, generated with the addition of a single basis function Eq. (28). In the second row of Fig. 1, we plot snapshots of dataset 2, generated with five parametrised basis functions Eq. (29). In the third row of Fig. 1 we plot snapshots of dataset 3, generated with a single basis function Eq. (28) and a Stratonovich-Itô correction drift Eq. (30). In the fourth row of Fig. 1 we plot snapshots of dataset 4, generated with a single basis function Eq. (28) and a pre-prescribed drift function Eq. (31) representing physical unresolved processes. Not plotted are an additional 1000 hidden testing/validation datasets per the above dataset.

4.3 Results and Discussion

We apply the SVD approaches based on Method 3.1, to datasets 1,2,3,4 in a context of a twin experiment for the re-simulation of data. We apply the ensemble backpropagation Method 3.2 to only datasets 1,2, as we have not described the extension of the Methods 3.6 and 3.2 to actively include explicit drift parameterisation. We then compute ensemble forecast verification metrics on the hidden test data, to see if the underlying distribution is well represented.

4.3.1 Results: Dataset 1

Twin Experiment Synthetic dataset 1 $\{\{x_i(t_n)\}_{i \in \wedge_0^i(t_n)}, \{u^m(x_d, t_n)\}_{x_d \in \wedge_d}\}_{n=0}^{n=n_t}$ described in Sect. 4.2 was generated using a single basis of noise (Eq. (28)), whose snapshot (at $t = 0, 16, 32$) of vortex positions is shown in the first row of Fig. 1.

Figure 2a contains a plot of the basis vector-field $\theta_1 \xi_1(x_d)$, used (in combination with the forward model) to generate synthetic dataset 1. Over 99.99 percent of covariance was explained by one basis function. Figure 2b contains the time average velocity field drift from the data. Figure 2c contains the recovered vector field $\sigma_1 \tilde{\xi}_1$ by Method 3.1. Visually Fig. 2a and c appear similar, and have agreement to 0.0467482 in the relative L^2 norm. The next recovered basis $\sigma_2 \tilde{\xi}_2$ has machine precision magnitude plotted in Fig. 2d. Figure 2e contains a plot of the recovered

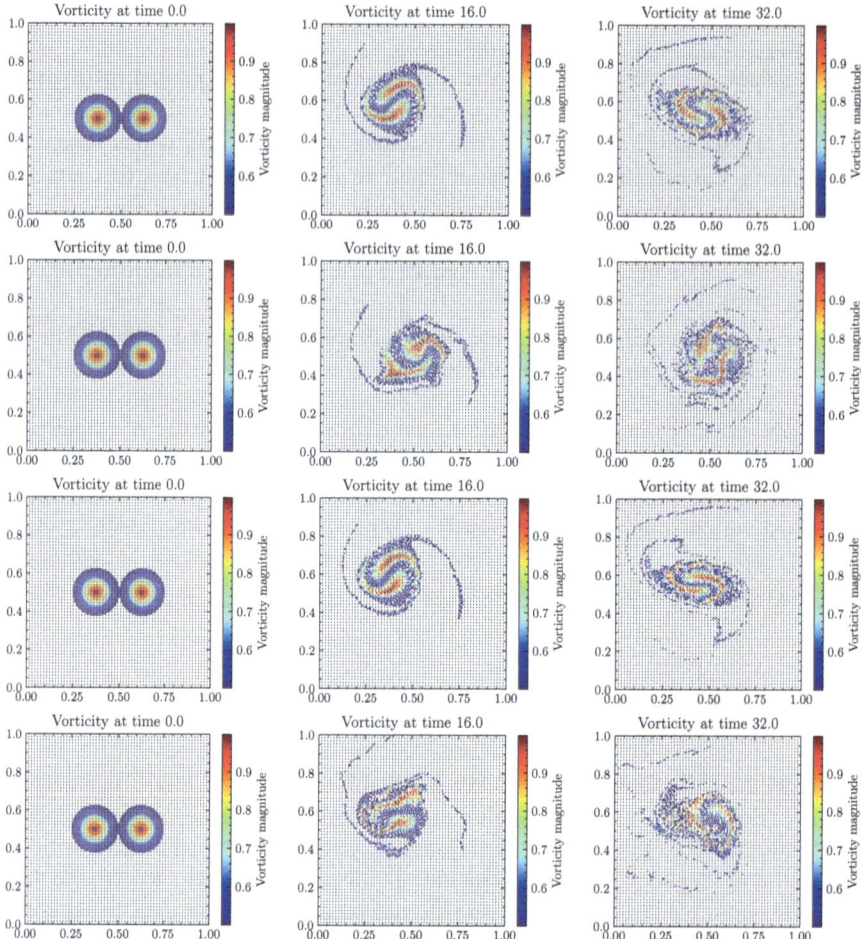

Fig. 1 Shown in each row is a scatter plot of point vortex positions whose vortex strength is indicated by the non-perceptually uniform colourmap jet, as to highlight finer structures in the flow. Black square dots denote weather stations from which data is collected. In row one we plot snapshots of the point vortex (at t = 0,16,32) for dataset 1 generated with one basis function. Row two corresponds to dataset 2 driven with five basis functions. Row three corresponds to dataset 3, generated with one basis function and an additional Stratonovich-Itô drift. Row four corresponds to dataset 4, generated with one basis function and a prescribed deterministic drift, representing additional unresolved processes. These datasets contain the effect of different prescribed physical processes (Stochastic diffusion and drift terms). This work aims to extract and parameterise the effects of these hidden processes, for use with a stochastic ensemble forward model

time increments ΔW_1^r, and the driving signal increments ΔW_1^*, ΔW_1^r and ΔW_1^* are nearly indistinguishable apart from a small difference in magnitude. They differ in the relative L^2-norm by 0.03818701. With the addition of the time mean it is possible to recover the data anomaly matrix to 2.00982e-13 using only one

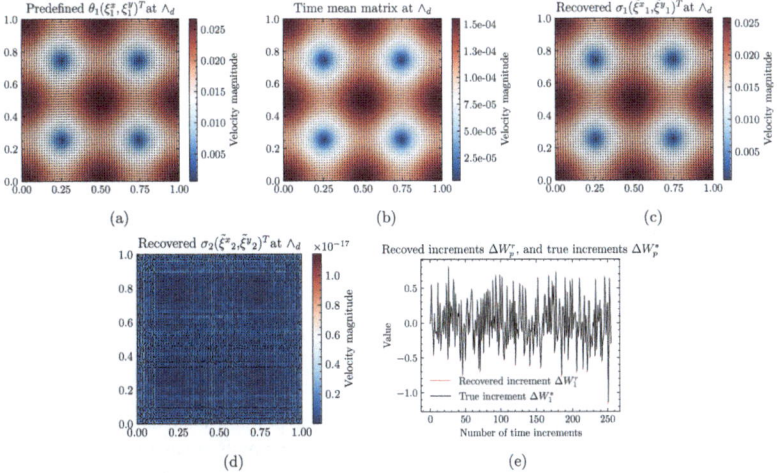

Fig. 2 Predefined basis of noise (**a**), recovered time mean (**b**), recovered first basis (**c**), second recovered basis (**d**), recovered first increment (**e**), after Method 3.1 is applied on dataset 1

singular value, one singular vector and the recovered driving signal. Note the SVD decomposition could equivalently result in an opposite signed $\boldsymbol{\xi}_1$, and opposite signed ΔW_1^r increments.

We have presented evidence indicating that the methodology in Method 3.1 identified a Data anomaly matrix related to the effect of $\theta_1 \boldsymbol{\xi}_1 (\boldsymbol{x}_d)$ from the synthetic data. We speculate the small difference between $\sigma_1 \tilde{\boldsymbol{\xi}}_1$ and $\theta_1 \boldsymbol{\xi}_1 (\boldsymbol{x}_d)$, is in part related to the removal of the time average, specific to the realisation of Brownian motion ΔW_1^* used in the synthetic data. This motivates the next test where we shall drive the solution with the recovered increments, recovered basis functions, and with or without the recovered time mean drift, all in comparison to dataset 1. We shall call this the re-simulation of data.

Figure 3 contains the solution of the SPDE when driven by the recovered signal ΔW_1^r with recovered basis function $\sigma_1 \tilde{\boldsymbol{\xi}}_1$ as compared with the original data using ΔW_1^* and $\theta_1^* \boldsymbol{\xi}_1^*$. In Fig. 3 the first 3 images in row one do not have the addition of the time mean, where as row two contains the time mean drift velocity. The relative (L2-spacetime) error of the reproduced time mean included solution is 0.01265 the relative error of the recovered time mean not included solution is 0.05499. As seen in Fig. 3 and measured by relative error of the reproduced solution the inclusion of the time mean was found to be significantly helpful in the re-simulation of the synthetic data in the context of a twin experiment. The remaining non-zero difference in relative error norm 0.01265 could be speculated to be caused by many things, such as small inaccuracies in Fourier interpolation growing over the course of the simulation run.

Verification of Learning During Training We train the ensemble back-propagation approach Method 3.2 using a $E_o \in \{1, 5\}$ sized member ensemble

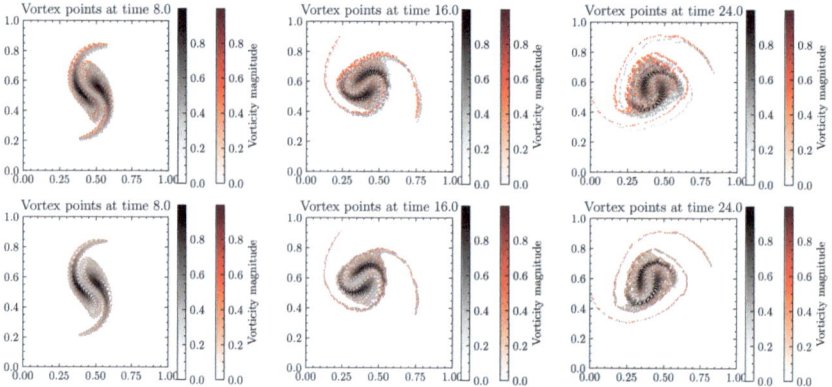

Fig. 3 Dataset 1 ($P = 1$), reconstruction of data. Grey data points are the data. The red points are the recovered increment driven system. In the first row, the red points are evolved without using the additional time mean contribution using Method 3.4. The second row contains the same experiment but with the additional time mean drift using Method 3.5. We observe that the time mean velocity drift has notable effect, in the context of this re-simulation of data twin experiment

forecast over a smaller time window $\mathcal{T}_o = [8]/8$ such that the forward model creates a E_o sized ensemble in which 8 SSP33 Runge Kutta steps are taken for each ensemble member. The CRPS-loss of the E_o sized ensemble forecast over the time window $\mathcal{T}_o$ is denoted crps$_o$. The parameters in the gradient descent algorithm used, are learning rate 1e-7, acceleration = true, max iterations 500, tolerance 1e-15, implemented using the "jaxopt.GradientDescent" algorithm. We take the initial guess of parameters to be $\theta_1^g = 10^{-12} e_p$, such that the initial CRPS score before training is essentially a measure of the forecast skill of an ensemble with E_o deterministic models.

In Fig. 4 we plot the Relative improvement in CRPSS$_o$ over the initial Deterministic ensemble forecast (30–40% improvement), and the relative error in parameter estimation magnitude, for $E_o \in \{1, 5\}$ respectively. For $E_o \in \{1, 5\}$, the true parameter value was $\theta_1^* \in \{0.003, 0.003\}$, the initial guess is $\theta_1^g = 10^{-12}$, after 500 iterations we have learned parameter value $\theta_1^r \in \{-0.00208943, 0.00360890\}$ respectively. The initial CRPS estimate is crps$_o = \{0.0101590, 0.0101590\}$, the after training the final CRPS estimate over the time interval $\mathcal{T}_o$ is crps$_o \in \{0.00809122, 0.00608491\}$ respectively, this CRPS is computed on a subset of the training data, $\mathcal{T}_o$.

Forecast Verification for Underlying Distribution For dataset 1, an additional 1000 hidden validation/testing datasets were created, each testing dataset is the exact stochastic model used for the production of dataset 1, but run forwards with new normally distributed sampled increments. We then compute a 30-member ensemble forecast using a stochastic forward model trained/calibrated on dataset 1. Per hidden dataset, we estimate the CRPS over all space-time observations (using Eq. (18)) representing the likelihood of observing the test dataset from the ensemble forecast.

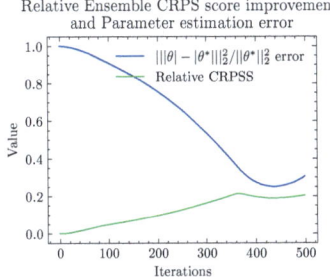 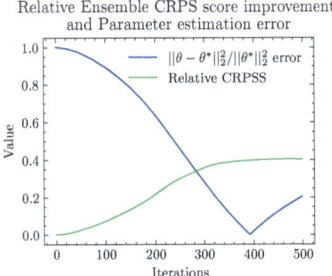

Fig. 4 The green line is the relative CRPSS improvement as compared to the initial deterministic proposed ensemble. The blue line is the relative error associated with the magnitude in parameter estimation on training data. The figures are plotted for $E_o = 1$, $E_o = 5$, respectively. Both figures show a 30–40% increase in forecast skill over the deterministic proposed ensemble. Learned parameters are closer in magnitude to the true parameters used for the generation of synthetic data. However, there are situations in which the true parameter does not necessarily give a better CRPSS

We then take the mean CRPS over the entire 1000 test datasets. A lower mean CRPS value informally indicates a better likelihood that the hidden testing dataset comes from the trained ensemble methods 30-member ensemble forecast prediction.

The raw averaged CRPS scores are displayed in the first column of Table 5, and the percent relative improvements in average CRPSS (calculable as $100(1 - \text{CRPS(model 1)}/\text{CRPS(model 2)}))$ are displayed in Table 1, we observe the following. The Without-time-mean ensemble (calibrated with Method 3.1) outperformed the perfect model by 0.04%. The Perfect model outperformed the With-time-mean model by 2.039%. The With-time-mean model outperformed randomised initial conditions by 8.071%. The Randomised initial condition outperformed persistence by 49.44%.

We conclude that the underlying distribution is best represented by the Perfect model and the Without-time-mean model performed similar in CRPS score. The With-time-mean model performed marginally worse, we speculate that this is because the observed drift was a sampling error rather than a systematic modelling error. Correcting for a sampling error, biased the solution towards the reference training data as seen in Fig. 3, but was not helpful for representing (on average) the hidden 1000 testing datasets i.e. the underlying distribution.

In summary, Tables 1 and 5 indicate that using the time mean drift is not helpful in representing the underlying distribution when the underlying distribution (e.g. hidden SPDE/SDE model) does not have an explicit time mean drift. This is fairly transparent as the time mean drift occurs only as a result of sampling data from the underlying distribution, and adding in the small observed drift biases towards the training dataset rather than compensating for a model-data drift mismatch. However, in practice one may not be able to tell the difference between model error and statistical sampling error, as one normally cannot resample from the underlying distribution. Biasing the model towards matching observed data with a time mean drift could be seen as a valid modelling assumption to make, and only decreased

Table 1 Hidden dataset 1, "Percent improvement" table of row scheme over the column scheme for representing the distribution, i.e. hidden datasets. Readable as follows, the With-time-mean ensemble is 8.071% better at representing the underlying distribution than the RIC ensemble. In practice this is the relative skill, hidden dataset averaged, observation averaged CRPS estimate, of the row scheme against the column scheme multiplied by a factor of 100

Scheme	RIC	With-time-mean	Without-time-mean	Perfect	Persistence
RIC	0	-8.779	-11.09	-11.04	49.44
With-time-mean	8.071	0	-2.123	-2.081	53.52
Without-time-mean	9.982	2.079	0	0.0413	54.48
Perfect	9.945	2.039	-0.04132	0	54.47
Persistence	-97.77	-115.1	-119.7	-119.6	0

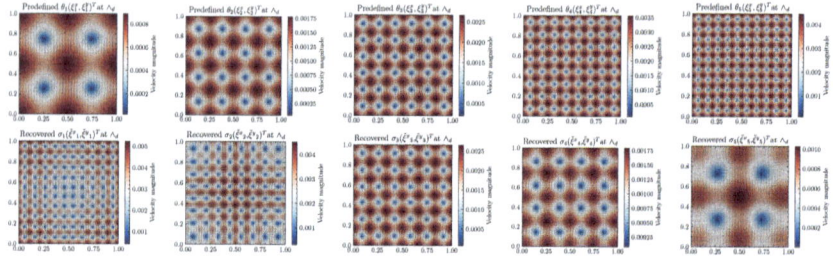

Fig. 5 The first row contains the vector fields and magnitude of the predefined basis $\{\theta_p^* \xi_p\}_{p=1}^{p=5}$ used in the generation of dataset 2, when evaluated at the weather stations $x_i \in \wedge_d$. The second row shows $\{\sigma_p(\tilde{\xi}_p^x, \tilde{\xi}_p^y)^T\}_{p=1}^{p=5}$ the recovered basis from truncated SVD at weather stations, generated by the algorithm described in Method 3.1, not shown here is the 6th recovered principle component, a machine precision zero vector-field similar to Fig. 2d

relative CRPSS by about 2% for this case. It should be noted that a sensible random perturbation of the initial condition performed remarkably well (10% worse CRPSS) at representing SALT-type data. Both Methods 3.5 and 3.4 (with and without the time mean) outperformed both RIC and Persistence, and approached the CRPS score of the perfect model.

4.3.2   Results: Dataset 2

Twin Experiment For synthetic dataset 2 described in Sect. 4.2 we used $P = 5$ velocity fields as a basis of noise Eq. (29) for the generation of data, whose snapshot (at $t = 0, 16, 32$) of vortex blobs is shown in the second row of Fig. 1.

In the first row of Fig. 5 we display the five basis functions used to generate the synthetic dataset 2, while in the second row, we plot the recovered 5 basis functions obtained through Method 3.1. Approximately 99.99% of covariance was explained by these five basis functions, with the sixth having machine precision magnitude. The recovered basis was not found unique or ordered (SVD is non-unique), but the vectorfields recovered exhibit some similarity in magnitude and shape to those used

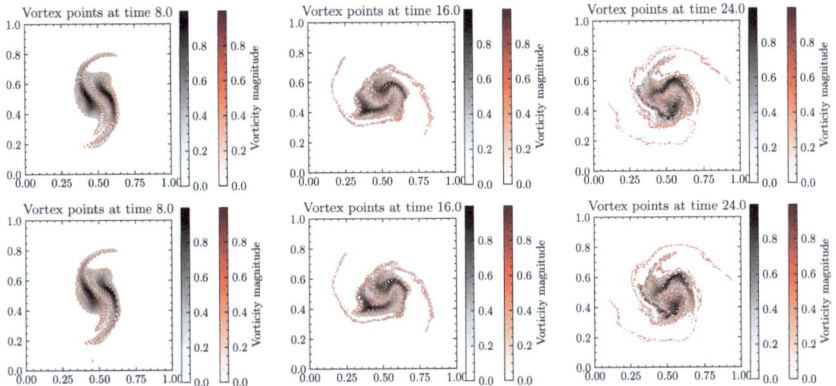

Fig. 6 Dataset 2 ($P = 5$). In the first row, the red points are evolved using recovered increment and recovered basis functions from Method 3.1 without the time mean contribution, and the grey points are the data. The second row contains the same experiment but with the additional time mean drift in

to generate dataset 2. Figure 6 contains the SPDE driven by the recovered signal $\{\Delta W_p^r\}_{p \in [5]}$ using the Fourier interpolated recovered 5 components $\{\sigma_p \tilde{\xi}_p\}_{p \in [5]}$, as compared with the original data. The relative L^2 error of the recovered increment time mean included driven solution from the data is 0.0531869, the relative L^2 relative error of the recovered time mean not included solution is 0.0629821. This indicates an improvement in the re-simulation of data by the inclusion of the time mean.

We test that recovered increments are normal. The Shapiro-Wilk test [42] tests gives a score of 0.99863, and p-value of 0.421952. The two-sample goodness of fit Kolmogorov-Smirnov test ([8]) between ΔW^* and, $U/\Delta t = \Delta W^r$ gives a test statistic of 0.0390625, with p-value 0.990. This is not below the threshold of 0.05, so we cannot reject the null hypothesis that this sample is distributed according to the standard normal with confidence level 95%. We perform the Anderson test, the value of the test statistic was 0.317 and doesn't exceed the critical values [0.574, 0.654, 0.785, 0.915, 1.089], indicating the null hypothesis of normality cannot be rejected at the associated percent significance levels [15, 10, 5, 2.5, 1.]. The recovered noise has moments displayed in the second column of Table 6. Hypothesis testing indicates that the recovered increments are likely sampled from a normal distribution, justifying the basis $\{\sigma_p \tilde{\xi}\}_{p \in [P]}$ as a reasonable choice for modelling new normal increments.

Verification of Learning We observe that the CRPS loss on training data decreases drastically (initially), indicating improved forecast skill in comparison to a deterministic forecast for training dataset 2. The proposed initial parameter guess is $10^{-12} e_5$ (essentially a deterministic proposal), the synthetic data was generated with "true" parameter values $\theta^* = 10^{-5} e_5$. After 500 iterations of gradient descent, the estimated parameter values are

Table 2 Hidden dataset 2, "Percent improvement" table of row scheme over the column scheme for representing the distribution, i.e. hidden datasets

Scheme	RIC	With-time-mean	Without-time-mean	Perfect	Persistence
RIC	0	−2.921	−5.631	−6.704	81.46
With-time-mean	2.838	0	−2.634	−3.676	81.98
Without-time-mean	5.331	2.566	0	−1.016	82.45
Perfect	6.283	3.546	1.006	0	82.62
Persistence	−439.3	−455.1	−469.7	−475.5	0

[0.0001621, −0.00001877, −0.00002568, 0.00003455, 0.0001050]   indicating identification of the rough sizes of the parameters, but not sign or exact size. We found examples of new parameters not equalling the true parameters used to generate the training data, but giving improved CRPS skill scores during training.

Forecast Verification for Underlying Distribution For dataset 2, an additional 1000 hidden testing datasets were created, with the same $P = 5$ basis functions with same parameter values $\theta^* = 10^{-5} e_5$, but with different driving increments. We tabulate the CRPS average scores in the second column of Table 5. We note that the backpropagation approach ensemble Method 3.6 had an unusually low CRPS average indicating good forecast skill. We tabulate the relative CRPSS scores in Table 2. From which we conclude that in terms of relative CRPSS. The perfect model ensemble on average outperformed the Without-time mean ensemble by 1.0%. The Without-time-mean ensemble on average outperformed the With-time-mean ensemble by 2.5%. The With-time-mean ensemble outperformed the randomised initial condition ensemble by 2.8%. The randomised initial condition ensemble outperformed the Persistence ensemble by 81.5%.

From Table 2, and Fig. 6 we conclude that in the instance when the data comes from a model without a time mean drift. Adding on the observed time mean drift was not helpful in representing the hidden model, despite predicting the training data more accurately. We speculate this occurs because the observed drift is a statistical error associated with sampling a unknown distribution rather than a systematic modelling error.

4.3.3 Results: Dataset 3

Twin Experiment Synthetic dataset 3 described in Sect. 4.2 was generated using a single basis of noise (Eq. (28)) and a predefined drift Eq. (30) (Fig. 7a) mimicking unresolved small scale drift dynamics, the snapshot (at $t = 0, 16, 32$) of vortex positions is shown in the third row of Fig. 1.

Using Method 3.1, we recover a time mean drift (plotted in Fig. 7b), a basis function (not plotted indistinguishable to Fig. 2c) and a driving signal, such that the DAM can be reconstructed to $1.90113e − 13$. The recovered ΔW^r gives a Shapiro-Wilk test W score of 0.989228 with p-value 0.0538488, and a Shapiro statistic of

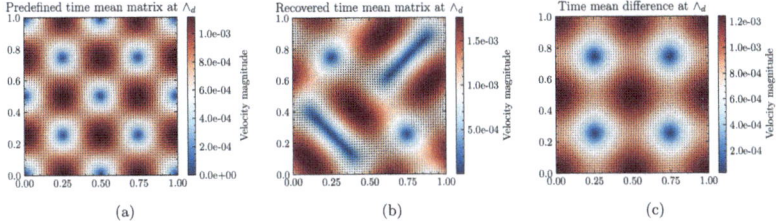

Fig. 7 Predefined Itô-Stratonovich drift (**a**), recovered time mean drift (**b**), attained by Method 3.1 applied on dataset 3. The difference is plotted in (**c**), indicating the recovered time mean drift is a contribution from both the predefined difference from the proposed forward model and the data (Itô-Stratonovich correction) as well as the specific Brownian motion realisation associated with sampling from underlying hidden data distribution

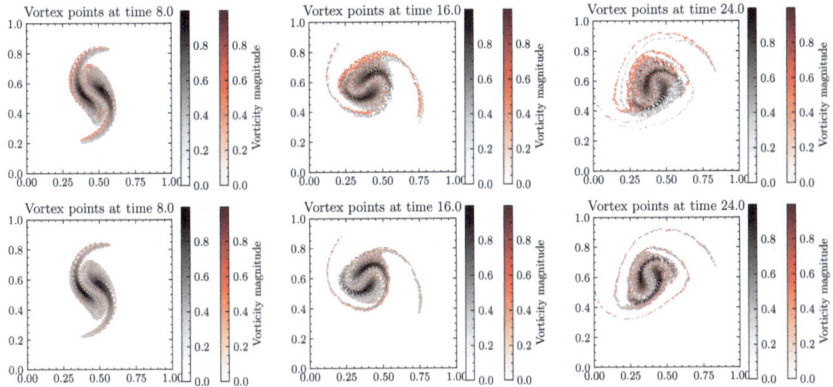

Fig. 8 Dataset 3, reconstruction of data. Grey data points are the data. The red points are the recovered increment driven system. In the first row, the red points are evolved without using the additional time mean contribution using Method 3.4. The second row contains the same experiment but with the additional time mean drift using Method 3.5. We observe that the time mean velocity drift has a significant effect, in the context of this re-simulation of data in this twin experiment

0.989228 with p value 0.0538488. The two-sample goodness of fit Kolmogorov-Smirnov test gives a KS statistic of 0.03515625, with p-value 0.997513 not below the threshold of 0.05. This indicates some evidence that the recovered increments are normal and the basis function is appropriate for use with different normal increments.

In Fig. 8 we plot the re-simulation of data with the recovered increments ΔW^r, and the recovered basis function $\tilde{\xi}_1$ with and without the inclusion of the recovered time mean drift. The relative (spacetime L2) error of the time mean included solution is 0.0160130, whereas the relative error of the recovered time mean not included solution is 0.0864620. We observe in both the relative L2 spacetime error and Fig. 8 that the recovered time mean is significantly helpful in the re-simulation of the observed dataset 3 training data. Demonstrating that the observed time mean drift can compensate for both the Itô-Stratonovich modelling deficiency and the

Table 3 Hidden dataset 3, "Percent improvement" table of row scheme over the column scheme for representing the distribution, i.e. hidden datasets

Scheme	RIC	With-time-mean	Without-time-mean	Perfect	Persistence
RIC	0	−10.29	−11.51	−12.52	47.9
With-time-mean	9.326	0	−1.11	−2.029	52.76
Without-time-mean	10.32	1.098	0	−0.9087	53.28
Perfect	11.13	1.989	0.9005	0	53.7
Persistence	−91.96	−111.7	−114.1	−116	0

unphysical drift observed in the DAM from statistical sampling error, in the re-simulation of data.

The interesting feature of dataset 3, is that the recovered drift (plotted in Fig. 7), is visibly affected by model inadequacy by missing an Ito-Stratonovich correction drift term Fig. 7a, and also by statistical sampling error (associated with the specific Brownian motion realisation used for data). This is highlighted in Fig. 7c, where the difference between the predefined time mean drift and recovered time mean drift, appears to be the same shape as the basis function Eq. (28) for noise. This motivates computing the CRPSS on hidden data, to see whether the time mean drift is important in terms of representing the Itô-Stratonovich model deficiency despite the addition of an unphysical statistical drift bias observed by the data anomaly matrix.

Forecast Verification for Underlying Distribution In the third column of Table 5 we tabulate the averaged CRPS score over the hidden datasets. We turn this into the improvement in relative CRPSS displayed in Table 3. From this, we conclude that the Perfect ensemble model on average represented the hidden 1000 datasets 0.9% better than the Without-time-mean ensemble. The Without-time-mean ensemble on average represented the 1000 hidden datasets 1.1% better than the With-time-mean ensemble. The With-time-mean ensemble on average represented the 1000 hidden datasets 9.3% better than the RIC ensemble. The RIC ensemble outperformed the persistence ensemble by 47%.

We conclude that the inclusion of a drift term from data, even when well moti-vated from modelling deficiencies may not necessarily improve the calibrated model in terms of CRPS score. We speculate (based on the previous two experiments) that this occurs because the finite realisation of the ξdW term in the generation of the synthetic training data resulted in an unphysical time mean drift observed in the DAM. Whose inclusion in a calibrated model with drift (e.g. ensemble Method 3.5) can dominate the potential benefit of modelling a small but well-motivated drift.

4.3.4   Results: Dataset 4

Twin Experiment Synthetic dataset 4 described in Sect. 4.2 was generated using a single basis of noise (Eq. (28)) and a predefined drift Eq. (31) (Fig. 9a) mimicking

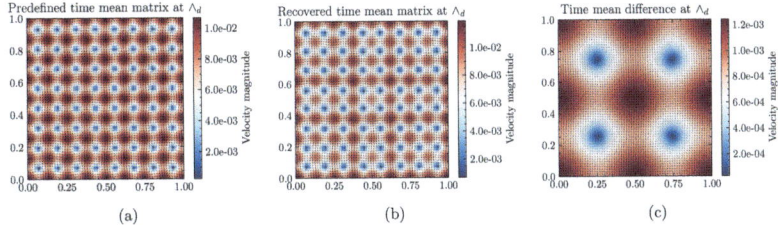

Fig. 9 Predefined time mean drift (**a**), recovered time mean drift (**b**), after method Method 3.1 is applied on dataset 4. The difference is plotted in (**c**), indicating the recovered time mean matrix is a contribution from both the physical predefined difference from the proposed forward model and specific Brownian motion realisation associated with sampling from the underlying hidden data distribution

the effect of unresolved small scale drift dynamics, the snapshot (at $t = 0, 16, 32$) of vortex positions is shown in the fourth row of Fig. 1.

Using Method 3.1, we recover a time mean drift (plotted in Fig. 9b), a basis function and a driving signal, such that the DAM can be reconstructed to 1.98225e-13. On the recovered increments we perform the Shapiro-Wilk test to evaluate the null hypothesis that the data was drawn from a normal distribution and get a score of 0.989228, and p-value of 0.0538488. We perform the two-sample goodness of fit Kolmogorov-Smirnov test for the recovered increments to test the null hypothesis that the recovered increments are distributed according to the appropriately scaled normal distribution. The p-value of 0.997513 is not below the threshold of 0.05, so we cannot reject the null hypothesis that this sample is distributed according to the standard normal with a confidence level 95%. Giving evidence that $\tilde{\xi}_1$ is an appropriate basis for stochastic parametrisation.

The interesting feature of dataset 4 is that the recovered time mean drift is made up from both real model inadequacies from missing a physical drift term in the underlying model (plotted in Fig. 9a) and specific sampling error. This is highlighted in Fig. 9c, where the difference between the recovered and predefined drift is plotted and appears to be in the same shape as the stochastic basis velocity Eq. (28).

Snapshots (at t=8,16,24) of the re-simulation of data with and without the time mean drift are plotted in the first and second row of Fig. 10 respectively. The relative spacetime error of re-simulating data with the time mean included is 0.0672719 whereas the relative error without the time mean included solution is 0.214116. We conclude that the inclusion of the time mean drift term is helpful in the re-simulation of the training dataset 4. Since this model-data mismatch in drift is larger in magnitude than the Itô-Stratonovich correction, it is well motivated to consider whether the observed time mean drift can be used in an ensemble forecast and improve the forecast skill. Despite the potential for a statistical bias associated with the sampling of the Brownian motion.

Forecast Verification of Underlying Distribution In the fourth column of Table 5 we tabulate the averaged CRPS score of each model over the hidden data. This

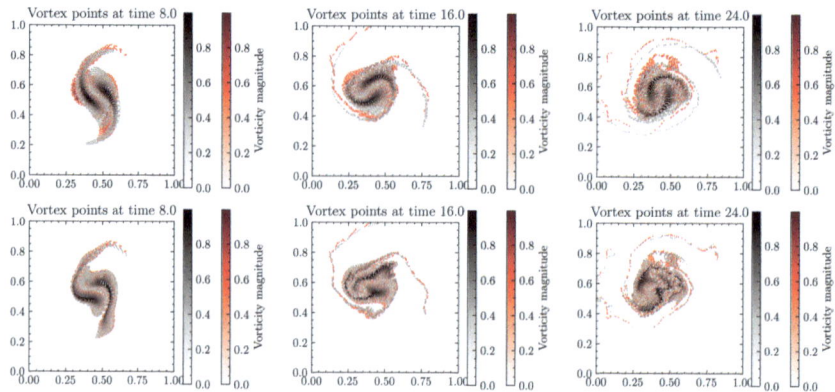

Fig. 10 Dataset 4, reconstruction of data. Grey data points are the data. The red points are the recovered increment driven system. In the first row, the red points are evolved without using the additional time mean contribution using Method 3.4. The second row contains the same experiment but with the additional time mean drift using Method 3.5. We observe that the time mean velocity drift has significant effect, in the context of this re-simulation of data in this twin experiment

Table 4 Hidden dataset 4, "Percent improvement" table of row scheme over the column scheme for representing the distribution, i.e. hidden datasets

Scheme	RIC	With-time-mean	Without-time-mean	Perfect	Persistence
RIC	0	−16.96	−12.46	−19.04	**42.41**
With-time-mean	14.5	0	**3.849**	−1.776	50.76
Without-time-mean	**11.08**	−4.003	0	−5.85	48.79
Perfect	15.99	**1.745**	5.527	0	51.62
Persistence	−73.64	−103.1	−95.28	−106.7	0

is presented in terms of a relative improvement in average CRPSS in Table 4. We conclude that in terms of representing the 1000 hidden realisations from the underlying SPDE/SDE, as compared by relative space-time averaged CRPSS. The RIC ensemble was on average 42.41% better than Persistence. Without-time-mean was on average 11.08% better than RIC. With-time-mean was on average 3.849% better than Without-time-mean. The Perfect model was on average 1.745% better than With-time-mean. These are indicated with bold values in Table 4.

We make the following important conclusion from dataset 4. If the data is generated with a model with a notable time mean drift, using Method 3.1 the time mean can be captured. Furthermore the inclusion of the measured drift (ensemble Method 3.5) improved the re-simulation of data as in Fig. 10. The inclusion of the measured drift also improved the average CRPS of hidden testing datasets as seen in Tables 4 and 5, even in the presence of statistical sampling error.

We hypothesise that observed time mean drifts are likely to be significant and expected in realistic modelling scenarios, there are likely unresolved drift processes between the forward model and observed data. In which case including the observed time mean as in ensemble Method 3.5, can be seen as an essential modelling step

Table 5 Each column contains the average CRPS scores associated with attaining hidden test datasets, from an ensemble(denoted in a row) calibrated on the corresponding training dataset

Scheme	Average CRPS Dataset 1.	Average CRPS Dataset 2.	Average CRPS Dataset 3.	Average CRPS Dataset 4.
Persistence	1.501e-01	1.223e-01	1.509e-01	1.520e-01
RIC	7.591e-02	2.268e-02	7.859e-02	8.755e-02
Perfect	6.836e-02	2.126e-02	6.985e-02	7.354e-02
Without-time-mean	6.834e-02	2.147e-02	7.048e-02	7.785e-02
With-time-mean	6.979e-02	2.204e-02	7.126e-02	7.485e-02
Learned forward model: $E_o = 5$	6.944e-02	1.789e-02	NA	NA

Table 6 Moments of the recovered increments

Moment	Dataset 1	Dataset 2	Expected value
$\mathbb{E}(\Delta W_r)$	$-6.349e-16$	$-2.325e-16$	0
$\mathbb{E}(\Delta W_r^2)/\Delta t$	$1.000e+00$	$1.000e+00$	1
$\mathbb{E}(\Delta W_r^3)$	$8.875e-04$	$-1.478e-03$	0
$\mathbb{E}(\Delta W_r^4)/\Delta t^2$	$2.727e+00$	$2.878e+00$	3
$\mathbb{E}(\Delta W_r^5)$	$-5.641e-03$	$-2.375e-03$	0
$\mathbb{E}(\Delta W_r^6)/\Delta t^3$	$1.224e+01$	$1.336e+01$	15

unless one has access to diagnostics tools capable of ruling out the data observed drift as a modelling error and classifying it as a statistical error. Such situations are unlikely, one does not necessarily get to resample from the underlying distribution from which the data was generated as we have in this idealised testing scenario.

4.4 Summary of Results

Under the assumption that the data comes from an SPDE realisation, the new SVD calibration technique Method 3.1 captures the same number of basis functions used to generate the data for $P = 1$ and for $P = 5$. The basis recovered are in some objectionable sense reasonable in magnitude and shape in comparison to the basis used to generate the data. The recovered noise increments have passed several hypothesis tests indicating normality. In the specific instance of one basis function, the recovered basis function was shown to be in agreement with the basis function used to generate the data, both visually and agreeing to 10^{-2} in the relative error norm.

We speculated that the 10^{-2} relative L^2 discrepancy in both recovered path (Table 6) and basis function in part came from the time mean removal in the SVD decomposition, and proposed adding this term back in as a deterministic drift velocity in the equation without violating the geometric structure of the model. To test the addition of this term we proposed the resimulation of data, by driving the SPDE with recovered increments, and recovered basis functions. In the resimulation of data, driving the solution of the model could more accurately represent the training dataset by the inclusion of the time mean drift velocity, for all datasets.

We also estimated the Continuous Rank Probability Score (CRPS) for each model, the persistence forecast (ensemble Method 3.1), the new SVD algorithm with the time mean (ensemble Method 3.5), and without it (ensemble Method 3.4), the perfect model (ensemble Method 3.3), the and the model rerun with learned parameter values (ensemble Method 3.6). This was done by estimating CRPS over all time and all state space values for a 30-member ensemble forming a global estimate of the CRPS score to quantify how likely the observations come from the ensemble. This is performed on 1000 hidden datasets and averaged to help distinguish the sampling error associated with sampling from the data distribution.

From which we concluded. Should data come from an SPDE/SDE realisation with a physical drift term, using the time mean from the DAM to evolve the ensemble is an important modelling step for improving the skill of the forecast. The potential drawback of adding a time mean drift term is that, if the data arises from a SPDE/SDE realisation with a small or insignificant drift term, sampling from the data distribution may result in an observed non-physical drift in the DAM matrix. The appearance of a nonphysical drift is typically small in magnitude (arising from the statistical error of sampling Brownian motion not having mean zero) and may justify ignoring small drifts such as the higher order (typically smaller) Itô-Stratonovich correction in the context of calibration. However, the possibility of a nonphysical drift in the DAM does not justify neglecting a drift of larger magnitude arising from model-data mismatches. Overall both SVD approaches ensemble Method 3.5 and ensemble Method 3.4 (with and without the time mean) outperformed both RIC and Persistence, and approached the CRPS score of the perfect model.

Regarding Method 3.2, evidence points towards a decreasing CRPS score with increased training time, and showed a 40% CRPSS improvement from the initial deterministic proposed ensemble. The lack of interpolation lead to a drastic improvement in compute speed in the forward model. With a fixed number of basis functions the SVD approaches produced ensemble forward models that took approximately 1 hour to run, whilst the parameter estimated model took approximately 1 minute to run, this scaling gets more drastic with the more weather-stations used in the model. It is also worth remarking the backpropagation approach Method 3.2 did not use any Eulerian weather station data in the training, and only trained on approximately 3% of available data Lagrangian path data, so direct comparison to SVD methodology may not be appropriate.

There are many free parameters involved, $n_c, E_o, T_o, n_v, P, n_t \ldots$, in which the effectiveness of both calibration methods could be studied. For example, we plot the CRPSS and parameter magnitude estimation error for $E_o = \{1, 10, 100\}$, trained over the entire time window $\mathcal{T}_o = [256]/32$ in Fig. 12, for a smaller dataset to illustrate the effect of E_o size during training. Back-propagation through an ensemble costs vastly more than an SVD approach. Nevertheless, in the context of offline approaches to data assimilation, it may be beneficial to incur the cost of the offline training, to improve the forward ensemble model speed (due to the lack of interpolation in the forward model).

5 Conclusion and Future Outlook

A methodology for the calibration of stochastic vector-fields was proposed using the combination of the Biot-Savart kernel and vortex positional data in combination with a SVD approach Method 3.1. The Eulerian vector-fields recovered are demonstrated relevant to the stochastic forward model proposed and we provided evidence Method 3.1 parameterised the difference between the proposed forward model and the synthetic data. The methodology has been shown consistent by the use of a twin experiment and using the CRPS skill score as compared relative to several benchmarks including the "perfect" model (ensemble Method 3.3).

The inclusion of a time mean drift velocity was motivated and shown important in the context of a twin experiment. This term was shown consistent with the geometric modelling assumptions (Kelvin theorem, coadjoint action section A, Hamiltonian and Poisson structure section B). The forecast verification results presented here suggest the addition of the time mean drift term is an advantageous modelling step, except in the setting when the data does not have a large difference in drift from the model. Realistic data may have a significantly larger time mean drift velocity from the proposed forward model than in the idealised experiments presented here, making this arguably an essential modelling step.

Using the Biot-Savart kernel to create a data anomaly matrix is not necessarily restricted to the setting of inviscid vortex methods, using the Biot-Savart kernel is likely applicable for the calibration of other fluid mechanics models. For the point of testing and to distinguish between model error data sample error and calibration error, we have biased all tests towards SPDE/SDE realisation data for which we know the parameters. The application of the calibration methodology proposed in this chapter could be adapted in the context of less idealised (and less testable) synthetic data. For example, the vorticity equation can be solved (Eq. (4), Example 2.1), using a finite volume discretisation, with passive tracer drifters (with an initially measured known vorticity) in the flow shown in Fig. 11. These tracers could be treated as analogues of the point vortices in the inviscid vortex method, and vector fields could be calibrated using Method 3.1. One could equally treat this as a reference dataset to calibrate an inviscid vortex method, serving as another example of stochastic coarse-grained model reduction.

Using a tangent linear model approach to auto differentiate through the ensemble forecast, as to minimise the CRPS Skill score Method 3.2. We were able to consistently propose basis functions with an improved relative CRPS skill score

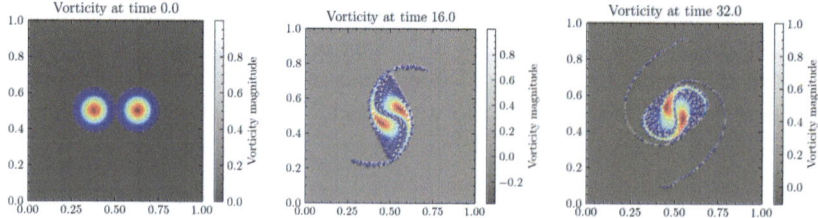

Fig. 11 Passive tracers carried by the Fourier reconstructed velocity field within a unlimited upwind bias high order flux form mimetic finite volume method (plotted grey) (solving strategy uses ideas from [1, 19, 28, 45]), the passive particle tracer are coloured by the initial vorticity (plotted jet)

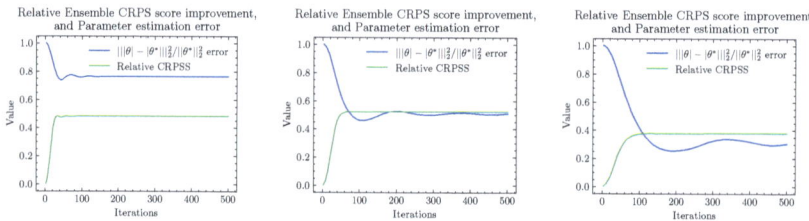

Fig. 12 On a different dataset generated with fewer dimensions (using much less points), we were able to run the ensemble training over the full 256 time intervals, with 1,10,100 ensemble members respectively. The effect of increasing the ensemble size appeared to improve the parameter estimation error

as compared to a near deterministic forecast, and did not require weatherstation data. The minimisation of the CRPS of an ensemble forecast did go someway to propose reasonable estimation of the true parameters magnitude (better than the initial guess). However, we found cases in which a decreases in CRPS did not necessarily lead to a more accurate estimation of parameter values. Parameters converged to roughly the correct magnitude of the proposed parameters, but not the right sign or exact same number (Fig. 12).

The inconsistency in the calibration problem as posed as minimisation of a CRPS score and as a parameter estimation problem, could be a barrier for reliable methodology but also a potential opportunity for model-specific calibration. One could foresee calibration vector fields chosen to produce a good (in a probabilistic sense) forecast of a specific event. Providing an event-informed approach to the choice of vector fields. One could potentially achieve such a task using an approach similar to Method 3.2 but with the minimisation of a Brier skill score of a single observed important event, e.g. hurricane hitting a specific location, sea surface height under a satellite track.

Acknowledgments I would like to acknowledge Darryl Holm for continued support and insight. I would like to acknowledge Wei Pan, Ruiao Hu for valuable insights into existing calibration methodology. I would like to acknowledge Theo Diamantakis and Darryl Holm regarding various geometric insights into point vortices and the SALT framework in the variational principle. I would like to acknowledge Wei Pan, Oliver Street, Alex Lobbe for interesting discussions in weather forecast verification techniques. I would like to acknowledge discussions regarding numerical methods with Ruiao Hu, Wei Pan as well as Aythami Bethencourt de Leon, and James Micheal Leahy regarding specifics in the JAX coding environment. I would also like to acknowledge two particularly helpful and thorough anonymous reviewers for comments leading to the improvement of this manuscript.

The work of JW is supported by the European Research Council (ERC) Synergy grant "Stochastic Transport in Upper Ocean Dynamics' (STUOD) – DLV-856408.

Appendix 1: A Geometric Structure

It may be worth noting that although Eq. (25) appears to have an additional drift not present in the original Stochastic Advection by Lie Transport(SALT) chapter [25]. The modelling remains faithful to the geometric framework in the following way. An additional deterministic drift velocity acting at the level of the particle trajectory map, gives a stochastic Kelvin theorem akin to [25] but with the additional feature of a data informed deterministic drift moving the loop

$$d \oint_{C(udt+\bar{v}dt+\xi_p \circ dW^P)} \boldsymbol{u} \cdot d\boldsymbol{x} = 0. \tag{32}$$

An Euler Poincaré equation(see [26]) with a modification to the Lie algebra resulting in a coadjoint operator of the form $\mathrm{ad}^*_{udt+\bar{v}dt+\xi_p \circ dW^P}$. In the Euler equation an additional deterministic drift term appears in the velocity of the Lie derivative operator, in 2D this appears as additional transport of vorticity by a time mean drift velocity as follows

$$d\omega_t + (\boldsymbol{u} + \bar{\boldsymbol{v}} \cdot \nabla)\omega_t dt + \sum_{p=1}^{P} (\theta_p \boldsymbol{\xi}_p \cdot \nabla)\omega_t \circ dW^P = 0. \tag{33}$$

Appendix 2: Hamiltonian and Poisson Structure

The finite dimensional regularised Stratonovich system has a "Hamiltonian" structure, when the vectorfield basis is assumed to have a streamfunction representation $\xi_p = -\nabla^\perp \psi_p$, $\forall p \in [P]$, and when the time mean drift has a streamfunction representation $\bar{v} = -\nabla^\perp \bar{\psi}$. We first define the operators $\nabla_i^\perp := (-\partial_{y_i}, \partial_{x_i})$, $\nabla_i := (\partial_{x_i}, \partial_{y_i})$, denoting derivatives with respect to specific particle positions. Then the Hamiltonian system can be written as the following

$$\Gamma_i d\boldsymbol{x}_i = -\nabla_i^\perp \left(H_\delta dt + \sum_{i\in\wedge_0} \Gamma_i \bar{\psi}(\boldsymbol{x}_i) dt + \sum_{i\in\wedge_0} \sum_{p=1}^{P} \Gamma_i \theta_p \psi_p(\boldsymbol{x}_i) \circ dW^p \right),$$

$$\forall i \in \wedge_0, \tag{34}$$

where H_δ is the deterministic regularised Kirchhoff Hamiltonian

$$H_\delta = \sum_{i,j\in\wedge_0^i, i\neq j} \Gamma_i \Gamma_j G_\delta(\boldsymbol{x}_i - \boldsymbol{x}_j). \tag{35}$$

For a function of particle positions $F(t, \{\boldsymbol{x}_i\}_{i\in\wedge_0^i})$, the time derivative reveals the following Poisson structure

$$0 = \partial_t F + \{F, H_\delta\}dt + \{F, \bar{\psi}\}dt + \sum_{p=1}^{P} \{F, \theta_p \psi_p\} \circ dW^p,$$

$$\{F, H_\delta\} := \sum_{i\in\wedge_0^i} \frac{1}{\Gamma_i}(\nabla_i F) \cdot (-\nabla_i^\perp H_\delta). \tag{36}$$

Example B.1 (Surface Quasi-Geostrophic) Surface Quasi-Geostrophic on $\mathbb{R}^2$ has the following differential relationship between deterministic stream function and vorticity, $\psi = (-\Delta)^{-1/2}\omega$, with Greens function given by $G(\boldsymbol{x}) = -(2\pi)^{-1}||\boldsymbol{x}||_2^{-1}$, and kernel by $K(\boldsymbol{x}) = (4\pi)^{-1}\boldsymbol{x}^\perp||\boldsymbol{x}||^{-3}$. This two dimensional model bears analogy to the three dimensional Euler Equation and kernel.

Example B.2 (Quasi-Geostrophic Shallow Water) Quasi-Geostrophic Shallow Water on $\mathbb{R}^2$ has the following differential relationship between deterministic stream function and vorticity, $\psi = (\Delta - \lambda^2)^{-1}\omega$, with Greens function given by $G(\boldsymbol{x}) = -(2\pi)^{-1}K_0(\lambda||\boldsymbol{x}||)$, and kernel by $K(\boldsymbol{x}) = \lambda(2\pi)^{-1}K_1(\lambda||\boldsymbol{x}||)$, where K_n is the modified Bessel function of second kind, and satisfies $K_0' = -K_1$. Numerically, the modified Bessel function of the second kind requires approximation, typically done via numerical integration either through trapesium rule [27] or peicewise Chebyshev quadrature, additional computational time or computational resources are required.

Example B.3 (Euler-α (Model of Turbulence)) Euler-α (model of turbulence)on $\mathbb{R}^2$ has the following differential relationship $\psi = (1 - \alpha^2\Delta)^{-1}(-\Delta)^{-1}\omega$. Using the fundamental solutions to the Helmholtz and Laplace operator D. D. Holm, M. Nitsche and V. Putkaradze [27] deduce the following Greens function $G(\boldsymbol{x}) = \frac{-1}{2\pi}(\log(||\boldsymbol{x}||_2) + K_0(||\boldsymbol{x}||_2/\alpha))$ and Biot-Savart kernel $K(\boldsymbol{x}) = (2\pi)^{-1}\boldsymbol{x}^\perp||\boldsymbol{x}||_2^{-2}(1 - ||\boldsymbol{x}||_2/\alpha K_1(||\boldsymbol{x}||_2/\alpha))$. Here one observes a regularisation of the Euler kernel, taking a similar but distinct form to that of the regularised vortex blob approximations. There is exponential decay for large values of $||\boldsymbol{x}||_2$, but unlike the vortex blob method the Greens function remains unbounded at the origin. The reconstructed velocity is bounded, but the vorticity is not.

Remark B.1 (Different Domains) Various modifications to the Greens functions have been proposed to deal with different domains. Point vortices have been proposed on a singularly periodic strip [32, 39] using the method of images, a doubly periodic square see [44] [36], for toroidal see [41], for gaps in walls see [14], for inclusion of islands see [29]. Milne-Thompson theorem for a cylinder can be used to take into account the introduction of a cylinder [10], and the conformal Schwarz–Christoffel mapping could be proposed more generally. Embedded closed surfaces conformal to the unit sphere are considered in [16], and for additional literature review see [5] [2].

Remark B.2 (CRPS) The CRPS is a continuous version of the Rank probability score, and can be interpreted as a Brier score but integrated over all possible thresholds. The CRPS is a strictly proper scoring rule [35]. The CRPS measures differences between the predicted and occurred cumulative density functions and gives a score indicating how good the ensemble is at matching observations. 0 is accurate, and 1 is inaccurate. For a deterministic forecast, the CRPS reduces to the mean absolute error. We use an estimator of the representation (pg8 eq 11 in [21]) in the loss function with an additional fairness modification see [17, 46], see same reference for the equivalence to other formulations of the CRPS. For discussion, literature review and details regarding the Continuous Rank Probability Score, and relationship to other skill scores (such as Breir skill score [9], and RPS) and other diagnostic tools such as (Rank Histogram/Talagrand diagram) see [21, 24], and [20]. For further insights into the Candille–Talagrand (2005), Brier score, Quantile score, Hersbach and more decompositions of the CRPS score see [3]. In the context of postprocessing of ensemble forecasting in the forecast verification community, the CRPS score has been found to produce sharper better-calibrated forecasts than with maximum likelihood estimation [18].

Remark B.3 (SVD) The SVD of the real $m \times n$ matrix M is a decomposition into two orthogonal matrices U, V and a diagonal matrix Σ, such that $M = U \Sigma V^T$, for $U, \Sigma, V^T \in \mathbb{R}^{m \times m}, \mathbb{R}^{m \times n}, \mathbb{R}^{n \times n}$. The diagonal entries $\sigma_i = \Sigma_{ii}$ of the diagonal matrix $\Sigma \in \mathbb{R}^{m \times n}$, are known as the singular values of M, they are uniquely defined by M up to ordering and there are $rank(M)$ of them. U, V are orthonormal (rotation) matrices, interpreted as forming an orthonormal basis from columns or rows. The SVD is typically ordered such that the singular values are decreasing in size, while the matrix Σ is unique in this ordering, matrices U and V are not unique. The columns of U and the columns of V(rows of V^T) are called the left-singular vectors and right-singular vectors of M. A left singular value of M is a non-negative real number σ, such that $Mv = \sigma u$ for $v \in \mathbb{R}^n, u \in \mathbb{R}^m$, a right singular value of M as a non-negative real number σ, such that $M^T v = \sigma v$ for $v \in \mathbb{R}^n, u \in \mathbb{R}^m$. We are interested in a truncated SVD, where we retain only the first t singular values along with their corresponding t column vectors in U, and t row vectors in V^T. Such that $M_t = U_t \Sigma_t V_t^T$, for $U_t, \Sigma_t, V_t^T \in \mathbb{R}^{m \times t}, \mathbb{R}^{t \times t}, \mathbb{R}^{t \times n}$ approximates M, we use the Truncated SVD algorithm in Halko [23].

Remark B.4 (Fourier Interpolation)   Fourier interpolation of an unstructured grid of points using a structured 2d regular grid of data on $[0, 1] \times [0, 1]$ (weather stations) can be performed in the following manner. Let $q \in \mathbb{R}^{n_c \times n_c}$, denote the discrete field at $\wedge_d$, where n_c is even, and $\hat{q} = \mathcal{F}_x \mathcal{F}_y(q)$ its discrete 2d Fourier transform. Then $q(x_l, y_l) = \text{Re}(\sum_{-n_c/2 \leq j,k \leq n_c/2} \hat{q} \exp(2\pi i j x_l / \Delta x) \exp(2\pi i k y_l / \Delta y))$, $\forall l$, is Fourier interpolation at (x_l, y_l). This is performed on the x, and y components of (deterministic and stochastic) velocity fields, at discrete time evolving points $x_l \in \wedge_0(t)$ in the methods Methods 3.5 and 3.4. Typically the error of Fourier interpolation is related to the continuity of the field. The reconstruction and evaluation of interpolating polynomials or other basis functions from weather stations has an inherent error and computational cost. Two dimensional Fourier interpolation from $n_c \times n_c$ weather stations to n_v points has roughly the following computational cost, in floating point operations $CPn_c^4 2\log(n_c)n_v$.

Remark B.5 (Notation, Initialisation and Weather Stations)   Let $\wedge_{[x_{\min}, x_{\max}] \times [y_{\min}, y_{\max}]}^{m,n} = X_{[x_{\min}, x_{\max}]}^m \times Y_{[y_{\min}, y_{\max}]}^n$, denote the $m \times n$ Cartesian product mesh defined from the following two sets $X_{[x_{\min}, x_{\max}]}^m := \{x_i | x_i = x_{\min} + h_x(i + 1/2), \forall i \in \{0, \ldots, m - 1\}\}$, $h_x := (x_{\max} - x_{\min})/m$, $Y_{[y_{\min}, y_{\max}]}^n := \{y_j | y_j = y_{\min} + h_y(j + 1/2), \forall j \in \{0, \ldots, n - 1\}\}$, $h_y := (y_{\max} - y_{\min})/n$ of equally spaced cell center points. Such that the (i, j)-th element $(x_{i,j}, y_{i,j}) = (x_i, y_j) \in \wedge_{[x_{\min}, x_{\max}] \times [y_{\min}, y_{\max}]}^{m,n}$, denotes the (i, j)-th cell center position of a Cartesian product mesh on $[x_{\min}, x_{\max}] \times [y_{\min}, y_{\max}]$. We denote the vectorisation of an arbitrary $A \in \mathbb{R}^{m \times n}$ matrix by

$$\text{vec}(A) = \left[a_{1,1}, \ldots, a_{m,1}, a_{1,2}, \ldots, a_{m,2}, \ldots, a_{1,n}, \ldots, a_{m,n}\right]^{\text{T}} \in \mathbb{R}^{nm \times 1}.$$

We denote the outer product by $\otimes$. We denote the vector of ones length m by e_m. We denote the set of natural numbers from 1 to M by $[M] := \{1, \ldots, M\}$. We denote the vector $x_m \in \mathbb{R}^m$, as the vector of increasing points in $X_{[x_{\min}, x_{\max}]}^m$, and $y_n \in \mathbb{R}^n$ as the vector of increasing points in $Y_{[y_{\min}, y_{\max}]}^n$. Such that we can construct coordinate matrices in the following manner $x_{i,j} = (e_n \otimes x_m) \in \mathbb{R}^{n \times m}$, $y_{i,j} = (y_n \otimes e_m) \in \mathbb{R}^{n \times m}$. And evaluate the initial vorticity ω as a function of coordinate matrices in the following manner $\Gamma_{i,j} = \omega(x_{i,j}, x_{i,j})h_x h_y$, $\forall (i, j) \in [m] \times [n]$. We consider a $n \times n$ uniform Cartesian meshgrid $\wedge_{[x_{\min}, x_{\max}] \times [y_{\min}, y_{\max}]}^{n,n}$, of initial points in a closed subdomain of $\mathbb{R}^2$. With initial condition $\omega_0(x, y)$, evaluated at the positions of this uniform mesh such that $\Gamma_{i,j} = \omega_0(x_i, y_j)h_x h_y$ $\forall (i, j) \in \{1, \ldots, n\} \times \{1, \ldots, m\}$. We define the following set of dynamically evolving points $\wedge_0 := \{(x_{i,j}, y_{i,j}) | (X_{i,j}, X_{i,j}) = (X_i, X_j) \in \wedge^{m,n}, \omega_0(X_i, X_j) \neq 0\}$, as those with nonzero initial vorticity on the initial mesh, assumed $n_v \leq nm$ dimensional. Note that we have defined the initial condition mesh of points on the dual mesh to that in [7, 34]. We consider a $n_c \times n_c$ Cartesian meshgrid $\wedge_d := \wedge_{[x_{\min,d}, x_{\max,d}] \times [y_{\min,d}, y_{\max,d}]}^{n_c, n_c}$, of fixed weather stations in a closed subdomain of $\mathbb{R}^2$ for all time. Where the subscript d denotes "data", and indicates that this is a weather-station position in which (velocity) data can be collected.

References

1. Akio Arakawa and Vivian R Lamb. A potential enstrophy and energy conserving scheme for the shallow water equations. *Monthly Weather Review*, 109(1):18–36, 1981.
2. Hassan Aref, James B Kadtke, Ireneusz Zawadzki, Laurence J Campbell, and Bruno Eckhardt. Point vortex dynamics: recent results and open problems. *Fluid Dynamics Research*, 3(1–4):63, 1988.
3. Sebastian Arnold, Eva-Maria Walz, Johanna Ziegel, and Tilmann Gneiting. Decompositions of the mean continuous ranked probability score. *arXiv preprint arXiv:2311.14122*, 2023.
4. Vladimir Arnold. Sur la géométrie différentielle des groupes de lie de dimension infinie et ses applications à l'hydrodynamique des fluides parfaits. In *Annales de l'institut Fourier*, volume 16, pages 319–361, 1966.
5. WWR Ball. Point-vortex dynamics. 2006.
6. J Thomas Beale and Andrew Majda. Vortex methods. ii. higher order accuracy in two and three dimensions. *Mathematics of Computation*, 39(159):29–52, 1982.
7. J Thomas Beale and Andrew Majda. High order accurate vortex methods with explicit velocity kernels. *Journal of Computational Physics*, 58(2):188–208, 1985.
8. Vance W Berger and YanYan Zhou. Kolmogorov–smirnov test: Overview. *Wiley statsref: Statistics reference online*, 2014.
9. Glenn W Brier. Verification of forecasts expressed in terms of probability. *Monthly weather review*, 78(1):1–3, 1950.
10. Alexandre Joel Chorin. Numerical study of slightly viscous flow. *Journal of fluid mechanics*, 57(4):785–796, 1973.
11. Colin Cotter, Dan Crisan, Darryl D Holm, Wei Pan, and Igor Shevchenko. Numerically modeling stochastic lie transport in fluid dynamics. *Multiscale Modeling & Simulation*, 17(1):192–232, 2019.
12. Dan Crisan, Darryl D Holm, James-Michael Leahy, and Torstein Nilssen. Variational principles for fluid dynamics on rough paths. *Advances in Mathematics*, 404:108409, 2022.
13. Dan Crisan, Oana Lang, Alexander Lobbe, Peter Jan van Leeuwen, and Roland Potthast. Noise calibration for the stochastic rotating shallow water model. *arXiv preprint arXiv:2305.03548*, 2023.
14. Darren Crowdy and Jonathan Marshall. The motion of a point vortex through gaps in walls. *Journal of Fluid Mechanics*, 551:31–48, 2006.
15. Theo Diamantakis and James Woodfield. L\'evy areas, wong zakai anomalies in diffusive limits of deterministic lagrangian multi-time dynamics. *arXiv preprint arXiv:2402.03026*, 2024.
16. David Gerard Dritschel and Stefanella Boatto. The motion of point vortices on closed surfaces. *Proceedings of the Royal Society A: Mathematical, Physical and Engineering Sciences*, 471(2176):20140890, 2015.
17. CAT Ferro. Fair scores for ensemble forecasts. *Quarterly Journal of the Royal Meteorological Society*, 140(683):1917–1923, 2014.
18. Manuel Gebetsberger, Jakob W Messner, Georg J Mayr, and Achim Zeileis. Estimation methods for nonhomogeneous regression models: Minimum continuous ranked probability score versus maximum likelihood. *Monthly Weather Review*, 146(12):4323–4338, 2018.
19. Amir Gholami, Dhairya Malhotra, Hari Sundar, and George Biros. Fft, fmm, or multigrid? a comparative study of state-of-the-art poisson solvers for uniform and nonuniform grids in the unit cube. *SIAM Journal on Scientific Computing*, 38(3):C280–C306, 2016.
20. Tilmann Gneiting and Adrian E Raftery. Strictly proper scoring rules, prediction, and estimation. *Journal of the American statistical Association*, 102(477):359–378, 2007.
21. Tilmann Gneiting and Roopesh Ranjan. Comparing density forecasts using threshold-and quantile-weighted scoring rules. *Journal of Business & Economic Statistics*, 29(3):411–422, 2011.

22. Ole H Hald. Convergence of vortex methods for euler's equations. ii. *SIAM Journal on Numerical Analysis*, 16(5):726–755, 1979.
23. Nathan Halko, Per-Gunnar Martinsson, and Joel A Tropp. Finding structure with randomness: Probabilistic algorithms for constructing approximate matrix decompositions. *SIAM review*, 53(2):217–288, 2011.
24. Hans Hersbach. Decomposition of the continuous ranked probability score for ensemble prediction systems. *Weather and Forecasting*, 15(5):559–570, 2000.
25. Darryl D Holm. Variational principles for stochastic fluid dynamics. *Proceedings of the Royal Society A: Mathematical, Physical and Engineering Sciences*, 471(2176):20140963, 2015.
26. Darryl D Holm, Jerrold E Marsden, and Tudor S Ratiu. The euler–poincaré equations and semidirect products with applications to continuum theories. *Advances in Mathematics*, 137(1):1–81, 1998.
27. Darryl D Holm, Monika Nitsche, and Vakhtang Putkaradze. Euler-alpha and vortex blob regularization of vortex filament and vortex sheet motion. *Journal of Fluid Mechanics*, 555:149–176, 2006.
28. Willem Hundsdorfer, Barry Koren, JG Verwer, et al. A positive finite-difference advection scheme. *Journal of computational physics*, 117(1):35–46, 1995.
29. ER Johnson and N Robb McDonald. The point island approximation in vortex dynamics. *Geophysical & Astrophysical Fluid Dynamics*, 99(1):49–60, 2005.
30. Ioannis Karatzas and Steven Shreve. *Brownian motion and stochastic calculus*, volume 113. Springer Science & Business Media, 2012.
31. Peter E Kloeden, Eckhard Platen, Peter E Kloeden, and Eckhard Platen. *Stochastic differential equations*. Springer, 1992.
32. Robert Krasny. A study of singularity formation in a vortex sheet by the point-vortex approximation. *Journal of Fluid Mechanics*, 167:65–93, 1986.
33. Ding-Gwo Long. Convergence of the random vortex method in two dimensions. *Journal of the American Mathematical Society*, 1(4):779–804, 1988.
34. Andrew J Majda, Andrea L Bertozzi, and A Ogawa. Vorticity and incompressible flow. cambridge texts in applied mathematics. *Appl. Mech. Rev.*, 55(4):B77–B78, 2002.
35. James E Matheson and Robert L Winkler. Scoring rules for continuous probability distributions. *Management science*, 22(10):1087–1096, 1976.
36. Kevin A O'Neil. On the hamiltonian dynamics of vortex lattices. *Journal of mathematical physics*, 30(6):1373–1379, 1989.
37. Valentin Resseguier, Long Li, Gabriel Jouan, Pierre Dérian, Etienne Mémin, and Bertrand Chapron. New trends in ensemble forecast strategy: uncertainty quantification for coarse-grid computational fluid dynamics. *Archives of Computational Methods in Engineering*, 28:215–261, 2021.
38. Valentin Resseguier, Wei Pan, and Baylor Fox-Kemper. Data-driven versus self-similar parameterizations for stochastic advection by lie transport and location uncertainty. *Nonlinear Processes in Geophysics*, 27(2):209–234, 2020.
39. Louis Rosenhead. The formation of vortices from a surface of discontinuity. *Proceedings of the Royal Society of London. Series A, Containing Papers of a Mathematical and Physical Character*, 134(823):170–192, 1931.
40. W Rüemelin. Numerical treatment of stochastic differential equations. *SIAM Journal on Numerical Analysis*, 19(3):604–613, 1982.
41. Takashi Sakajo and Yuuki Shimizu. Point vortex interactions on a toroidal surface. *Proceedings of the Royal Society A: Mathematical, Physical and Engineering Sciences*, 472(2191):20160271, 2016.
42. Samuel Sanford Shapiro and Martin B Wilk. An analysis of variance test for normality (complete samples). *Biometrika*, 52(3/4):591–611, 1965.
43. Chi-Wang Shu and Stanley Osher. Efficient implementation of essentially non-oscillatory shock-capturing schemes. *Journal of computational physics*, 77(2):439–471, 1988.
44. Jeffrey B Weiss and James C McWilliams. Nonergodicity of point vortices. *Physics of Fluids A: Fluid Dynamics*, 3(5):835–844, 1991.

45. James Woodfield, Hilary Weller, and Colin J Cotter. New limiter regions for multidimensional flows. *Available at SSRN 4668131*, 2023.
46. Michaël Zamo and Philippe Naveau. Estimation of the continuous ranked probability score with limited information and applications to ensemble weather forecasts. *Mathematical Geosciences*, 50(2):209–234, 2018.

Index

© The Editor(s) (if applicable) and The Author(s) 2025
B. Chapron et al. (eds.), *Stochastic Transport in Upper Ocean Dynamics III*,
Mathematics of Planet Earth 13, https://doi.org/10.1007/978-3-031-70660-8

Lecture Notes in Computer Science 7501

Commenced Publication in 1973
Founding and Former Series Editors:
Gerhard Goos, Juris Hartmanis, and Jan van Leeuwen

Advanced Research in Computing and Software Science
Subline of Lectures Notes in Computer Science